THE 'TEN-MILE' MAPS

OF THE

ORDNANCE SURVEYS

by

ROGER HELLYER

THE CHARLES CLOSE SOCIETY
FOR THE STUDY OF ORDNANCE SURVEY MAPS
1992

First published November 1992

by

The Charles Close Society for the Study of Ordnance Survey Maps

c/o the Map Library, British Library Reference Division, Great Russell Street, London WC1 3DG

All rights reserved.
No part of this work my be reproduced, stored in a retrieval system or transmitted,
in any form or by any means, electronic, mechanical, photographic, electrostatic, recorded or otherwise,
without the prior written permission of the publisher

Introductory essays and editorial matter Copyright (c) Roger Hellyer, 1992

Projection essay Copyright (c) Brian Adams, 1992

ISBN 1 870598 12 1

British Library Cataloguing-in-Publication Data
A catalogue record for this book is available from the British Library

Printed and bound for the Charles Close Society
by Dotesios Limited, Kennet House, Kennet Way, Trowbridge, Wiltshire BA14 8RN

The Charles Close Society for the Study of Ordnance Survey Maps

FOREWORD BY YOLANDE HODSON

Ten miles to one inch....is the smallest scale which really helps the curious. (Winterbotham 1936)

The curious will find this book a fascinating tale of the development of a mere index map into a multi-purpose cartographic database which is familiar to us all today in its metric guise of the 1:625,000 Routeplanner. Over a period of many decades, the ten-mile map has been the basis for a wide variety of thematic overprints, the subject matter of which touches so many aspects of public life that this scale became the obvious choice for the projected post war national atlas, whose story, among many others, is recounted here.

The story of the ten-mile map begins in the library of St John's College, Cambridge, where the only known copy of the earliest (c.1817) index to the Old Series survives. Such elusiveness is a feature of many of the ten-mile maps, but is a particular characteristic of the index map which, over the years, has conventionally been regarded as an ephemeral cartographic artefact. Its uncertain status is nicely summed up by the British Library Map Catalogue sub-heading of "cartographic map".

Indeed, when the production of a series of Charles Close Society guides to Ordnance Survey maps was first proposed in January 1982, the suggestion that a study of the index maps to the Old Series One-inch maps would form a guide in itself was thrown in as an aside. Now, ten years later, Dr Hellyer has given substance to a chance remark and has expanded the subject in a work which breaks new ground with every chapter. His elucidation, for example, of the Ordnance Survey involvement in the production of the ten-mile base maps of Griffith and Harness in nineteenth-century Ireland is a notable expansion of our knowledge for this period.

It is probably true to say that most map readers pay scant attention to, or never consider, the mathematical base of the map they are looking at, save to note its scale. The elements of map construction are easily overlooked, or taken for granted, by many map users. The detailed account of these matters, as they relate to the ten-mile map, in this volume written by Brian Adams (a mathematician who served in the geodetic and specialist branches of the Hydrographic Department), is therefore an essential contribution to our understanding of the map as a whole.

Dr Hellyer continues the longstanding tradition which has produced outstanding contributions by the amateur map historian to the field of the history of cartography. A musicologist, and bassoonist with the Royal Shakespeare Company, he has contributed to *The New Grove Dictionary of Music and Musicians*, and has written widely on Harmoniemusik. He received in 1990 the British Cartographic Society's Survey and General Instrument Company Award for his work on the archaeological maps of the Ordnance Survey, and he brings to this book the same measure of precise recording married to a fluent and scholarly prose which makes it an engrossing chronicle of one of Ordnance Survey's longest enduring maps.

Yolande Hodson, Chairman, Charles Close Society.
August 1992.

CONTENTS

Foreword by Yolande Hodson		iii
Contents		iv
Illustrations		vi
Acknowledgements		vii
Abbreviations and column numbers		viii
Introduction		x

I. TEN-MILE MAPS OF GREAT BRITAIN IN THE NINETEENTH CENTURY — 1

Text	1. Old Series Index	1
	2. New Series Index	6
Specification table	Comparative specifications of all Ordnance Survey ten-mile maps	7
Summary list of states	1. Old Series Index	14
	2. New Series Index	15
Cartobibliography	1. South sheet, ?1817	15
	2. Middle sheet, ?1824	34
	3. North sheet, ?1881	49
	Supplement 1. Special issues based on the Index	50
	Supplement 2. Rivers and their Catchment Basins	51
	Supplement 3. Military issues	52

II. TEN-MILE MAPS OF IRELAND IN THE NINETEENTH CENTURY — 54

Text	1. Thomas Larcom's map	54
	2. The projects of Sir Henry James	56
Cartobibliography	1. Larcom's map, 1838	59
	2. James's map, 1868	63
	Supplement 1. Rivers and their Catchment Basins	69
	Supplement 2. Military issues	71

III. THE JOHNSTON TEN-MILE MAPS — 73

Text	1. Great Britain	73
	2. Ireland	77
Standard attributes	1. Great Britain. 2. Ireland	81
Variable attributes	Both maps	82
Cartobibliography	1. Great Britain, 1903	82
	Supplement 1. Special sheets	84
	Supplement 2. Military issues	86
	2. Ireland, 1905	87
	Supplement 1. Skeleton maps	90
	Supplement 2. Province maps and other special sheets	91
	Supplement 3. Military issues	92
	3. Ireland maps at 1:625,000, 1960	92

Contents v

IV.	THE TEN-MILE MAP BETWEEN THE WARS AND ITS METRIC CONVERSION			93
	Text	1.	The three-sheet map	93
		2.	The two-sheet map (including the Road Map (IV.2R))	98
		3.	Ministry of Town and Country Planning Series	100
		4.	Smaller scales	112
	Standard attributes	1.	1:633,600 Great Britain (three sheets)	114
		2.	1:633,600 Great Britain (two sheets)	115
		3.	1:625,000 Great Britain	116
		4.	1:1,250,000 Great Britain, to "D" edition	117
	Variable attributes		All four maps	117
	Cartobibliography	1.	1:633,600 Great Britain (three sheets), 1926	119
			Supplement. Special sheets	122
		2.	1:633,600 Great Britain (two sheets), 1932 and 1937	123
			Supplement 1. Editions published by the Land Utilisation Survey of Britain	124
			Supplement 2. Military issues	125
			Supplement 3. Royal Air Force issues	126
			Supplement 4. County indexes	127
		3.	1:625,000 Great Britain, 1942	128
			Supplement 1. Special sheets	139
			Supplement 2. Military issues	139
		4.	1:1,250,000 Great Britain, to "D" edition, 1947	139
			Supplement 1. Military issue	143
			Supplement 2. Editions not published by the Ordnance Survey	143
			Supplement 3. Skeleton maps published by the Ordnance Survey and others	144
		5.	1:1,900,800 Great Britain, 1954	144
		6.	1:2,500,000 Great Britain	144
V.	TEN-MILE MAPS SINCE 1955			145
	Text	1.	"Ten Mile" Map of Great Britain	145
		2.	Physical Map of Great Britain	148
		3.	Route Planning Map	148
		4.	1:1,250,000 Great Britain, "E" edition	151
	Standard attributes	1.	"Ten Mile" Map of Great Britain. 2. Physical Map of Great Britain	152
		3.	Route Planning Map from "B" edition	153
		4.	1:1,250,000 Great Britain, "E" edition	154
	Variable attributes		All four maps	155
	Cartobibliography	1.	"Ten Mile" Map of Great Britain, 1955	156
		2.	Physical Map of Great Britain, 1957	161
		3.	Route Planning Map from "B" edition, 1965	162
			Supplement 1. Military issues, published by D Survey, Ministry of Defence	171
			Supplement 2. Editions not published by the Ordnance Survey	171
			Supplement 3. Special sheets	172
		4.	1:1,250,000 Great Britain, "E" edition, 1975	173
	Appendices	1.	Projections of the Ordnance Survey Ten-mile Maps (contributed by Brian Adams)	176
		2.	Sir Henry James's Map of the World	184
		3.	County and other indexes	189
		4.	Unique numbers	192
		5.	Planning Maps: Explanatory Texts	194
		6.	Unpublished maps at 1:625,000	195
			Bibliography	197
			Documentary sources	200
			Index	201

ILLUSTRATIONS

1.	Characteristic symbols for nineteenth-century Ordnance Survey ten-mile maps (drawn by Richard Dean)	8
2.	Styles of lettering used on nineteenth-century Ordnance Survey ten-mile maps	14
3.	Extract from the earliest known Ordnance Survey ten-mile map, c.1817	16
4.	North Lincolnshire, c.1830	35
5.	The Kendal area on States M-P9, c.1858 and M-P10, c.1859	39
6.	**Ordnance Survey Characteristic Sheet for the Revised Ten-Mile Map of Great Britain**, 1900	75
7.	Intended sheet lines and numbers for the 1905 ten-mile map of Ireland	78
8.	Legends on Ordnance Survey ten-mile maps, 1903-1946	80
9.	Unused Ordnance Survey map covers: Jerrard's hand-lettered **Monastic Britain** and Martin's **Ten-mile Map**	94
10.	Martin's covers with the additional word **(Scotland)**	95
11.	Aviation symbols used on air editions of Ordnance Survey ten-mile maps, 1930 and 1934	97
12.	The northern half of the unpublished 1:2,500,000 map, 1942	113
13.	Index map (drawn by Brian Adams)	174
14.	Index map (drawn by Brian Adams)	175
15.	Extract from **Index Shewing the State of the Ordnance Survey of the Shetland and Orkney Islands**	190
16.	The legend from plate 15	191

Acknowledgement is gratefully made to Mr R.Dean (Plate 1); St John's College, Cambridge (Plate 3); Royal Geographical Society (Plate 4); Mr P.K.Clark (Plate 5a); Ordnance Survey (Plates 8,12); Mr R.A.Jerrard (Plate 9a); Dr T.R.Nicholson (Plate 9b); Mr B.W.Adams (Plates 13,14); Edinburgh University (Plates 15,16)

Most covers are not illustrated here, but are identified by the number sequence given in my list "The Covers" in J.P.Browne **Map Cover Art** (Southampton, Ordnance Survey, 1991, 122-144), where illustrations of them may be found. The number sequence was extended in **Supplement to Sheetlines** 31: 19-23.

ACKNOWLEDGEMENTS

Many friends and colleagues have assisted in the writing of this book, but I would like to give particular thanks to:

Brian Adams, whose careful analysis of the scanty evidence has thrown much new light on the vexed question of the construction of the Ordnance Survey ten-mile maps. Brian writes from long experience in the Hydrographic Department, many of whose charts were copper-based up until 1981; after 1981 service in the marine science branches alternated with practical compilation of charts, starting in the days of original drawings on tracing paper. His comparative indexes are worthy of the closest scrutiny. He has also found the time in his busy life to read this book from end to end. He has guided my untutored technical thinking, and saved me from many embarrassing lapses in language and fact. Any that remain, nonetheless, are my responsibility alone.

Professor John Andrews, who opened doors for me throughout Dublin, most generously allowed me unrestricted use of his transcripts of OSI documents, and kindly weeded errors from the Irish sections of the book.

David Archer, who was unfailingly prepared to allow what is after all his livelihood to become a research library, and who offered much sensible advice on the needs of the user of a catalogue such as this.

Dr Christopher Board, generous with time, expertise, and collections both private and professional.

John Paddy Browne, always a welcoming face beyond the portals of Maybush, for his willingness to make accessible to me the resources of his particular OS Department, not to mention the countless arrangements he made with the Record Map Library. Without him, many sources listed would simply have been missing.

Peter Clark, first Chairman of the Charles Close Society and until recently Keeper of Maps at the Royal Geographical Society, who actively interested himself in this project from the beginning, and offered much sound advice which led to the location of many new sources.

Richard Dean, whose detailed illustrations of nineteenth century characteristic symbols grace these pages.

Paul Ferguson, whose organisation at very short notice saved me from wasting many precious hours in Dublin.

Richard Haworth, for an unending stream of expert advice and personal favours during my visits to Dublin, and for so assiduously undertaking supplementary work in the libraries there. But for him, many Irish sources would be missing.

Yolande Hodson, wise counsellor who taught me how to evaluate the evidence in the earliest states of the ten-mile map, for generous hospitality, and for her words as Chairman of the Charles Close Society in the foreword to this book.

Leonard Hynes, for the time and trouble he took during my first visit to Dublin to make available to me surviving OSI ten-mile map materials.

Ian Mumford, who went to great lengths to satisfy my curiosity about the ten-mile map resources at MCE Tolworth, and for generously allowing me access to his notes on Sir Henry James and his Map of the World.

Dr Tim Nicholson, generous as always with access to a private collection of rare quality, and unselfish as always in sharing new research discoveries.

Ian O'Brien, for kindly making available to me the resources of his collection and expertise. It was his memory of events forty years ago that finally led to the discovery of what is probably the first OS ten-mile map.

Dr Richard Oliver, ready to offer advice and informed opinion at every juncture, and who unasked made over to me his transcripts of many valuable source documents.

Ordnance Survey, for permitting me access to much unpublished information, in particular the many files in the Public Record Office which remain closed under the thirty year rule.

Ordnance Survey of Ireland, for similar access to unpublished material in Dublin.

David Webb, for his photographic skills which illustrate these pages.

Dr Christie Willatts, O.B.E., who shared with me several rare source materials, who masterminded the making of a whole generation of ten-mile maps, and who never failed to respond to my many and often exasperating enquiries about that period of his life.

Many other members of the Charles Close Society, and others, who assisted in many ways, often by permitting me access to their private collections: in particular Michael Ashworth, John Beer, Andrew Bonar Law, John Coombes, Louis Curl, Edward Dahl, Richard Evans, the late Norman Gillard, Peter Hall, Maria Mealey, Guy Messenger, Forbes Robertson, John Symons and Phillip Tyler.

The staff at every library and archive listed, who have always been willing to assist with what must often have been irritating, not to mention laborious, requests, and if I single out Betty Fathers and her colleagues at the Bodleian Library in Oxford for particular thanks, it is only because they have seen a great deal more of me than anyone else.

ABBREVIATIONS AND COLUMN NUMBERS

This reference system for ten-mile map families is maintained throughout the book. It appears at the top of each page:-
I. Index to the one-inch maps: 1. South sheet, c.1817. 2. Middle sheet, c.1824. 3. North sheet, c.1881
II. 1. Ireland, 1838. 2. Ireland, 1868
III. 1. Great Britain, 1903. 2. Ireland, 1905
IV. 1. Three-sheet map, 1926. 2. Two-sheet map, 1932,1937. 3. 1:625,000, 1942. 4. 1:1,250,000, 1947, to "D"
V. 1. "Ten Mile" Map, 1955. 2. Physical Map, 1957. 3. RPM, from "B", 1965. 4. 1:1,250,000, "E", 1975
Appendix 1 to 6. (Appendix 2 concerns James's Map of the World. Appendix 3 concerns Indexes)

Titles as given here may be taken from different areas of the map: these elements are separated by /. Subdivisions of a title, if expedient, are shown by :. Overprinted title elements are marked ■: overprinting always ceases at /, and, if necessary elsewhere, with >. + titles are members of the Planning Series. Upper case and italic writing have been ignored except in specification tables and lists of changes where they may be relevant to specification change

Abbreviations

Map libraries. Capital letter abbreviations for places: **AB**: Aberystwyth, **B**: Birmingham, **BF**: Belfast, **BS**: Bristol, **C**: Cambridge, **D**: Dublin, **DR**: Durham, **E**: Edinburgh, **EX**: Exeter, **G**: Glasgow, **L**: London, **LD**: Leeds, **LE**: Leicester, **LV**: Liverpool, **M**: Manchester, **N**: Nottingham, **NT**: Newcastle upon Tyne, **O**: Oxford, **R**: Reading, **S**: Sheffield, **SO**: Southampton, **Y**: York; also **CPT**: copyright collection, **PC**: private collection, **RH**: the author's collection (reprints only listed)

Combined with lower case letters for collections, thus: **BFpro**: Public Record Office, **BFq**: The Queen's University; **Cjc**: St John's College, Cambridge; **Da**: Royal Irish Academy, **Dki**: King's Inns Library, **Dna**: National Archives, **Do**: Oireachtas, **Dos**: Ordnance Survey of Ireland, **Dtc**: Trinity College, **Dtf**: Freeman Library, Trinity College, **Ebgs**: British Geological Survey, **Lbl**: British Library, **Lgh**: Guildhall Library, **Lkg**: King's College, **Lmd**: Ministry of Defence Map Library, M.C.E., Tolworth, **Lnmm**: National Maritime Museum, **Lpro**: Public Record Office, **Lraf**: Royal Air Force Museum, Hendon, **Lrgs**: Royal Geographical Society, **Lsa**: Society of Antiquaries, **Lse**: London School of Economics, **Ob**: Bodleian Library, **SOos**: Ordnance Survey, **SOrc**: Royal Commission on Historical Monuments (England), **SOrm**: Record Map Library, Ordnance Survey. With **g**: university or college geography department, **n**: national (ie **ABn** of Wales, **Dn** of Ireland, **En** of Scotland, **C-On**: Ottawa, Canada), **p**: public, **u**: university

Official organisations: **BGS**: British Geological Survey, **DAS**: Department of Agriculture for Scotland, **DEP**: Department of Employment and Productivity, **DHS**: Department of Health for Scotland, **DOE**: Department of the Environment, **GS**: Geological Survey (Great Britain), **GSGS**: Geographical Section, General Staff, **GSGS (AM)**: Geographical Section, General Staff (Air Ministry), **GSI**: Geological Survey (Ireland), **IBWO, IDWO**: Intelligence Branch (later Department, then Division), War Office, **IGS**: Institute of Geological Sciences, **LUS**: Land Utilisation Survey of Britain, **MAF[F]**: Ministry of Agriculture [and] Fisheries [and Food], **MFP**: Ministry of Fuel and Power, **MHLG**: Ministry of Housing and Local Government, **MOS**: Ministry of Supply, **MOT**: Ministry of Transport, **MOW**: Ministry of Works, **MTCP**: Ministry of Town and Country Planning, **MWT**: Ministry of War Transport, **OS**: Ordnance Survey (Great Britain), **OSI**: Ordnance Survey (Ireland), **QDWO**: Quartermaster General's Department, War Office, **SDD**: Scottish Development Department, **TDWO**: Topographical and Statistical Department, War Office, **TSGS**: Topographical Section, General Staff, **WO**: War Office

Railway Companies: **B&D Jcn**: Birmingham & Derby Jcn, **CR**: Caledonian, **EC**: Eastern Counties, **EU**: Eastern Union, **GE**: Great Eastern, **GW,GWR**: Great Western, **I&B**: Ipswich & Bury, **L&B**: London & Birmingham, **L&C**: London & Croydon, **L&M**: Liverpool & Manchester, **L&NW**: London & North Western, **L&SW**: London & South Western, **L&Y**: Lancashire & Yorkshire, **LM&SR**: London, Midland & Scottish, **LT&S**: London, Tilbury & Southend, **M&C**: Maryport & Carlisle, **MD**: Metropolitan District, **MR**: Midland, **MS&L**: Manchester, Sheffield & Lincolnshire, **N&E**: Northern & Eastern, **NB**: North British, **SE**: South Eastern, **TV**: Taff Vale, **VN**: Vale of Neath, **WH&FR**: Welsh Highland & Festiniog, **WHR**: Welsh Highland, **WL**: West London, **WM&C**: Wilsontown, Morningside & Coltness, **WM&CQ**: Wrexham, Mold & Connah's Quay, **Y&N**: Yarmouth & Norwich; also **Ry**: Railway, **Ty**: Tramway, **Br**: Branch, **Ln**: Line, **Jcn,Jn,J**: Junction

Watermarks: **APSL**: A.P.& S.London, **APSLB**: A.P.& S.London Bodleian, **HC**: Hodgkinson & Co, **JW**: J.Whatman, **JWTM**: J.Whatman Turkey Mill, **MK**: Monckton Kent, **RT**: Ruse & Turners, **TEW**: T.Edmonds Wycombe, **TEWNB**: T.Edmonds Wycombe Not Bleached, **THSL**: Thos H.Saunders London, **TJH**: T.& J.Hollingsworth, **TJHK**: T.& J.H.Kent

Others: **anl,bnl,lnl,rnl**: above, below, left of, right of neat line; **tl,tc,tr,bl,bc,br**: top (or bottom) left, centre, right; **m**-onth, **q**-uarter, **y**-ear; **ARRR**: All rights of reproduction reserved, **CCR**: Crown Copyright Reserved; **gsm**: grams per square metre; **i**-nitialled **d**-ate; **NA**: Notice to Airmen. Bibliographical and documentary abbreviations are on p.197

Colour codes appear in columns 11 and 12. It is unfortunate that brown, black and blue all begin with "b": this has enforced some alternative abbreviations. The following are used throughout the cartobibliography: **b**-lue, **g**-rey (**g*** see IV.3), **j**-et (= black), **k**-rystal (or crystal) black (earlier broken black), **o**-range, **r**-ed, **s**-ienna (= brown), **w**-hite, **y**-ellow. These last two in effect mean the colour of the paper, and are noted when no sea spray colour is used.

With developments in printing colour combinations and screens, the number of overprint colours actually used is ever more difficult to assess until the general introduction of the trichromatic process in 1978, after which all colours have probably been derived from magenta, yellow and cyan, with black. It is now known as four-colour process printing.

Column numbers: as far as possible I have retained a common function for each column in lists throughout the book. No edition of the ten-mile map has entries in all columns. Sometimes they are irrelevant, and some would list unchanging features which are better noted as "Standard attributes". A few types of information change columns for convenience of organisation. The most important of these is the sheet price. Overprinted matter is marked ■. This includes most entries in columns 13 and 14: details in other columns occasionally supplied on overprint plates are here marked ■, and preceded by ■ in the lists. For further details, see p.82 (III), p.117 (IV), p.155 (V).

1. **Sheet number**, or **N**-orth, **S**-outh, **NS** back to back, **n**-orth of England strip map, **E**-ngland & **W**-ales sheet (some ■)
2. **State** (I): of **S**-outh, **M**-iddle, or **N**-orth sheet, **P**-artially completed state, **N**-ew **S**-eries
   Print code (III,IV,V). Reprint code (III), or OSI re-publication date (III.2). Print code (IV,V), in the form: base map:■overprint elements, ■overprinted code, or **p**-roof (perhaps with proof number or letter, or print code)
3. **Publication date.** Dates in I,II are estimated. III,IV,V: printed publication date of base map, or **n**-o **d**-ate. The place by default is S-outhampton to 1946 and from 1968, C-hessington 1948-1967, Dublin on Ireland maps. Exceptions are preceded **S** or **C**. For ■IV.2 dates included here, see col.13. V: **P**-rinted and published was used until 1958, **M**-ade and published from 1958. 1958 dates and exceptions are preceded **p** or **m**
   Legend codes. A double letter system is used. See p.82 for **a** (III) codes, p.117 for **b**, **c** (IV.1), **d** (IV.2), **q** (IV.2R), **e**, **f**, **X** (IV.3), **s**, **t** (IV.4) codes, p.155 for **g** (V.1), **p** (V.2), **h**, **j**, **k**, **r** (V.3), **u** (V.4) codes
   Sheet price, or **n**-o **p**-rice (IV,V). *NB*: I,III price is in col.8
4. **Railway revision:** see also col.5
5. **Reservoir revision.** *NB*: IV.3 information appears in col.4. Reservoir labels were deleted following a security directive in 1957: those in V.1 which precede or survive it are shown as + in col.8
6. **Minor correction date** (III,IV.1,V)
   Map revision (IV.3,4): see p.109 and p.140 for details
   Graticule intersections: + when present (IV.3), - when lacking (IV.3,V)
7. **E**-mbossed **P**-rinting **D**-ate (I: the earliest one known is quoted)
   Printed in Southampton date (III): derived from **f**-our **m**-ile, or **q**-uarter **i**-nch (given as ¼-inch) map
8. **Copyright statement** (I,III,IV.4,V). IV,V: **C**-rown **C**-opyright **R**-eserved, or **C**-rown **C**-opyright **19xx**. This is ignored for IV.1,2,3. *NB*: ■IV.3 appear in col.13. V: ■ dates are quoted if later than the base map CC date
   Sheet price, or **n**-o **p**-rice (I,III). III: + or - the relevant altitude statement. V: **u**-nivers lettering, or **d**-igitally produced sheets from data stored on computer; additional + or - are explained in cols.5,6. *NB*: IV,V price is in col.3
9. **Grid** information: **A**-lpha **N**-umerical squares, **g**-ra-**t**-icule, **I**-rish **G**-rid, (War Office) **C**-assini **G**-rid, **CGg**: CG+ graticule, **N**-ational **G**-rid, **NGg**: NG+graticule, **NGm**-ilitary system, **NGmg**: NGm+graticule, National **Y**-ard **G**-rid, **YGan**: YG with marginal AN system, (some ■). Grid information is implicit in col.3 codes, so is omitted from IV.3,4,V. On IV.3 maps printed in 1942 the National Grid was referred to as the **O**-rdnance **S**-urvey **G**-rid
10. **Magnetic variation date** (some ■). V.1 Sheet 1: + corrected figures in "Difference from Grid North" panel
11. **Colour of base map.** Second letters (if used): on **f**-ilm, **h**-eavy (usually 152gsm) paper, **t**-racing paper, or the colour of grid if different from the base map: an upper case letter implies a grid in 100km squares. III,IV: additional symbols (+,-,x) concern the water plate if uncharacteristic: see p.119, and p.viii on colour codes
12. **Depiction of height: H**-ill **S**-haded, **O**-utline with **H**-ills, **O**-utline, **R**-elief map, or number of coloured layers, the <u>number</u> underlined if there are uncoloured layers below
    Colour of sea. This may be uncoloured paper. See note on colour codes on p.viii
    ■**Number of overprint colours.** IV: + or - contours, if uncharacteristic: see p.119 for details
■13. **Overprint publication date. Price.** See col.3 for details and abbreviations, which may also be relevant here. IV.3 with **c** are ■copyright dates. Maps not for publication may be marked **f**-or **i**-nternal (or **o**-fficial) **u**-se **o**-nly. *NB*: ■III.1 and ■IV.2 details are given in col.3
■14. **Date of overprinted information**
15. **Print run. Date.** The information is taken as far as possible from the job files, and by default gives the number of good copies printed, and the date of printing. Less positive evidence comes from **E**-stimate, **R**-equest, History **C**-ard (issue date), **D**-ispatched information in job files, and PRO files, usually OS 1/432, 1/433, 1/999. An approximation is of course often implicit in the print code
16. **Cover type** by reference to my lists in Browne (1991) and Sheetlines Supplement 31, **x** if in unlisted cover
17. **O**-rdnance (or **G**-eological) **S**-urvey **P**-ublication **R**-eport (yyyy/q [quarterly], m/yy [monthly] reports). Since 1988 OSPR publication has postdated that of the maps: map publication dates are given. GSPR reports suffixed **g**
18. **Location of copies** (see list of abbreviations: "on sale" as in 1992)
19. **Notes**

INTRODUCTION

This book appears just a year too late to join those celebrating the Ordnance Survey's (OS)[1] bicentenary. The delay has in part been caused by attempts to trace as many current sources as possible, in order to bring the list of maps right up to date. This was, of course, always a forlorn hope, with the constant revision and replacement of current editions of the ten-mile[2] maps that remain an important feature of the OS Catalogue (OSC), and the other official maps for which it provides a base. So, if in no other way, this is likely to be a "first" among Charles Close Society (CCS) publications in that it will be out of date as soon as it appears. New maps indeed seem imminent. The scale is convenient for national summary maps, and the 1991 Heseltine announcement of yet more Local Government and boundary reorganisation is likely to lead to a whole crop of new administrative maps, a type that has been remarkably quiet at the 1:625,000 scale for a decade or more. The **Routeplanner** is now digitally produced, and though the current popularity of "floppy" road atlases seems to have put paid to the annual editions it enjoyed until 1988, it is still very popular, and remains of significant importance both to the Ministry of Transport (MOT) and the Ministry of Defence (MOD). Perhaps there is never a right time to publish a book about a current map, but publication in 1992, if not celebrating anything as grand as a bicentenary, at least marks the more mundane yet more relevant fiftieth anniversary of the change of scale from 1:633,600 to 1:625,000.

If one overlooks this almost imperceptible change from imperial to metric measurement, the OS ten-mile map has been in existence longer than any other, with the sole and honourable exception of the one-inch itself. This longevity has naturally witnessed wholesale changes in specification, and, in sum, we are concerned in this book with at least four different generations of map in Great Britain and three in Ireland. These are as interesting for their differences as for their similarities. In the cartobibliographies that follow, I have attempted to draw attention to both aspects in the development of maps at the ten-mile scale. Attributes that are standard to one family of maps may vary with another, so each family demands the listing of different features. Since it is obviously pointless to list details which never changed, each list of maps is preceded by a section describing its standard attributes, before describing and listing the variable ones. In order to achieve this with some degree of overall consistency, I have adopted a total of nineteen columns for the listing of variable attributes. There is no list that includes all nineteen since they are never all relevant, but, by maintaining as far as possible a fixed purpose for each column number throughout the book, I hope the common numbering will make their function more memorable. Although what follows undoubtedly (and unashamedly) owes much to the excellent models created by Guy Messenger and Richard Oliver, these probably had influence more on my methods as compiler than anything users will find on the page.

One detail that is almost entirely ignored, which in previous cartobibliographies has been of cardinal importance, is the revision date of a map. To the ten-mile, usually a by-product of one-inch, then quarter-inch rev-

---

1. This book deals with OS maps produced in Great Britain and Ireland. In order to avoid confusion, the abbreviation OS will always refer to London or Southampton, and OSI to Dublin, both before and after independence in 1922.
2. 'Ten-mile' is not an accurate definition of the scale, and it would be more precise, if cumbersome, to use the phrase 'ten miles to one inch'. But I will follow OS practice and use the short form, without quotation marks, even in relation to the metric 1:625,000 equivalent. Expressions such as 'imperial ten-mile' and 'metric ten-mile' adequately distinguish, where necessary, the 1:633,600 from the 1:625,000. I will reserve quotation marks for the specific case of the 1955 generation of maps, which was entitled **Ordnance Survey "Ten Mile" Map of Great Britain**.

Introduction

ision, this is effectively irrelevant, so much so that only two states of one map even bother to mention it. Even minor correction dates are very much the exception, not the rule. This is not to say that ten-mile maps were left unrevised: they were not, but to locate change in the topographical information, it is usually necessary to undertake a detailed inspection of the content. I have usually only noted such changes in this book where it was necessary in order to identify different map states. For the nineteenth century map a comprehensive analysis of topographical, and especially railway, revision, none of which was itemised even by so much as a print code, proved vital. But such analysis is usually unnecessary with the twentieth century map, with all the paraphernalia of print codes, minor correction and railway revision dates, price changes, *et al*. I undertook no analysis of that kind at all with the Johnston maps, where any minor corrections and railway revisions were theoretically noted by date code. Since railway revision on the next generation of maps beginning in 1926 seemed to evolve into two specific railway states from 1942 onwards, it became my ambition to identify the various stages that led to this, but I felt it unnecessary to provide a complete list of changes at each stage. Reservoir revision, on the other hand, was a compact enough subject to deal with in a comprehensive fashion, and I have attempted to note all changes leading to the enforced elimination of all references to them following a security directive in 1957. From 1955 onwards, I confess to overlooking topographical change almost entirely, since modern day print codes, in theory at least, control all such change. Thus topographical development of the twentieth century ten-mile map remains very much a field open to systematic research. Much more important in this book than the content and date of any revision to the base map is the detail of any overprint.

Overprinting has been the lifeblood of maps at the ten-mile scale. All editions have supported overlays or overprints, and in more than one case it has been more or less its sole purpose. Since the overprint is usually of more importance than the base map, some anachronistic pairings have been the result: hence the use of the 1903 map on First World War Official History maps dated 1935, or even today the 1990 **Administrative Areas** map which continues to use the 1974-75 base and ignores the more recent 1980 outline map. Considerable effort has been taken to locate as many overprinted ten-mile maps as possible, in books, in Royal Commission reports, etc, although, in my pessimistic moments, I am convinced that there must be hundreds more examples waiting to be found in the more abstruse state papers and elsewhere. But to scan all possibilities would require endless resources in time, and free access to shelves containing them, which is unlikely. Catalogues are rarely of use: usually they list the books and not the maps - even the tempting "(with map)" tells nothing - not even scale, never mind publisher. It is a rare catalogue indeed that specifically lists maps in books, and rarer still one that describes them. I can merely hope that those listed here will encourage researchers to report any others they find for inclusion in later supplements.

The sheer quantity of overprinted ten-mile maps leads quickly to the greyer areas of OS responsibility. Nobody would dispute the inclusion in these pages of maps "Published OS", but there is immediate disagreement over the "Printed OS", "Prepared OS", editions published by others on the OS base (eg GSGS), maps which failed to reach publication, or which were reproduced by dyeline and other such methods etc. In the end, there is no clear cut-off point, and any compiler of a list such as this will have to find his own, and the rationale for his choice. It would have been inconceivable to me to omit the pre-war **Land Utilisation** and **Types of Farming** if only because without them the origin of the post-war OS issues would have been inexplicable. Yet their only connection to the OS is the base map: they were prepared and published by the Land Utilisation Survey and printed by Bacon. I have also decided to include some early Irish maps which are very important sources, even though, beyond the original drawing and no doubt the results of a good deal of research at the disposal of the map makers, they owe nothing to the OSI at all.

Some would have War Office (WO) maps omitted, but would this mean also the omission of civil aviation maps with War Office print codes? Many civil aviation maps form part of common print runs with RAF and even military versions: it seems a nonsense to me to list the one and omit the others, just because of the technicality that their publisher was not the OS - worse, in fact, because their omission would have distorted the overall picture. It was a late decision to include here James's ten-mile Map of the World, but I felt justified since the specification of the Ireland map of 1868 derives directly from the North American sheets. It seems to me valid that provided one does not attempt to offer up everything as an OS map, but describes its connections to the OS for what they are, and ensures that the rightful publisher is so acknowledged, a place can be found in a listing such as this for such supplementary material. But I have decided to omit OS published maps that were then dyelined, or put through some other similar photographic process, and then overprinted with new information. Such methods have been adopted for all manner of official purposes, such as to show hydrometric areas, forestry commission land, and mineral resources. Though the Maps Office of the Ministry of Town and Country Planning (MTCP) was responsible from 1942 for the production of more than thirty different printed pairs of ten-mile maps (not all of them published), they produced far more than this by less permanent processes, usually for internal purposes only. Some of these, especially those which in other circumstances would have proceeded to printed map status, appear in an appendix.

It has also been a matter of some debate as to what printings should appear in cartobibliographies. Should only those reprints which show new information, identifiable by print code or revised dates of some kind, be included, or also facsimile reprints with no readily identifiable features at all? One can supplement such a list with proofs and pulls for internal OS use. To some an ideal catalogue would be one that lists all recorded states of each map, including facsimile reprints, proof states, special pulls for experimental purposes, printings to satisfy government and military requirements, or for OS internal use, such as of the Archaeology Department. But this is ultimately only possible when such states have been recorded, and copies located, identified and analysed. To do otherwise would be imbalancing and probably misinformative, since it is dangerous to catalogue the details of a source one has not seen. It is axiomatic that map states not intended for publication or the wide dissemination of military maps in time of war are rare, and likely to be idiosyncratic. They will probably be unlisted, only be located by chance, and in most cases no longer survive at all. Many published maps are rare enough, and copies of some of these I have yet to locate. My criteria for what to list have not remained immutable throughout this book. I have followed traditional methods on engraved maps and sought to record all those ways in which alterations were made to the plates, then to order the sequence of those changes in a series of known states. A few suspected states are also provisionally included. Other states will emerge of which I know nothing. It may be interesting to know that ten impressions were taken of the Index on 4 September 1820, and on 5 January, 21 June and 29 November 1821, but in the end it teaches us more about production methods than the map itself. The evidence is that they would all have been the same state of the map, and I have chosen not to list such printings separately. It would anyway at this stage be impossible to relate a surviving copy to any specific date.

But for more modern printed maps, I have decided to include all printings for publication, including facsimile reprints, of which I have knowledge, whether I have seen them or not. Where job files survive, it is possible to do this with some degree of confidence. Facsimile reprints are usually identifiable by nothing more than the difference in the overlay of one colour plate upon another, but if that is all there is, at least it is possible to tell them apart (if you can get them together to make the comparison) even if it is virtually impossible to be precise about the sequence of such reprints, copyright copies excepted. Such differences have not been described here, and it may be assumed, if other identification points have not been noted, that this is the only clue noted to their existence. Some so-called "facsimile" reprints prove not to be so (in itself a good reason for recording them), and in spite of common print codes in fact may have different topographical detail. Again only detailed comparison will reveal these alterations. But without the evidence of job files or other documents (surviving in any quantity only since the war), one is fortunate indeed even to be aware of some facsimile reprints, so there must be many others which still have to be recorded.

Beyond published maps, my general principle has been to list only printed maps on the OS ten-mile base that I have located, whether they be repayment service maps or unpublished printings for Government departments. A few such are listed even when unlocated if their content seems to be of particular interest, or when they are relevant to the published history of a sheet - perhaps by filling in otherwise missing print codes. But most of the special printings of base maps for MTCP have been ignored, or merely noted: it would be fortunate indeed that any copy located could be related to a specific printing, since they occurred so often. Naturally, all maps went through a proof stage, and many of these were preceded by models or one kind or another. Models have usually been ignored, and proofs only recorded where they are known to survive. A perusal of the printings of the 1942 **Physical Map** - admittedly an unusual case - in Hellyer (1992a) will demonstrate how far a cataloguer could pursue the principle of listing everything, including all proof states, and even that list has ignored some known pulls for internal OS purposes. Entries for that map have been reduced more or less to normal here, and with one exception only sources actually located have been listed.

I conclude this introduction with an observation to those who would like to read this account truly from the beginning. It is forgotten all too often by laymen that a map originates long before the drawing of its topography, in the mathematics of its projection. Therefore bypass the next five chapters and turn to Appendix 1, which is Brian Adam's account of the projections and sheet co-ordinates that brought the ten-mile maps into being. The non-mathematical need not be dissuaded by his lists of co-ordinates: these can be circumvented in pursuit of his essay which clarifies many of the problems faced by the practical cartographer.

CHAPTER I

TEN-MILE MAPS OF GREAT BRITAIN IN THE NINETEENTH CENTURY

1. Old Series Index

1. South sheet

There already exist Indexes engraved for both England and Scotland on this scale; but they are merely Index diagrams to shew the area of the 1-Inch maps of the two countries, that is, they have never been published as maps; while the details on them represent in England surveys from 50 to 100, and in Scotland from 20 to 50 years old. They are therefore, if judged as maps, entirely obsolete.[1]

So wrote J.H.Elliot, Secretary to the Board of Agriculture, on 29 December 1898, in evidence supporting the OS application to the Treasury to make a new ten-mile map. Such things had been heard before. The Index to the Old Series one-inch map was some seventy-five years old when the Dorington Committee met in 1892, yet it was still its fate in evidence to be condemned as "*only an index map, that is all it is*".[2] Director General Sir Charles Wilson ignored its existence altogether in his spoken evidence. It remains a common opinion even within the OS that an index is not a map, and perhaps that is true of those which are at best diagrams. But it is hard to see this as an objective view of the Old Series ten-mile Index, which was an admirable small-scale topographical map carrying, among other things, sandbanks, mudbanks, woodland, heathland, marshland, parkland, Anglican churches, rivers, canals, roads, county boundaries and hundreds of place and geographical names. Railways were subject to the most constant revision and became much the most prominent feature. Sometimes new symbols were introduced, such as for mineral tramways and underground railways. But for all that it is easy to see why the OS ignored it as a topographical map. An index to the one-inch map it was, first and foremost, and it grew largely in parallel with its parent map. Thus the south sheet was not published in complete form until c.1841, shortly before the last of its constituent sheets appeared in print. It was 1881 before the middle sheet was completed,[3] and for the north sheet 1884. By such dates, topographical detail printed early could, as with the parent one-inch sheets, be absurdly out of date.

Had no other use for it been found other than as an index, perhaps Henry Tipping Crook's appraisal of the map before the Dorington Committee quoted above would not have been unfair. But long before this time, it had proved its value in a variety of other ways. The War Office had found it an immensely useful base map, either in outline or even in complete form, for plotting military information on a national basis which even by the early 1890s had run into a score or more different editions. As an 1892 War Office Committee reported:[4]

It is very useful as the basis of special maps to show military and regimental districts, headquarters of volunteers, Post Office districts, main telegraph lines, etc...The Committee think that this map should be kept revised by the Ordnance Survey Department in regard to railways and first class roads.

---

1. PRO T1/9335c.
2. *Report of the Departmental Committee...* q.1446.
3. In preferring the term "middle" to "central", I am following OSC 1890's practice. In fact, until the northern Scotland sheet came to be made, the North England sheet was known as the north sheet. But in order to avoid any confusion I will always refer to it as the middle sheet. The sheets themselves never carried such labels, though, on completion of the north sheet, they were numbered 1 to 3 from north to south.
4. War Office (1892).

The OS also had used it as a base map, in 1858 for marking the course of an eclipse of the sun, and in 1861 for a plan of the catchment basins of the rivers of England and Wales, though, admittedly, these were subjects which rendered outdated topographical detail unimportant.

No record has been traced of its origins, and there can be no certainty that the earliest known state of the map was indeed the first. In some form it may even have served a military function during the Napoleonic War, or perhaps derived in part from a military source. Evidence to support this lies in the remarkable number of military sites that appear on the first known state, irrelevant in an index, which were never supplemented later. These include martello towers, fortifications, barracks and what were probably temporary encampments. Dartmoor Prison, well known for its origins as home to French prisoners of war, is also marked. But as an index, the date usually quoted for the first appearance of the map is c.1817. There is a variety of evidence for this, all of it circumstantial. Between 1811 and 1816 the sale of the one-inch map to the public had been banned on security grounds.[5] Once the ban was lifted it was presumably considered desirable to do something about publicising the maps, both by issuing leaflets and by engraving indexes. Leaflets were printed to advertise the availability of the one-inch map, sold at the time in "Parts" of the General Survey of Great Britain. The earliest known is one listing the first five parts, and is dated 25 October 1816.[6] A later one, advertising Part VI as well (publication completed by 14 August 1817), has in addition information about an index:

> Mr. Baker will shew an Index Map of the Survey, and Gentlemen wishing to procure Maps of the County adjacent to their own Estates, may select any sheets they please for that purpose.[7]

This leaflet is undated, but it would indeed be perverse if it were issued later than mid-1817, since Part VII, though not mentioned on it, carries the same publication date as Part VI - 14 August 1817. An entry in the Board of Ordnance minutes of 27 June 1817 may provide further evidence.[8] This refers to a contemplated enterprise of John Cary's for using the Old Series as the basis for a smaller-scale map of Cornwall. William Mudge, then in charge of the Survey, suggested that the Ordnance make its own "reduced map", scale unspecified.

But even from the earliest times indexes of the Old Series one-inch map seem to have been printed in three versions: a small index showing only one of the one-inch "Parts", to be applied to covers of these maps when sold as sets, a small index of England and Wales on one sheet at scale 1:4,000,000, and the full-sheet version on the ten-mile scale, reflecting the publication progress of the one-inch map, which ultimately covered Great Britain in three sheets. We know from documentary evidence that fourteen impressions of an index were engraved at the Tower in June 1820,[9] and subsequent entries refer to the printing of proofs in August 1820 followed by further impressions of ten in September 1820, January, June and November 1821. These are quantities similar to those of one-inch sheets engraved at the same time. Such small numbers suggest the full sheet index, and are in marked contrast to the much higher print runs of 200 or 300 quoted for April, July and December 1820. To judge by the payments made to Cox Son & Barnet who did the work (6/- for 200 indexes, as opposed to £4 for 100 full one-inch sheets), these can only refer to one of the smaller indexes. And, while it may appear frivolous, it may not be irrelevant to point out that of the three types, only the ten-mile Index actually shows "*Gentlemen's Estates*". If for this reason alone, therefore, the likelihood is that the leaflet quoted above was referring to the Index at this scale. From 1820 onwards, a great deal more care was taken in the representation of estates, as they were named and further adorned by delicately engraved paling boundaries, even though these are hard to detect with the naked eye.

The internal evidence of the earliest known state of the ten-mile Index adds weight to an 1817 date. It shows in almost "complete" form all those sheets published in Parts I to VIII of the one-inch map, with some of the unpublished sheet 7, where turnpikes, rivers, country estates and clusters of buildings for towns are shown in skeletal form, with no names. Although the sheets comprising Part VIII would not be published for more than another year, they were in preparation in 1817 and their publication imminent, and of more compelling importance is the absence from the Index of the first sheet in Part IX (sheet 38), which was in the event ready for publication before them, early in 1818. Subsequent issues of the Index confirm a policy that not only sheets in print should be shown complete, but also others whose issue was imminent. One must assume that those shown in a partial state were at the time not completely surveyed. The policy is, however, difficult to justify with the 1820 Index (State S-P2), because among the complete sheets it includes are 36 and 37 (published 1833 and 1830 respectively), yet excluded are seven other sheets which were published 1824-1830. The answer probably lies in the expectation that publication of sheets 36 and 37 was indeed imminent in 1820: surveying the area had been undertaken c.1811, and proofs of both sheets were printed in July 1820. In the event further work on these sheets was delayed until additional field revision had been undertaken, by which time, of course,

---

5. Margary (1977) xxviii-xxix.
6. Illustrated in Margary (1977) xxxii.
7. Illustrated in Sheetlines 12 [1], and Owen and Pilbeam (1992) 21.
8. PRO WO 47/661.
9. Ordnance Survey Manuscript Letter Book, 1817-1822 (PRO OS 3/260).

the Lincolnshire sheets and others had jumped the publication queue. But this does not explain the even more curious feature about sheet 36, that it is more completely filled than any other sheet on the Index, in that its river and canal names are included, whereas elsewhere they are at this early stage almost all omitted.

Beyond the mapped one-inch sheets, the Index had a neat line, ticks outside this showing lines of latitude and longitude, a single word title **Index**, but otherwise it is blank. As originally drawn, the scored northern section of the neat line lay 551mm above the south, situated just north of where the Norfolk coast would lie, though not coincident with any one-inch sheet line. This line was certainly deleted by c.1830 on State S-P6, on which appear one-inch sheets 64,65,69 and 70. Traces remain where it would have been breached as its path passes through sheets 69 and 70. This would presumably have occurred in 1824 if a State S-P5 was made to accompany the arrival of the Lincolnshire sheets.

Even though the neat line had yet to be redrawn, this new northern limit of the Index south sheet may at the time have been viewed as final. The simultaneous publication of all eight Lincolnshire sheets is unlikely to have been greeted by their appearance on the index at different times, and a start was made on an index middle sheet when the four northern Lincolnshire sheets were initially engraved in a separate copper plate, certainly c.1830, and perhaps also in an earlier c.1824 version. Had as a consequence the row of one-inch sheets 77-84 formed a permanent part of such a sheet, it is interesting to speculate on the final format the index would have taken. The middle sheet could have covered England and Wales with more than 100mm spare above Berwick. A third sheet above that could encompass mainland Scotland, though not the northern islands, and its western border would cut through the Western Isles.

For its size, the Index south sheet was engraved in a copper plate of generous proportions, 960mm W-E by 662mm S-N. This should have provided ample margins all round, and indeed outside the neat line this is true. It is curious therefore that the land mass was squashed against the neat line in the south-west corner, leaving no room even for the full dimensions of either sheet 32 or 33 (in its original landscape form), let alone the Scilly Isles beyond them. Certainly there was enough space remaining on the plate, late in the 1830s, to dispense with the original neat line, and extend mapping beyond it. This elimination of border seems first to have affected State S-P8. Only one (defective) copy of this is known, trimmed even within the original neat line, but from later states it is quite clear that the southernmost of the fathom lines added to this state was for some reason extended to the edge of the plate. It may be seen breaking through the piano key border which was applied later. Such a curiosity would hardly have occurred with a border in place.

Deletion of the neat line presumably coincided with decisions on remapping in the south-west at the one-inch scale, and therefore the Index, showing new roads, new physical names, a completely redrawn coastline, and in particular a portrait version of sheet 33, published in 1839. More plate space was thus absorbed in the south, though still not enough to provide for a complete sheet 32, nor the Scilly Isles, and with enough room remaining on the copper in the north, the decision was taken to add an extra row of one-inch sheets (77-84). The policy behind this may have been that the Orkney and Shetland Islands could thus be positioned geographically within three sheets, but probably of more immediate importance was the obvious benefit that northern Wales would now fit on the south sheet, though on the first known state with these dimensions (S-P8), one-inch sheets in this area were still incomplete.

The very filling of sheets caused some revision on the Index as names at the one-inch sheet limits often had to be moved to more convenient places, to make space for others on adjacent sheets. Any more radical revision seems usually to have been the by-product of alterations made to the parent one-inch sheets, as in the south-west, Lincolnshire, or even north Essex. But once mapping on the Index was completed, little further topographical revision, except to railway information, took place. The date of completion is again difficult to establish conclusively, but the indications are that it was 1840, or soon after. One telling piece of evidence is that until this time the Index seems never been on sale to the public, though they may well have been presented to those clients who became subscribers. Hodson (1989, 91) points out that, because they did not appear in the accounts of the Survey's agent William Faden, they could hardly have been on sale until 1822, and one can with some confidence extend this a further eighteen years, as the following Board of Ordnance document dated 24 June 1840 reveals:

Index Sheet of the Map of England & Wales to be sold for 4/- each.

The Board having on the 19th instant submitted to the Master General a letter from Colonel Colby stating that the Southern Portion of the Ordnance Map of England & Wales as far as Sheet No 84 being nearly completed, an Index Map to serve as a guide to enable Purchasers of the Maps readily to refer to the Sheets they require, had been in process of engraving, & would be added to as the remainder of the Map is published, - and submitting that the Sheets of the Index Map might be sold by the Agents, in the same manner as the Maps themselves, and as the Index Sheets are only in Outline, that they might be sold at 4s./ p. Sheet: And the Master General having on the 20th instant approved thereof, Ordered accordingly.[10]

Certainly State S-1 is the earliest known to display a

---

10. PRO WO 47/1865 7829-30.

price, and the introduction of lists of plate prices and agents names can hardly be irrelevant. The railways shown are consistent with a c.1841 date. It was probably also the first with a piano key border. Once complete, the map was awarded a more definitive title, **Index to the Ordnance Survey, of England and Wales.** Thereafter the evidence of surviving examples demonstrates regular new printings during the 1840s and 1860s, with (unless one or more states remain undiscovered) a curious gap in the sequence in the early 1850s. Surviving copies become noticeably more common following the price reduction to 2/-, probably effective from 1848. Revision principally affected first the list of agents, then railways, which rather bulldozed their way across the map, being simply engraved on top of any detail in their way. Otherwise the Index was not much revised until the making of the first Anglo-Scottish state of the middle sheet c.1873, and this is discussed further below.

2. Middle sheet

Sweated on to a copy of State S-P6 in the Royal Geographical Society (RGS), referred to above, is a small additional map carrying in index format the four northerly Lincolnshire one-inch sheets, 83-86. It has been trimmed, probably all round, and certainly on the east side where some 13mm of sheets 84 and 85 are missing. One of its more curious features is the mapping of its component sheets, which is more complete than the other half of Lincolnshire printed on the Index south sheet - indeed it is remarkably complete in that river names, still lacking on the south sheet generally, are included. Whether an earlier more thinly mapped version, consistent with the southern Lincolnshire sheets on State S-P6, preceded it is speculative, but probable. It is impossible now to ascertain whether the copper used was a fragment which merely fulfilled a local need, or was a full plate intended for the complete Index middle sheet. But whatever the case, the northern Lincolnshire sheets were certainly engraved in it in isolation. Outside the west and north borders is 5mm of paper with no printing, and probably there was no printed matter beyond this. Furthermore the road casings are cut by the outer sheet lines, which proves the integrity of the map.

Sometime in the history of the Index, the Lincolnshire sheets were all re-engraved. Many topographical features and placenames were added, while others were deleted, altered or moved. Oddly, all water feature names, except "Mouth of the Humber", were deleted, so conforming with the practice still standard on the Index south sheet. This probably occurred c.1839 when the one-inch row 77-84 was transferred to the south sheet, leaving sheets 85 and 86 alone to constitute the first state of the revised middle sheet (State M-P3). It is impossible now to tell whether the original northern Lincolnshire plate was reused, or a new one begun. We see here sheets 85 and 86 in complete form, empty sheet lines for 87-90, the Delamere Meridian extending from top to bottom of the sheet, and the Preston to Hull line

extending side to side, upon which the grid of northern England one-inch sheets (91-110) would be built. The only known copy has been cropped, but it is quite likely that its original dimensions were the same as those of the few middle sheets to survive intact, which have plate measurements of 977mm W-E by 674mm S-N (from paper impressions). This width is greater than the south sheet, and in itself provides one good reason why so many copies were trimmed, in order to marry them with south sheets. Furthermore, it extends westwards 200mm beyond the Isle of Man sheet lines, where it could have no relevance to any English sheet. Therefore, given that this was not entirely wasteful, it can only mean that from the start the OS were intending this to be a combined England-Scotland sheet which would encompass Scottish sheets out to the sheet 42 column.

There is thus evidence that the Index middle sheet had a planned structure from the start in a way that the south sheet did not. New also seems to have been the engraving of sheet lines ahead of infill, which does not occur on any known state of the south sheet. State M-P3 shows the unquartered sheet lines of the original Old Series up to sheet 90. The next known state (M-P6), carries the complete framework of English sheets above the Preston to Hull line, including quarter sheets. It also carries a sparse outline of the northern part of England, depicting coastline, county and national boundaries, a few rivers and Ullswater (unnamed), and some placenames, principally market towns. Detail was added to this framework sheet by sheet as on the Index south sheet, though apparently even further ahead of publication, and probably at the same time as it was reduced from the six-inch for use on the one-inch map. This continued until sheets 91-96 and 97SE had been filled, probably by 1854. Some of these were not actually to be published until 1861.

But then this policy altered radically. Further topographical detail was added, both fully and in outline, without respect to sheet lines, but rather filling to county boundaries and in other less obvious topographical units. This seems to have been caused by the introduction in 1853 of surveying at the 1:2500 scale, and survey and publication by parishes. It would appear that smaller scale mapping even down to the ten-mile scale was now achieved by photo reduction (via the six-inch and the one-inch), and finished work seems to have been added to the Index wherever it occurred. The main part of Lancashire was quickly filled, County Durham was tackled from north and south simultaneously, and, with survey parties working outwards from Kelso, the area around Berwick was covered long before the remainder of the northern English counties. On at least one state, the approximate route of substantial lengths of road or railway were added along arbitrary corridors through relatively uncharted territory. This growth continued until completed c.1866, by which time all the component English one-inch sheets of the Index middle sheet (except sheet 100) were issued or in preparation.

From this date, unless there remain undiscovered states, the rate of revision to both south and middle sheets slowed. It is possible that this is in some way related to a comment in Winterbotham (1936, 79f): "*In 1867...the index was officially promoted to a place in the estimates and to the status of a map.*" This might even imply the authorisation of the OS topographical map at the ten-mile scale to which Wilson referred in written evidence to the Dorington Committee in 1892.[11] The date of authorisation was never established, and without documentary support further comment would again be speculative. But in purely practical terms a place in the estimates may well have meant, with mapping of England and Wales complete, at least as an index, that all available energies could now be turned to the ten-mile mapping of Scotland. There was if nothing else a military imperative for this, and it seems that as early as 1870 it had proved possible to provide an outline map as far north as Fife, and by 1875 of the entire country which temporarily satisfied military requirements, even if the completion of topographical detail was to take nearly a decade more.

The whole question of the origin of indexes to the one-inch map of Scotland is shrouded in mystery. If there were earlier Anglo-Scottish states of the ten-mile Index, they have not yet been located, and it was probably not until c.1873 that Scottish mapping was added to the English Index. By that time the one-inch map of Scotland had been in progress for seventeen years, certainly time enough to permit a series of ten-mile index sheets depicting this growth, but none have yet emerged. Diagrams at the thirty-mile scale date back at least as far as OSR 1857, but, other than these, the earliest known indexes of the Scotland one-inch map are in fact progress indexes which detail completion of the elements of work involved during the 1860s in the construction not only of the one-inch, but the six- and twentyfive-inch as well. A mixture of scales was used, including the ten-mile and even the seven-mile. But in any event, since these were almost certainly not available to the public, their existence is to this enquiry a red herring.[12]

The appearance of Scotland on the ten-mile Index began c.1869 on State M-P14, though this was more in principle than practice, since all that was shown were coastlines, to the west, and north as far as Fife, and main-line railways which reached half-way to Glasgow and all the way to Edinburgh. This feeble gesture towards mapping north of the border, nonetheless, persuaded the OS to delete the old title from the south sheet, and rather grandiloquently to engrave a new one in the middle sheet: **Index to the Ordnance Survey, of England and Wales, and Part of Scotland.** The two sheets were in the future evidently to be considered a pair, albeit a mismatched pair, with the width of the middle sheet, already greater than the south, now extended to 1072mm across the plate in order to accommodate the Western Isles of Scotland. A small addition also seems to have been made to the top of the plate. It cannot be confirmed whether these extensions affected State M-P14 (the one known copy has been trimmed), but State M-P15 was certainly so altered. Scotland *de facto* made its first appearance c.1873 on State M-P15 of the Index, uncomfortably united at the border where one-inch maps on Cassini's and Bonne's Projections meet. Mapping was temporarily confined to the mainland. The Isle of Man was added, mapped in Scottish style. Beyond the mainland is a full complement of unnumbered and empty sheet lines, which were left virtually untouched through several subsequent issues until final completion of the Index middle sheet c.1881 when the Western Isles were added. As to the projection lines, the Preston to Hull line had been deleted on State M-P15. But the Delamere Meridian survived longer. Indeed, it may have been mistaken for a sheet line, as it was extended to the lower border when these lines were re-entered on State M-P13. It was finally removed c.1881.

3. North sheet

Only one partial state of the north sheet seems to have been made. There is a reference to it in OSC 1881, that the sheet extended to the parallel of Duncansby Head. We also know that water names were being engraved on 4 October 1883.[13] With its completion in 1884, three years ahead of the one-inch maps appearing on it, the phrase **and Part of Scotland** was altered to **and Scotland** in the map title on the middle sheet. Since Scotland sheet lines were unaffected by changes further south, it was relatively unchanged with the almost immediate appearance of the New Series Index, and its further development is discussed in this essay at that point.

The demise of the Old Series Index

Thus in 1884, for the first time, there was a complete three-sheet Index to the one-inch maps of Great Britain. But, with the 1882 change from Old to New Series sheet numbering north of the Preston to Hull line, the Index was effectively obsolescent even before it was completed. Since c.1873, no further railway information had been added to either middle or south sheet.[14] Some topographical detail was occasionally revised, with, for instance, the addition of the outline of the French and Irish coasts, but the major revision of both railway and other topographical information that appears on issues

---

11. *Report of the Departmental Committee...* Q.1444.
12. See p.188. So too are the coastal outlines of Scotland on the Ireland maps of 1838 and 1868 (II.1,2).
13. PRO OS 1/144 70B.
14. Indeed the south sheet took a backward step in this regard, presumably when an electrotype was discarded and a return made to an earlier version of the matrix.

of the New Series Index never seems to have been applied to it. The cautionary tone of this remark is intentional, because the evidence is both contradictory and, with so few latter day source maps found, thin. On the one hand, OSC 1890 lists indexes for both Old and New Series with uniform sheet size 27ins by 40ins, which remained in Irish and Scottish versions of OSC until 1908. But no Old Series Index sheets of these dimensions have been located. Late copies of the south sheet reveal no such increase in their original width, nor uniform styling with the more northerly sheets. The earlier dimensions of the middle sheet also apparently remained unchanged, though the information it carried was rapidly overtaken by events. By March 1884, copies had been printed with an advisory note: "*For Sheets 91 to 110, refer to Index to the One Inch Map of England & Wales, New Series*". This left this Old Series Index middle sheet relevant to only one row of maps.

Engraving

The sequence of engraving the index appears not to follow precisely that of the contemporary one-inch map given by Harley and Hodson,[15] because index sheet lines had to be engraved in addition. These would have followed the outline and water features, but preceded the roads and boundaries. Great care was taken to clear the sheet lines from the places where they crossed the roads. Symbols and placenames came next, and railways last, usually engraved without deletion over any detail, especially names, in their path. The common failure to clear them where they should pass under roads may be a sign of engraving prior to opening.

---

2. New Series Index[16]

Even before the completion of the Index north sheet in 1884, work seems to have begun on modifying the south and middle sheets to show one-inch New Series sheet lines. Above the Preston to Hull line this entailed deleting some unused quarter-sheet lines and renumbering the rest, and below it redrawing the one-inch sheet lines completely. Matrices were made from which unnecessary sheet lines and numbers were scraped, then duplicate plates were made. We have noted above how by March 1884 Old Series Index middle sheets were already advising users of the change, and this would appear to confirm that a document dated 22 September 1884 refers specifically to the south sheet:

Preparation of an index to the 1-inch New Series, by carrying out alterations to a duplicate plate of the 10-mile, was approved by the DG.

No pre-1890 New Series Indexes have so far been discovered, but it is probable that the map revision applied in 1884-85 to both middle and south sheets (irrelevant, of course, in the case of the north sheet) would have been the same as that on contemporary Old Series Indexes. But it is possible that railway revision may have been applied to them that the obsolescent Old Series Index lacked. Post-1885 military ten-mile maps not only carry the same revised topography, albeit with different marginalia, but also updated railway information, and the remark that they were "Transferred from the Ordnance Survey Index Map Jany 1885" (ie the state that was to hand in January 1885, rather than made in January 1885). This new railway information was very inaccurately drawn, and was probably added not to the Index, but to the transfer. But until a New Series Index of this date emerges to confirm the railway state that appears there, this conclusion must remain uncertain.

But whatever took place in 1884-85, the OS considered the 1890 revision of the New Series Index important enough to warrant appearance both in the OS Publication Report (OSPR)[17] and OSC. There was a further increase in railway information - evidently independent of the 1885 revision - and revised topographical detail. The Index was now designed as a three-sheet map, with a common border provided to surround all three, if mounted together. Sheet widths were adjusted: that of the south sheet was increased to accommodate the Scilly Islands, and of the middle and north sheets slightly reduced to the same width. It was still able to encompass the Western Isles, though the western border now cut through Scotland sheets 58,68,78 and 88. And the joining line between middle and south sheets was adjusted, to accommodate New Series sheet lines. So, a thin strip of land, wider on the east coast than the west, was removed from the south to the middle sheet, so relocating Stretford and Saltfleet, for instance, on the middle sheet. For the first time, the date of the railway information was given. This was updated, it would appear, only once, in 1893. Further updating is unlikely since the map with this railway state remained on sale during the First World War, even though it had been removed from the OSC of Scotland and Ireland by 1909, and of England and Wales as early as 1905.

Postscript

"*The ten mile to the inch map was originally produced as an index for the one-inch and ¼-inch sheets prior to 1892*".[18] I do not know the basis of this statement, and I think it is erroneous. Certainly no quarter-inch index use of this map is known.

---

15. Margary (1977) xxvii.
16. Documentary evidence for this section comes from PRO OS 1/144 70B.
17. *Index to the One-Inch General Map of Great Britain (Scale 10 miles to 1 inch); in 3 sheets*, "New Series." Price 2s. each (OSPR 10/1890).
18. *Final Report.....* (1938) 8.

Specification table

Section A of this specification table deals only with the nineteenth century British map. Sections B and C attempt to draw together the characteristics of the basic ten-mile map families (I-V), as distinguished by the chapters of this book. Such a table can hardly be a perfect tool, if only because documentary evidence of a complete specification survives only for the 1955 map, so for all others the student is forced to analyse what he sees on the map itself. The 1900 **Characteristic Sheet** is a useful yardstick for what to place into what category, but for the nineteenth century map alternative treatment sometimes proved expedient. There may well be some disagreement, for instance, over features I have elected to classify topographical and what appears to be an OS euphemism - the "extensive district". It is also not always apparent quite what basis the OS used to distinguish certain types of writing. Many usages have changed, nowhere more so than when dealing with communities and designations for them, be they cities, towns of various sizes, market towns, county towns, villages, hamlets *et al*. The student is not assisted in this by the OS who until 1955 never published on the map the rules governing these population classifications, though OS practice in 1952 was to consider a medium town as one with 10,000-200,000 inhabitants, and small and large below and above these figures. But, all in all, I thought it of more practical use to pool this information in one place, rather than have it spread throughout the book, with the inevitable problems in comparative analysis that creates. The only exception to this is the modern map. The "M" column reflects the position as in the first issue of the map in 1955: I thought it better to refer to the many new symbols invented for the Route Planning Map (= RPM) in my discussion of it, rather than clutter this list unnecessarily.

The list deals with nineteenth century maps in most detail because there is no information elsewhere about their characteristics, whereas from 1903 legends appeared on the maps, and even characteristic sheets were provided, even if the information they offered was incomplete. Drawings by Richard Dean are provided on p.8 of the various symbols identified on the nineteenth century maps, coupled with facsimile examples of the lettering used. Most of the examples given in the lists are also drawn from those maps. The first state of the British map is recorded on which a new classification is used.

For the nineteenth century British maps (classes A-C), divisions other than sheet lines pertain:
A. England & Wales sheets 1-90 (I)
B. England & Wales sheets 91-99,101-110
C. England & Wales sheet 100, Scotland sheets 1-131
D. Ireland, 1838 (II.1)
E. North America, 1864, Ireland, 1868 (App.2,II.2)
   NB: E° references exclude Ireland
F. E&W Rivers and their Catchment Basins, [1868]
G. Ireland Rivers and their Catchment Basins, 1868
H. Scotland Rivers and their Catchment Basins, 1893
J. Great Britain, 1903, Ireland, 1905 (III.1,III.2)
K. Great Britain, 1926,1937,1942 (IV.1,IV.2,IV.3)
L. Road Map of Great Britain, 1932,37 (IV.2R)
M. "Ten-mile" Map of Great Britain, 1955 (V.1)
N. 1:1,250,000 Great Britain, 1947,1962,64,69 (IV.4)
P. 1:1,250,000 Great Britain, 1975 (V.4)

A "-" indicates a feature lacking in that category. A "+" indicates a feature that exists, but is given different treatment, so is listed elsewhere. Lower case letters indicate that there is more than one treatment of that feature. Unless noted otherwise, the examples are all drawn from nineteenth century maps. Names in brackets do not appear. Examples cited without "eg" are the only ones located. Examples are chosen to give some idea of the variety of name types assembled in a particular classification. Location of examples is assisted by the use of one-inch sheet numbers. Some categories in "E" (eg international boundary) naturally only occur in North America. Many American placenames have suffixes ("P.O." for Post Office, "C.H." for Court House, [Railway] Station) in the writing of the placename. These have been disregarded here. Specific types of symbols, other than modern ones, are grouped together, but overall the list reflects chronologically the order in which characteristics were introduced to the map.

For a map that was nearly seventy years in the making, one is struck more by the consistency of the characteristics in the British nineteenth century map than the maverick quality of the exceptions, some of which must have been errors (eg classifying Yaxley a city and Peterborough a market town). As one might expect, many of the conventions follow those on the parent one-inch sheets, especially as regards lettering, though in practice this was not always achieved with consistency. The origin of the specification of James's ten-mile Map of the World remains unknown, but it is clear that the 1868 Ireland map shared its characteristics, and furthermore that their existence influenced what came to be used when the British map reached Scotland. For instance, this would appear to be the origin of the altered railway symbol, which was quite probably adopted not only for its cleaner appearance but also for the considerable savings in engraving time.

Sources: Principally the maps themselves, and documents, especially PRO OS 1/789, 36/4, 36/5, 36/6 (for V.1 maps).
Ordnance Survey Characteristic Sheet for the Revised Ten-Mile Map of Great Britain. / All rights of reproduction reserved. Ordnance Survey Office, Southampton, July 1900. Price Sixpence (OSPR 8/1900). An engraved sheet, see p.75.
"*A sheet showing in full the Conventional Signs, Ornaments, Characters of Writing, etc., used on the "Ten-Mile" Maps is published. Price 6d.*" (DOSSSM [1937], 12). Not found.

CHARACTERISTIC SYMBOLS FOR 19TH CENTURY ORDNANCE SURVEY TEN-MILE MAPS

....	County boundary (B.1a)
—·—·—·—	National (B.1b) or State (B.1b [E]) border
— — — —	County boundary (B.1c [E])
—·—·—·—	Province border (B.1d [E])
—+—+—+	National border (B.1e [E])
Unfenced	Turnpike (B.2g)
Under construction	Road (B.2h [E]) Minor road (B.2h [D])
Unfenced	Minor road (B.2j) Main road (B.2j [D])
— — — —	Track (B.2k)
▭▭▭▭	Railway (B.2ll)
═══	Railway (B.2mm [D])
▬▬▬	Railway (B.2nn [C,E])
═══	Mineral tramway (B.2pp)
++++++	Underground railway (B.2qq)
....	Railway in tunnel (B.2rr)
::::::::	Railway in tunnel (B.2ss)
ooooooo	Railway in tunnel (B.2tt)
	Waterlining (B.4)
―	Low water line (B.5a)
	Sandbank (B.5b)
⊔	Breakwater (B.5d)
ᴀ ᴀ	Lighthouses (B.5e)
○	Lighthouse, light vessel (B.5f [D,E])
●	Lighthouse (B.5g [E])
⚓	Light vessel (B.5h)
⚔	Wreck (B.5j)
♭	Buoy (B.5k)
....	Fathom line (B.5l)
—·—·—	Fathom line (south-west England) (B.5m)
(⁺⁺)	Underwater rocks (B.5c) in danger area (B.5n)
(⁺)	Underwater rock in danger area (B.5p [E])
-------	Sailing route (B.5q [E])
—·—·—	Electric telegraph (B.5r [E])

～～	River (B.2a)
◯	Lake (B.2b)
── Tunnel ──	Canal, drainage dyke (B.2c)
▬▬▬	Canal (B.2d [D])
───	Ship canal (B.2e) Canal (B.2e [E])
⊶	Dock (B.2f)
🏠	Built-up area (B.2l)
⌐⌐⌐	Buildings (B.2m)
≋	Buildings blocked together in towns
+	Anglican or parish church (B.2n)
⊚	Large town (B.2p)
◎	County town (B.2q [E])
● ● ●	Medium towns (B.2r)
⊙ ⊙	Small towns (B.2s [E])
○	Village (B.2t) Town (B.2t [D])
≡≡≡	Barracks (B.2u)
▯▯▯▯	Barracks (B.2v [E])
◊ ⋈ ⌒ ∧ ·	Fortifications (B.2w)
∘∘∘∘	Mortello towers (B.2x)
ᴧ	Sea mark (B.2y)
∘	Prison (B.2z)
ᴧ	Tower (B.2aa)
∘	? Temporary encampment (B.2bb)
∘	? Temporary encampment (SW England) (B.2cc)
±	Cross (B.2dd)
□	Ancient site (B.2ee)
🌳	Parkland (B.2ff)
	Heathland, marshland (B.2gg)
♣ ♣ ♣	Woodland (B.2hh)
⌒	Sea defence embankment (B.2jj)
	Earthwork (B.2kk)
⬭	Contour (B.2uu [D])
⚔	Site of battle (B.2vv [E])
)(	Bridge (B.2ww [E])
ᴖᴖᴖᴖ	Cliff (B.2xx [E])

All the examples illustrated are shown approximately twice original size.

Plate 1. Drawn by Richard Dean.

Specification table

A. Marginalia (British nineteenth-century maps only)

A.1	Title (S-P1)		A.10	Electrotype notice: south sheet only (S-5)
A.2	Border (S-P1)		A.11	New Series index notice: middle sheet only (M-2)
A.3	Printer's notice: south sheet only (S-P3)		A.12	Cromartyshire notice: north sheet only (N-1)
A.4	Geology: south sheet only (S-P8)		A.13	Index sheet number (S-15)
A.5	Scale statement and scale-bar (S-1)		A.14	Copyright notice (NS-N-1,NS-M-2,NS-S-2)
A.6	Price statement (S-1)		A.15	Railway revision date (NS-N-1,NS-M-2,NS-S-2)
A.7	List of plates: south sheet only (S-1)		A.16	Isle of Man notice: middle sheet only (NS-M-2)
A.8	List of London agents: south sheet only (S-1)		A.17	Adjacent sheet numbers (NS-M-4)
A.9	List of country agents: south sheet only (S-5)			

B. Cartography

ABC DE FGH JKL MNP   State   Description and examples

NB: Map types in columns A-P are described on p.7. Sheet number references are to the nineteenth century Index, within a one-inch quarter sheet (48SE if the sheet was quartered, 9sw if it was not), or Scotland full sheet (S57).

B.1 Boundaries
- a ABC DE°FGH ++L +++   S-P1   County, detached part of county (eg 44,54NW,62SW), ridings of Yorkshire, city counties of Norwich, Lincoln, Hullshire, York (A,B,C,D,L). E° and many B dotted borders are in fact rouletted, provisionally engraved. River catchment area (F,G,H)
- b -BC -E°+-H J++ +++   M-P14   National (B,C,J). State (E°). Joint national-river catchment (H). Province (J)
- c +++ +E F-H JK+ MNP   County (E,J,K,M,N). National (F,H). County and region (P)
- d --- -E --- +-- ---   Province
- e -++ -E°+-+ +K+ MNP   National. *NB*: Pecks lacking along rivers (E°)
- f -++ -+ +-+ ++L +++   National (dot-cross sequence)

B.2 Topographic features.   *NB*: The coastline is assumed, and most skeleton forms are ignored
- a ABC DE FGH JKL MNP   S-P1   River, afon, water; creek (E°). *NB*: Pecked when unsurveyed (E°)
- b ABC DE FGH JKL MNP   S-P1   Lake, loch, lough, pond, pool, mere, water. Reservoir. *NB*: Inland open water in D,F has horizontal ruling
- c ABC ++ +-- JKL M-P   S-P1   Canal, drainage dyke. *NB*: Canals are pecked in tunnels: a tunnel in 71NW has portals
- d +++ D+ +-- +++ +-+   Canal
- e A+C +E F-- +++ +-+   S-P1   Ship canal (A,C): eg Caledonian Canal (S62). Canal (E,F): eg Lagan Navigation
- f A-- -- --- JK- M-P   S-P1   Dock: (West India Docks, Perrys Dock, 1sw)
- g ABC ++ --- +K+ MN+   S-P1   Turnpike, main road (A,B,C). Main (or MOT Trunk & Class 1, Class 2) road (K,M)
- h +++ DE --- J+L ++P   Road (E,L,P). Minor (or Second Class) road (D,J)
- j ABC D+ --- JK+ M--   S-P1   Minor (or other, other tarred) road (A,B,C,K,M). Main (or First Class) road (D,J)
- k A-- -E°--- J+- ---   S-P8   Track (A,E°), agger (A): eg 51NW,51SW,64se,66SW. Ford (J: Valencia, Holy Islands)
- l ABC -E F-- JK+ M++   S-P1   Built-up area (hatched lines run NW-SE on south sheet only)
- m ABC D- --- JKl MN+   S-P1   Building, often in parkland (A,B,C also blocked together in towns: l,m may be combined (eg Huntingdon), some (eg Mendlesham, 50SW) lack roads). *NB*: Black town infill (K,L,N). Coloured town infill on early coloured issues superseded by hatching (M)
- n ABC -- --- J-- ---   S-P1   Anglican or parish church, almost always with name of parish: m,n often combined
- p AB- -+ --- +++ +++   S-15   Large town (A,B: not in GB): Dublin
- q +-- -E°--- +++ +++   County town, sometimes combined with l
- r AB- -E FGH J++ +++   S-15   Medium town (A,B: not GB, E: GB only). Town of over 10,000 (F), 5000 (G) population
- s --- -E +++ JKL M++   Small town
- t A-- DE --- +KL MNP   S-15   Village (A,E,K,L; A: not GB). Town (D,N,P). Settlement (eg mission) (E°)
- u A-- -+ --- +-- ---   S-P1   Barracks: eg Hythe (4nw)
- v +-- -E --- J-- ---   Barracks: The Curragh Camp
- w A-- -E --- --- ---   S-P1   Fortification: eg (Sandham Fort, 10se), Battery, (Redoubt), Greencastle Fort (E)
- x A-- -- --- --- ---   S-P1   Martello towers: (5sw)
- y A-- -- --- --- ---   S-P1   Sea mark: on Ashey Down (10se)
- z A-- -- --- --- ---   S-P1   Prison: (Dartmoor Prison, 25se)

Specification table

	ABC	DE	FGH	JKL	MNP	State	Description and examples
aa	A--	--	---	---	---	S-P1	Tower: (Walton Tower, 48se)
bb	a--	--	---	---	---	S-P1	?Temporary encampment: eg especially 8
cc	a--	--	---	---	---	S-P1	?Temporary encampment in south-west England. *NB*: These quickly became invisible
dd	A--	--	---	---	---	S-P1	Cross: (Nuns Cross, 25se)
ee	A--	--	---	++-	---	S-P1	Ancient site: Dover Castle (3se), Willbury Hill (46NE)
ff	ABC	--	---	J--	---	S-P2	Parkland (A: mostly with paling boundaries)
gg	A--	--	---	---	---	S-P3	Heathland, marshland: eg New Forest (15se), Romney Marsh (4nw)
hh	A--	--	---	---	---	S-P3	Woodland, and for use in parkland
jj	A--	--	---	---	---	S-P8	Sea defence embankment: eg bordering the Wash (69sw)
kk	A--	--	---	---	---	S-P8	Earthwork: (Devil's Dyke, 51SW,51SE)
ll	AB+	++	---	++-	+++	S-P8	Railway, private mineral line mainly in North-East England
mm	+++	D+	---	++-	+++		Railway (1845): the double lines later filled to form a single pair
nn	++C	+E	---	JK-	MNP	M-P14	Railway
pp	A--	--	---	---	---	S-8	Mineral tramway: eg at Liskeard (25sw), Dowlais (36nw)
qq	A--	--	---	---	---	S-10	Underground railway: (Metropolitan Railway from Paddington to Farringdon, 7se)
rr	a+-	--	---	JK-	M+P	S-P8	Railway in tunnel: eg (Watford, 7ne), (Balcombe, 9ne), (Sapperton, 34nw)
ss	aB-	--	---	++-	+N+	M-P6	Railway in tunnel: eg (Crigglestone, 87NW), (Balcombe re-engraved)
tt	a+-	--	---	++-	+++	NS-S-2	Railway in tunnel: eg (Cymmer, NS248), (Severn re-engraved, NS250)
uu	---	D-	---	-K-	---		Contour (D: 1845); a coloured line with height (K: not IV.2,3)
vv	---	-E°	---	J--	---		Site of battle, with date
ww	---	-E	---	-K-	---		Bridge (E), (K: Forth, Tay rail bridges: conventional overhead view with buttresses)
xx	---	-E	---	J--	---		Cliff
yy	---	--	F--	---	---		Crow's foot, with number, for height above sea level
zz	---	--	---	J--	---		Triangulation pillar (triangle)
aaa	---	--	---	JK-	M--		Spot height (dot with height in feet)
bbb	---	--	---	J--	---		Cave (small infilled rectangle)
ccc	---	--	---	JK-	---		Railway station (bar across railway line)
ddd	---	--	---	J+-	---		Airport (runway layout): Dublin, Shannon (eg added on **Biological Subdivisions**)
eee	---	--	---	+K-	---		Aerodrome, aerodrome or seaplane station (dot in square) (not IV.2,3)
fff	---	--	---	JKL	M-P		Breakwater, pier (depicted): eg at Holyhead, Tynemouth, Kingstown
ggg	---	--	---	JK-	---		Ruined building (open building block)
hhh	+--	--	---	JK-	---		Ancient site (ring of hachures or dots): eg Devil's Dyke (Brighton), Stonehenge, Old Sarum, Maiden Castle, Castle Hill (south of Huddersfield), Moat of Urr, Kaims Castle
jjj	---	--	---	J--	---		Standing stone (depicted): Rufus's Stone (near Cadnam)
kkk	---	--	---	-K-	M--		Graticule intersection (large cross) (K: IV.3: 1944)
lll	---	--	---	+K-	---		Ford (double dotted line): to Holy Island
mmm	---	-+	---	+K-	---		Sand hill (black stipple cluster): eg (Braunton, Pendine, Rattray Head). See B.5b
nnn	---	--	---	--L	---		Mileage measurement (line through town or road junction)
ppp	---	--	---	--l	--P		Stipple built-up area infill
qqq	---	--	---	--L	M--		Telephone call box (dot with "R" (for RAC) or "A" (for AA)) (M: not till RPM)

B.3 Geology

    A-- -- --- --- --- S-P8 Geological areas marked on 20-27,29-33, and an associated key (see A.4)

B.4 Waterlining

    AB- -E --- --- --- S-P1 Round coast, sandbanks, islands, rocks, low water and (later) lakes (E: Ireland only)

B.5 Hydrographic features

a	ABC	--	---	---	---	S-P1	Low water line
b	AB-	DE	F--	JKL	M--	S-P1	Sandbank, mudbank (D: not stippled; J: black, K,L: blue, M: brown). Sandbanks sometimes extend onshore, eg west of Dunfanaghy (E), Braunton, Pembrey (J)
c	A--	-E	---	---	---	S-P1	Underwater rock, rock ledge: eg 9sw,16nw,24se,33sw
d	A--	--	---	J--	M-P	S-P1	Offshore breakwater: (Plymouth 24nw, Cherbourg)
e	A-C	++	F--	JK-	--+	S-P1	Offshore and coastal lighthouses: eg 28,33sw,37se,48NE,49SW,S49. *NB*: Some have smoke (A), none have smoke (C,E,J), all are alight (K)

Specification table 11

	ABC DE FGH JKL MNP	State	Description and examples
f	+-+ De +-- ++- --P		Lighthouse, light vessel (E: lighthouse usually with *L.H.*, *L.Ho.*, or *L.Ho^s*)
g	+-+ +e°+-- ++- --+		Lighthouse
h	A-- ++ --- JK- ---	S-P8	Light vessel: off 9,2sw,48SE,67NW, eg Nore (2sw)
j	A-- -- --- --- ---	S-P8	Wreck: (59SW)
k	A-- -- --- --- ---	S-P8	Buoy (the point of the cone at buoy position): (2,3,48,49,67)
l	a-- -- --- -+- ---	S-P8	Fathom line (NOT 38 round to 22)
m	a-- -- --- -+- ---	S-P8	Fathom line in south-west England: (38 round to 22 only, and beyond sheet lines)
n	A-- -+ --- --- ---	S-P8	Danger area, often surrounding underwater rock symbol (B.5c) (16sw,17se)
p	+-- -E°--- --- ---		Danger area, surrounding underwater rock symbol
q	--- -E --- JKL M--		Sailing route, ferry. *NB*: Marked as a ford across Dornoch Firth (K)
r	--- -E --- --- ---		Electric telegraph, Atlantic telegraph
s	--- -- --- JK- ---		Flat rocks: eg in the Mouth of the River Severn
t	+-- -- --- -K- ---		Fathom line (continuous line on blue plate)

C. Lettering

NB: Titles, lists and marginal writing are ignored here. Original spellings are quoted. Alphabets used on each map differ slightly, but these variations have rarely been noted. *NB*: All lettering in column "P" is sanserif

C.1 Roman capitals

	ABC DE FGH JKL MNP	State	Description
a	ABC ++ --- jK+ M+P	S-P1	County, large detached part of county (italic suffix "(det)")
b	a+C DE +GH JKL MNP	S-P1	City, county town. City and town (M), with population above 200,000 (N)
c	A-C DE FGH JKL MNP	S-P1	Large island: eg Isle of Wight, Isle of Man, Rum, Achill
d	AB- -- +-H JK- MNP	S-15	Country: France, Ireland
e	ABC ++ -+H J++ +++	N-1	Marine name: eg Atlantic Ocean
f	+++ D+ +++ +++ +++		Principal bay, harbour, haven, estuarial river, sea lough, inland lough
g	--- -e°--- +-- ---		Province. *NB*: Ireland in C.6e
h	--- -E°--- --- ---		State

C.2 Roman open capitals

	ABC DE FGH JKL MNP	Description
a	+++ DE -G+ +++ +++	Marine name
b	++- -- F-+ ++- +++	Country
c	--- -- --- --- -N+	County

C.3 Roman

	ABC DE FGH JKL MNP	State	Description
a	ABC -+ --- +K+ M-P	S-P1	Parish, village. *NB*: Usually with "+" (A,B,C), though Bowness (98NE) has buildings
b	ABC DE°++H JK+ MNP	S-P2	Minor island
c	a++ DE +G+ +++ +++	S-P2	Principal headland (D,E,G). Minor headland (A): eg Penbrush P^t, Linney Head (38,39)
d	a++ -E°+++ ++- +--	S-P2	Minor bay, harbour, basin: Broad Sound (38nw). *NB*: C.3c,d in England & Wales seem limited to 38 and 39 only, and probably should have been in C.9 style
e	A-C D+ ++- ++- --P	S-P2	Offshore lighthouse (A,C,P): Smalls (39sw), Eddystone (24sw), Inchcape or Bell Rock (S49). Lighthouse, light vessel (D)
f	a-- -- --- --- +--	S-P8	Supplementary part of market town name: eg *NEWARK* upon *TRENT* (70nw)
g	+b+ DE +GH +KL MNP	M-P6	Market town (B: on skeleton mapping,D,G,H). Small town (E,K,L,M: with dot in circle). Town (N,P. N has 2 categories: 20,000-100,000, 100,000-200,000 population)
h	+++ D+ --- ++- ---		Railway company name (1845)
j	--- D+ --- --- ---		Inland harbour (Shannon Harbour, Richmond Harbour)
k	+-- D+ --- --- ---		Canal
l	+-- D- --- ++- +-+		Sandbank
m	--- -E°--- +-- ---		Important port: eg harbour and basin in Quebec
n	--- -e --- --- ---		Fortification: eg Battery, Camden Fort, Carlisle Fort, Charles Fort, Greencastle Fort
p	--C -- --- --- ---	M-P14	"Church" (with "+"): unnamed parish, eg in Aberdeenshire, Banffshire, Nairnshire, S37
q	+-c -- --- J-- ---	N-1	Letter code to refer to small detached county areas: eg "c" for Cromartyshire
r	--C -- --- --- ---	N-1	Railway terminus: Strome Ferry Terminus (S81)

Specification table

	ABC DE FGH JKL MNP	State	Description and examples

C.4 Roman sloping capitals

- a   AbC ++ +++ JKL +++   S-P1   Market town (A,B,C). Large, medium, small town (J). Medium town (K,L)
- b   ABC +E FGH JKL MNP   S-P1   Principal bay, channel, harbour, estuarial or navigable river, sea loch, sea lough, basin, firth, sound: eg Bristol Channel, Luce Bay, Kilbrannan Sound, The Wash, Loch Linnhe, Lough Foyle, Mouth of the Shannon, Firth of Forth, Scapa Flow. Dock (24nw)
- c   A++ -E --- JK- M--   S-P1   Large forest, moor, marsh: eg Salisbury Plain, Dartmoor Forest, Romney Marsh, South Downs, The Peak (81NE); swamp (E°)
- d   ABC ++ F+H JK+ MNP   S-P1   Principal headland: eg Selsea Bill, Denge Ness, Start Point, The Needles, Lands End, Flamborough Head, Mull of Kintyre, Cape Wrath
- e   A-- -+ --- J-- ---   S-P2   Extensive district: eg Isle of Ely, Gower, Isle of Thanet, Sheppey Isle
- f   A-+ -- --- +++ ++-   S-P8   Large detached part of county: eg Worcs, Gloucs (44ne), Shropshire (54NW,62SW)
- g   aB+ ++ --- +++ +++   M-P7   City, county town: eg Lancaster, York
- h   -B+ +E FGH JKL MNP   M-P9   Principal lake, lough: Ullswater, Windermere, Lough Neagh. Reservoir (P)
- j   +++ ++ -++ +KL MNP   Marine name
- k   --- -- --- --- M--   Supplementary part of Boulogne *(SUR MER)*

C.5 Roman sloping

- a   +++ ++ +++ +KL +++   Middle range headland (K). Principal headland (L)
- b   +++ ++ +++ ++L +++   Minor island

C.6 Egyptian capitals

- a   A-- -- --- J+- ---   S-P8   Ancient site: Willbury Hill (46NE), Roman Fosse Way (63,71)
- b   a+C +E --- J+- ---   S-8   Railway company name (A: additions made on State S-8 only)
- c   --C ++ --- JKL M--   M-P14   Important range of hills (C,J,K,L,M): Paps (S28), Grampian Mountains (S46-65). Important peak (K,L: 1937,M): eg Ben Nevis
- d   +++ DE --- j++ +++   County (J: county names in margins only)
- e   --- -e --- J-- ---   Province (Ireland only). *NB*: Canada in C.1g (E)
- f   +++ +E°F++ +++ +++   County town (E°: usually with concentric circles symbol). Town (F)
- g   --- -E --- --- ---   Electric telegraph, Atlantic telegraph
- h   +++ ++ --- +KL +N+   With serifs: County (L). Riding of Yorkshire (K,N)

C.7 Egyptian

- a   --C +E --- +K+ M--   M-P14   Important range of hills (E). Important peak (C): Ben Nevis (S53). Minor peak (K,M)

C.8 Egyptian sloping capitals

- a   +-C +e --- ++- +--   M-P14   Ship canal (C): Caledonian Canal (S62), Glen Roy! (S63), Canal (S83). Canal (E): Royal Canal, Grand Canal, Ulster Canal, Newry Canal, Lagan Navigation
- b   --- -E°--- --- ---   Important mountain range: eg Blue Ridge Mountain
- c   +-- -E --- +-- ---   Extensive district: Connamara, Jar Connaught, Joyce's Country, Cloghan, The Rosses
- d   +++ ++ --- +K- ---   Railway company abbreviation

C.9 Italic

- a   a-- -- --- --- +--   S-P1   Supplementary part of placename, eg Withycombe *in the Moor* (25se), Newark *upon* Trent (70nw: State S-P6), Stoke *upon* Trent (72NW)
- b   ABC DE FGH JK- MNP   S-P2   River: eg Sirhowey River (36), Camddwr (57SE), Afon Rheidol (59SE), Ettrick Water (S16), Source of the Severn (60SW); source (E); creek (E°)
- c   A-- +e°--- JK- M--   S-P2   Canal: Royal Military (4), Paddington (7se), Basingstoke (8nw), Neath, Cardiff (36), Oakham (64nw), Birmingham and Liverpool Jcn (73), (Mkt Weighton) Canal (86nw)
- d   ABC -E -GH JKL M--   S-P2   Minor lake (A,B,E,G,H,J,K,L,M): eg Dozmare Pool (30se), Whittlesea Mere (64se), Bala Lake (74SW), Coniston Water, Bassenthwaite Lake; (L: 1937). All inland lochs (C)
- e   a-C -E -GH JKL M--   S-P2   Minor headland: eg Firebeacon Point, Tintagell Head (30ne)
- f   A-C +E FGH JKL MN+   S-P2   Small island, rock: eg I.of Dogs (1sw), Wolf Rock (33sw)
- g   A-C -E --- JKL MN+   S-P2   Hamlet, village (A,C,E,J,K,L,M): eg Hirwain (36nw), Goole (86nw), Lybster (S110). Town (N: population under 20,000). Settlement (eg mission) (E°)
- h   A-C -- --- JKL M--   S-P2   Isolated building: Gentleman's seat, many in parkland: eg Hampton Court Palace (8ne), Guidea Hall (1sw), Newick Place (5nw), Court Lodge (5ne), Tyrthegston Court (36),

Specification table

	ABC	DE	FGH	JKL	MNP	State	Description and examples
j	A--	--	---	JK-	M--	S-P2	Golden Grove (41), Balmoral Castle (S65), Guisachan House (S73). Religious house: eg Leeds Abbey (6sw), Erith Hermitage (51NW), Haverholme Priory (70ne), Monastery (J). Inn: eg Stonham Pie Inn (50SW), Kate Kearney's Cott. (P.H.) (J), Angler's Retreat (K) Parkland, deer park (50SE). *NB*: Many are also gentlemen's seats
k	A-+	+E	F--	JK-	---	S-P2	Coastal lighthouse: eg Mumbles (37se), Longships (33sw)
l	A-c	--	---	+--	---	S-P3	Small detached part of county: Part of Oxon (7nw)
m	a-C	-E	FGH	JK-	M--	S-P6	Minor bay, channel: eg Sunk Channel or Black Deeps (2), Mouth of the Ex (22nw), Start Bay (23nw), Coverack Cove (32nw), Wicca Pool (33nw), Milford Haven (38nw), Nicholas Gat (67SW), Stiffley Overfalls (68NW), Brancaster Roads (69ne), Fossdyke Wash (69sw), Kinsale Harbour. Strait, sound, inlet (E°). Westport Quay (J)
n	A--	--	---	JK-	M-P	S-P6	Drainage dyke: eg South Holland Drain (65nw)
p	A--	+-	---	JK-	M-P	S-P6	Sandbank: Roger Sand (69nw), Dogger Head (69ne), Sarn Badrig or Causeway (59NW), Barnstaple Bar (27sw), Sizewell Bank (49SW)
q	A-C	-E	---	JKL	M--	S-P6	Important valley, or lowland topographic feature: eg Cannock Chase (62NW), Charnwood Forest (63NW), Whaplode Fen (65ne), Holbeach Marsh (69sw), Dove Dale (72NE), Chat Moss (89SE), Wicklow Gap (E); swamp (E°), pass, glen (J), (L: 1937)
r	A--	--	---	+--	---	S-P6	Minor extensive district: East Holland (69nw), Bank Lands (69se)
s	A--	-E°	---	J--	---	S-P6	Road: The Watling Street (61,62), Littleworth Drove (64ne). Parallel roads (J)
t	aB+	++	---	++-	---	S-P8	Railway company name
u	A--	--	---	---	---	S-P8	River bank: The North Bank, The South Bank (of River Nene, 64,65)
v	A--	--	---	---	---	S-P8	Bridge: Ponds Bridge (64se), Fossdyke Bridge (69sw), Cavendish Br (71SW)
w	A--	--	---	--L	---	S-P8	Toll-bar (A): Causeway Bar (64ne). Toll (L)
x	A--	+E	-G-	JK-	---	S-P8	Ship: eg light vessel (2sw,48SE,67NW), wreck (59SW)
y	A--	--	---	---	---	S-P8	Fathom measurement (38-22)
z	---	D-	---	---	---		Characteristic of lighthouse, light vessel
aa	A--	--	---	JKL	---	S-2	Railway station: (Ammanford, 41se)
bb	-BC	D-	---	J+-	+--	M-P8	Minor range of hills, peak, ridge (B,C,J): Whernside (97SW), Snaefell (100), Cheviot Hills (108), Ben Lomond (S38). Important range (D: 1845): Slievebloom Mountains
cc	-B-	-E	---	J--	---	M-P9	Industrial location: eg colliery (103NE), silvermine (E), mill, salt work (E°)
dd	-B-	-E°	---	JKL	M-+	M-P9	Reservoir (Killington Reservoir, 98NE), (L: 1937)
ee	--C	--	---	JKL	M--	M-1	County to which island is attached
ff	--C	-E	---	JK-	---	N-1	Railway junction: Boat of Garten Junction (S74)
gg	---	-E°	---	---	---		Date (ie month) of battle
hh	---	-e	---	J--	---		Military establishment: eg U.S. Arsenal, The Curragh Camp, Battery
jj	---	-E°	---	---	---		Portage
kk	---	+E	---	J--	---		Inland harbour: Shannon Harbour
ll	---	-E	---	JK-	M--		Falls, rapids, spring, waterfall (eg The High Force)
mm	---	-E°	---	JKL	M--		Ferry
nn	---	-E°	---	JKL	M--		Railway tunnel. Road tunnel (K: 1935,L,M) (Mersey)
pp	---	-E	---	---	---		Sailing distance
qq	---	--	---	j--	---		Cave
rr	---	--	---	J--	---		Airport: Dublin, Shannon (eg added on **Biological Subdivisions**)
ss	---	--	---	JK-	---		Additional railway information: eg Disused Railway, Mineral Railway, Corrour Siding
tt	---	--	---	J--	---		Standing stone: Rufus's Stone (near Cadnam)
uu	---	--	---	-K-	---		"French landed in 1797" (Carreg Gwastad Point)

C.10 Gothic

| a | --C | -E | --- | JK- | --- | M-P14 | Antiquity: Blair Castle (S55), Fyvie Castle (S73), Dunrobin Castle (S103), John O' Groats House (S116), Seven Churches (Co. Wicklow) |
| b | --- | -- | --- | jK- | --- | | Cave: eg St Medan's Cave |

C.11 Numerals

a	ABC	--	---	---	---		Upright with serifs: sheet numbers
b	---	-E	---	J-L	---		Italic: date of battle, wreck, etc: eg (Naseby). "B" road numbers (L: some arrowed)
c	---	--	---	JKL	M--		Upright, sanserif: Spot heights. Contour values (K: IV.1 only). Fathom values (K). "A" road numbers (often arrowed) (L). Mileage (L)

Plate 2. Styles of lettering used on nineteenth-century Ordnance Survey ten-mile maps.

Style	Example
Roman capitals (C.1)	ORKNEY ISLANDS Ipswich
Roman open capitals (C.2)	IRISH SEA
Roman (C.3)	Fair Isle Naseby
Roman sloping capitals (C.4)	DARTMOOR FOREST Olney
Egyptian capitals (C.6)	PAPS WILLBURY HILL
Egyptian (C.7)	Ben Nevis
Egyptian sloping capitals (C.8)	CALEDONIAN CANAL
Italic (C.9)	Berwick Head Papa Sound R. Lea
Gothic (C.10)	John o'Groats House

NB: C.1, C.2 and C.4 are at actual size, C.3, C.6, C.7, C.8, C.9 and C.10 at double size

Summary list of states

Column numbers: 1. Sheet number. 2. Edition code. 3. Publication date. 4. Date of railways. 7. Ramshaw's printings, or earliest known "Embossed Printing Date". 8. Copyright statement (A-ll rights of reproduction reserved): price. 17. OSPR. 18. Location of copies

1. Old Series Index

	1	2	3	7	8	18
		N-P1	(?1881)			not found
?1		N-1	(1884)	5.84	np	Cu,Lgh
		?M-P1	(?1824)	-	np	not found
		M-P2	(?1830)	-	np	Lrgs
		M-P3	(?1839)	-	np	Lbl
		?M-P4	(?1844)	-		not found
		?M-P5	(?1847)	-		not found
		M-P6	(?1852)		np	Lpro
		M-P7	(?1854)	8.9.56	np	Lrgs
		M-P8	(?1857)	24.9.57	np	Cu,Ob,PC
		M-P9	(?1858)		np	Mp,PC
		M-P10	(?1859)		np	RH,PC
		M-P11	(?1862)	7.62	np	Cu,Dtc,PC
		M-P12	(?1864)		np	Lrgs
		M-P13	(?1866)	4.66	np	Cjc,Lmd,PC
		M-P14	(?1869)		np	Lmd
		M-P15	(?1873)	2.75	np	Lpro
		M-P16	(?1876)	3.76	np	ABn,Lpro,RH
		M-P17	(?1878)	1.79	np	Lsa,Ob,PC

1	2	3	7	8	18
	M-1	(?1881)	8.81	np	PC
	M-2	(?1883)		np	Lpro
2	M-3	(?1885)	6.87	2/-	Cu,Lgh,Lsa
	S-P1	(?1817)	-	np	Cjc
	S-P2	(1820)	-	np	Lbl,Og,RH
	S-P3	(?1823)	-	np	Og,PC
	S-P4	(?1823)	Ramshaw	np	ABn,Lrgs
	?S-P5	(?1824)			not found
	S-P6	(?1830)	Ramshaw	np	Lrgs
	?S-P7	(?1837)			not found
	S-P8	(?1839)	-	np	Lbl
	S-1	(?1841)	-	4/-	Lbl,SOrm
	S-2	(?1843)	-	4/-	SOos
	S-3	(?1845)	-	4/-	PC
	S-4	(?1847)	-	4/-	Og
	S-5	(?1849)	22.12.51	2/-	Lbl,Lkg,Lrgs,Ob,RH
	S-6	(?1857)		2/-	Cu,Lrgs,Mp
	S-7	(?1859)	10.59	2/-	Ob,Og,RH
	?S-8	(?1861)			not found
	S-9	(?1863)	3.63	2/-	Cu,Dtc,PC
	S-10	(?1865)		2/-	Cjc,Lmd,PC
	S-11	(?1869)		np	Lmd
	S-12	(?1873)		np	PC
	S-13	(?1874)	4.76	np	ABn,LVg
	S-14	(?1876)	1876	np	Lpro,Lsa,Ob,RH,PC
3	S-15	(?1885)	2.85	2/-	Cu,Lgh,Lpro
3	S-16			2/-	EXg

I. Summary list of states

2. New Series Index

1	2	3	4	7	8	17	18
	NS-N-1	(1890)	1890	10.90	A:2/-	10/90	CPT
1	NS-N-2	(?1893)	1893		A:2/-	-	not found
1	NS-N-3	(?1900)	1893	6.05	A:2/-	-	Lpro,Mg
?2	NS-M-1	(?1884)				-	not found
2	NS-M-2	(1890)	1890	10.90	A:2/-	10/90	CPT
2	NS-M-3	(?1892)	189		A:2/-	-	Lpro
2	NS-M-4	(?1893)	1893		A:2/-	-	not found
2	NS-M-5	(?1900)	1893	1.09	A:2/-	-	Lpro,Mg
?3	NS-S-1	(?1885)				-	not found
3	NS-S-2	(1890)	1889	10.90	A:2/-	10/90	CPT,Mg
3	NS-S-3	(?1892)	1889		A:2/-	-	Lpro,SOrm
3	NS-S-4	(?1893)	?1893	5.05	A:2/-	-	Lpro,Mg

Cartobibliography

NB: In the lists of Index states that follow, additions to one-inch mapping in either complete or skeleton form are described separately, not as a changes. Roads and rivers were not always made good to sheet limits until the completion of neighbouring one-inch sheets, and some names near sheet borders were moved at the same time: changes such as these have not been noted. County names added later are noted as changes. These lists are representative: no claim is made to completeness.

Comprehensive lists of railway changes until 1893 have been included, both for their intrinsic value to railway historians, and because they are much the most obvious, and sometimes the only, subject of revision on the Index. They will also prove the most valuable aid when testing for new intermediate map states. How their appearance here parallels that on the one-inch map to which this is an index is a matter of further investigation. Ordering railway lists is unavoidably complex, not least because, at this scale, some of the routes as engraved are generalised. But some observations should simplify their use. They are organised essentially by railway company. Primary routes are entered to the left of each column, roughly in clockwise sequence on the south sheet from the River Thames, and in anti-clockwise sequence on the middle sheet from Piel, south of Barrow-in-Furness. Midland Railway primary routes radiate from Derby. Branches of these primary routes, and subsidiary branches, are displaced in turn one step further right. All branches are listed in sequence away from the point of origin. So any entry is a branch off the last entry one step to its left, however much higher up the list this may be. It should thus be possible to locate any line by means of reference as far back towards its primary route as the user finds necessary. Once recorded, no entry reappears unless it is modified, or required as part of the chain to another unopened route.

In the early years especially, lines were mapped prior to opening according to railway acts. Some were not built, or were built along different alignments. These are often subject to notes. Corrections to these alignments may be connected with ".....". Attempts have been made generally to note errors in railway depiction where these occur. The opening dates given are initial openings to commercial traffic, not necessarily passenger openings. Reversals (eg Tilbury, Alnwick) have been disregarded. Newly opened routes are printed in capital letters, and extensions from or to an already opened piece of line have the new destination in capital letters. The "=" symbol means that the next entry of that route will appear at a difference place in the sequence, perhaps because it will by then be joined to a route also being built from the other end. Some routes as mapped only go to a one-inch sheet line. These are indicated by the sheet number to the edge of which the line is completed. The haphazard nature of the development of the middle sheet especially leaves many routes incomplete. Where there are gaps, "//" has been used, until the gap is filled. Sometimes, on one or other side of the divide, only one name is given, when there are no obvious terminals.

Placenames in italics imply the designated terminus of a section of line which is not yet mapped, or, if in italic capitals, the approximate location reached by an uncompleted stretch of line. As to brackets: "[]" embrace termini which are on neighbouring sheets, and "()" embrace additional information, normally the locality of listed junctions, though those within half a mile of a station have not normally been named. A date followed by "=" implies a section of line opened well before its appearance on the Index, and which can thus have no significance as to the its date. Finally, it should be noted that railways were never drawn on the ten-mile Index through Birmingham, Leeds, Manchester, Glasgow and Edinburgh, and are often sketchy elsewhere (eg Exeter, Liverpool, London and Newcastle). These railway lists assume continuity of routes through these conurbations.

I.1. South sheet, ?1817

State S-P1 (?1817) (?first state)

Dating evidence: 38 (1818) absent, 3,6 (1819) present. For further discussion, see p.2

Dimensions and visible constructional lines:
    Paper: 1013mm (1018mm at top) W-E by 681mm S-N
    Plate: 960mm W-E by 662mm S-N
    Neat line: 891mm W-E by 551mm S-N
Distances between neat line and edge of plate:
    34mm to west, 33mm to south, 35mm to east
    There are latitude and longitude ticks outside the neat line. The side neat lines fade at the top where the plate was not inked.
One-inch sheets completed:
    1-6,8-12,14-27,29-33,47-48 (= Parts I-VIII complete)

INDEX

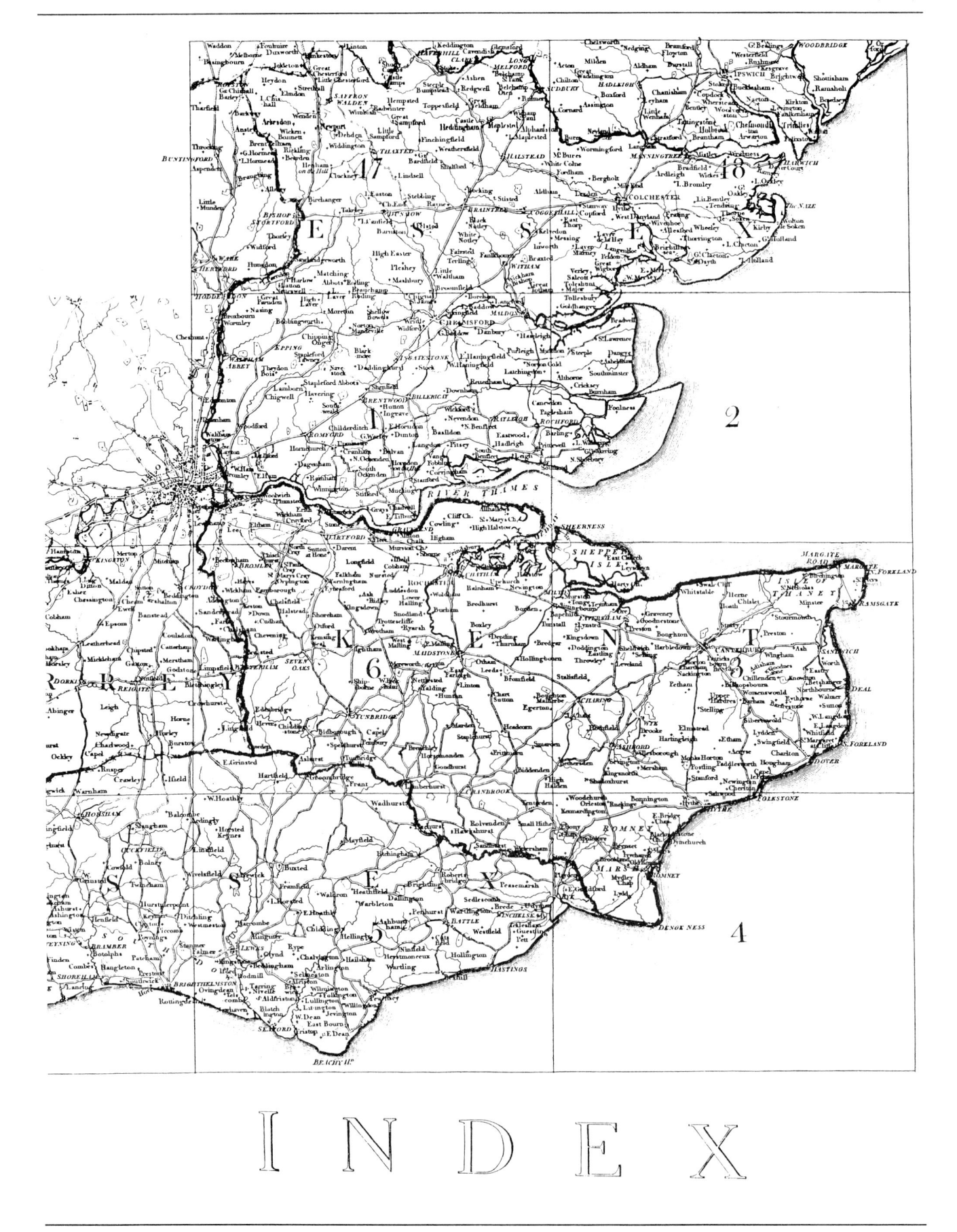

Plate 3. Extract from the earliest known Ordnance Survey ten-mile map, c.1817, reduced to 82.1%. The title has been moved. Reproduced by permission of the Master and Fellows of St John's College, Cambridge.

I.1 (S-P1)

In skeletal form: 7 (unnumbered) - showing the east side, the south west Roman road and Windsor Park. There is no sheet line on the west side; a constructional line is shown double for half the length of the north side.

NB: The south sheet lines of 9,5,4 were engraved 7mm south of their true position to include detail mapped on the extrusions on 9,5.

Other matter still not completely mapped:
No names on Wales and Bristol Channel islands (20)
Many roads on 1,2,3,6 in Kent, and two roads west of Launceston (25nw,30ne) are shown in outline only
Parkland has blocks if buildings are present, but no names, and outline boundaries: Richmond Park alone on this state has a paling boundary
Hurstpierpoint church and building detail (9se)
The engraver's guidelines above and below names are still largely visible
Stane Street to north edge of 9 (never extended)

Standard features present
A.1 Title: **Index**
A.2 Border: neat line, visible west, south and east
B.1 Boundaries
  a County boundary
B.2 Topographic features
  a River
  b Lake: Dozmare (Dozmary) Pool (30se)
  c Canal
  e Ship canal: across the Isle of Dogs (1sw)
  f Dock: West India Docks, Perrys Dock (1sw)
  g Turnpike
  j Minor road
  l Built-up area: eg London
  m Building in parkland, blocked together in towns
  n Anglican church, usually with name of parish
  u Barracks: eg Hythe (4nw)
  w Fortification: eg Sandham Fort (10se), Gosport
  x Martello towers (5sw)
  y Sea mark: on Ashey Down (10se)
  z Prison: Dartmoor Prison (25se)
  aa Tower: Wolton (Walton) Tower (48se)
  bb ?Temporary encampment: eg especially 8
  cc ?Temporary encampment in south-west England
  dd Cross: Nuns Cross on Dartmoor (25se)
  ee Ancient site: eg Dover Castle (3se)
B.4 Waterlining: complete around coast, sandbanks, islands and rocks, but not yet lakes
B.5 Hydrographic features
  a Low water line
  b Sandbank, mudbank: eg Maplin Sands (2sw)
  c Underwater rock: eg off Bolt Head (24se)
  d Offshore breakwater: Plymouth (24nw)
  e Lighthouse: eg Longships Lighthouse (33sw)
C.1 Roman capitals
  a County
  b City, county town: including Colchester
  c Large island: Isle of Wight

C.3 Roman
  a Parish (with "+")
C.4 Roman sloping capitals
  a Market town
  b Principal bay, channel, estuarial river
  c Large forest, moor, marsh: eg Romney Marsh (4), Salisbury Plain (14) Dartmoor Forest (25)
  d Principal headland: eg Selsea Bill (9sw)
C.9 Italic
  a Supplementary part of parish name: eg Withycombe *in the Moor* (25se)

Cjc. This remarkably well preserved copy has no watermark. It is flat, with excess paper beyond the plate.

State S-P2 (1820)

One-inch sheets added:
28,38-40,58 (Part IX, 1818-20), 36 (1833), 37 (1830)
In skeletal form: 7 (full area),13,41 (all numbered)
NB: The north 7,13 intersection is shown stepped: as published the one-inch maps reached a common point here.

Introduction of new standard features
B.2 Topographic feature
  ff Parkland (now with infill and paling boundaries)
C.3 Roman
  b Minor island: eg those off Pembrokeshire
  c Minor headland: eg St Anns Head (38nw)
  d Minor bay: Broad Sound (38nw)
    NB: C.3c,d seem limited to 38 and 39 only
  e Offshore lighthouse: Smalls Lighthouse (39sw)
C.4 Roman sloping capitals
  e Extensive district: Gower (37)
C.9 Italic
  b River, especially on 36
  c Canal: eg Neath (36nw), Royal Military (4nw)
  d Minor lake: eg Dozmare Pool (30se)
  e Minor headland: eg Whitford Point (37nw)
  f Small island, rock: eg Monkstone (20ne)
  g Hamlet, especially on 36
  h Isolated building: Religious house: eg Leeds Abbey (6sw); Gentleman's seat: eg Hill Hall (1nw)
  j Parkland: eg Longleat Park (19se)
  k Coastal lighthouse: Mumbles Lighthouse (37se)

Changes
1. Latitude and longitude ticks enlarged
2. Absent Roman Road below Greenwich completed (1sw)
3. *WORTHING* (9se), Goathurst (20se) added
4. Names added to Wales and islands on 20
5. County names added: BERKSHIRE, MIDDLESEX
6. Waterlining added to lakes: eg *Fleet Pond* (8nw)
Matter developed from outline to final state:
7. Some roads in Kent on 1,3,6 (but not on Sheppey)
8. Two roads west of Launceston (25nw,30ne)
9. Hurstpierpoint church and building detail (9se)
10. Parkland (see above)

NB: No new waterlining, so not on 28,36,37,38,39,58
NB: *Fishguard* misspelled *Fisguard* (never corrected)

Lbl Maps 148 e27; Og A/2/1; RH
Documentary source: PRO OS 3/260

State S-P3 (?1823)

Skeletal one-inch sheet completed: 7 (1822)

Introduction of new standard features
B.2   Topographic features
  gg   Heathland, marshland: eg New Forest (15se), Romney Marsh (4nw)
  hh   Woodland: eg in New Forest, and in parkland
C.9   Italic
  l   Small detached "Part of Oxon" (7nw)

Changes
1. Sheppey road from outline to final state (2sw,3nw)
2. "+" symbols added at Chigwell (1sw), Hailsham (5sw), Cranley (8sw), Ewhurst, Ockley, Capel, Leigh (8se), Morested (11nw), Shaw, Linkenholt (12nw), Leckford (12sw), Lassham, Shaldon (12se), N.Wotton (19sw)
3. Beer Ferrers (northern one) to Beer (25sw)
4. Waterlining added on 28,36,37,38,39,58
5. Eddystone Lighthouse added (24sw)
6. *R.Thames*, *I.of Dogs* added in London (1sw)

Og (in library office) (TEW 1823 [?NB]); PC (2 copies)

State S-P4 (?1823) - Ramshaw printing

Introduction of new standard feature
A.3   *Printed by Ramshaw*: bottom centre, 25mm bnl

ABn PB 6186, Lrgs 1 (both TEW 1823 NB)

State S-P5 suspected (?1824 - ?Ramshaw printing)

Evidence: none, but 64,65,69,70 published 1 March 1824
Not found.

State S-P6 (?1830) - Ramshaw printing

One-inch sheets added:
   34 (1828), 35 (1830), 44 (1828)
   Skeletal sheets completed: 13 (1830), 41 (1831)
   Thinly mapped sheets: 64,65,69,70 (1824). With fewer placenames, incomplete road and river networks. Natural watercourses unnamed, man-made ones named. Many "+" are missing, many present without placenames. Most parkland areas are misshaped. Several categories of topographic feature were named on this issue which were deleted on the next. The river west-east through Brandon is shown as a road! Traces of the original north horizontal of the neat line are visible across 69,70.
   NB: Sheet 54 quarters published 1831, before 41

Introduction of new standard features
B.2   Topographic feature
  c   Additional use: Drainage dyke
C.9   Italic
  m   Minor bay, channel: eg Fossdyke Wash (69sw)
  n   Drainage dyke: eg South Holland Drain (65nw)
  p   Sandbank: eg Roger Sand (69nw)
  q   Lowland topographic feature: eg Holbeach Marsh, East Fen (69w), Rockingham Forest (64sw)
  r   Minor extensive district: East Holland (69nw)
  s   Road name: Littleworth Drove (64ne)

Change
1. County names added: CAERMARTHENSHIRE, MONMOUTHSHIRE, GLOUCESTERSHIRE, [Buckingham]SHIRE, [Oxford]SHIRE, [Lincoln]SHIRE, RUTLAND (but not Hertfordshire)

Lrgs 2 (TEW 182? NB). *NB*: Last figure cut off

State S-P7 suspected (?1837)

Evidence: hydrographic features around east coast probably added to 49,67,68 when published (1837-38), and to 48 when republished in quarters (1837-38)
Not found.

State S-P8 (?1839)

Dating evidence: 71NW,71NE published 1.7.39

One-inch sheets complete: 1-70,71SW,71SE,72,73,74,83,84
   NB: Still blank: 71NW,71NE,75,76,77,78,79,80,81,82
   NB: 64,65,69,70 redrawn to full specification
   NB: 83,84 redrawn and added to south sheet
   NB: Very faint quarter sheet lines, as though not yet properly engraved: 42,43,53,54,55,56,57
   NB: Not quartered: 45,46,48,49,50,51,52,59,60,61,62, 63,64,65,66,67,68,68E,69,70,71,72,73,74 (*NB*: 71)
   NB: A few rivers names on 42,43,50,54,61,67

Introduction of new standard features
A.4   Geological classification in south west, listed in a separate legend between 24 and neat line
B.2   Topographic features
  k   Track, or agger: eg 51NW,51SW,64se,66SW
  jj   Sea defence embankment: eg 69sw
  kk   Earthwork: Devil's Dyke (51SW,51SE)
  ll   Railway
  rr   Railway in tunnel (8ne)
B.3   Geological areas marked on 20-27,29-33, and an associated key between 24 and neat line
B.5   Hydrographic features
  h   Light vessel (off 9,2sw,48SE,67NW)
  j   Wreck (59SW)
  k   Buoy (2,3,48,49,67)
  l   Fathom lines (not 38 round to 22)
  m   Fathom lines on 38-22, and beyond sheet lines. The southernmost extends west to the plate edge

I.1 (S-P8)

          n   Danger area (eg 9sw,9se,16sw,17se, outside 9sw)
    C.4   Roman sloping capitals
          f   Large detached part of county (44,54NW,62SW)
    C.6   Egyptian capitals
          a   Ancient site: Willbury Hill (46NE), Roman Fosse
               Way (63,71)
    C.9   Italic
          h   Isolated building: additional uses: Stonham Pie
               Inn (50SW); Erith Hermitage (51NW)
          t   Railway company name
          u   River bank: River Nene North, South Banks (65sw)
          v   Bridge: at Fossdyke (69sw), Ponds (64se)
          w   Toll-bar: Causeway Bar (64ne)
          x   Ship: eg light vessel, wreck (2sw,59SW)
          y   Fathom measurement (38-22)

Changes
1. Neat line probably deleted (see p.3)
2. Map dimensions revised with the addition of 77-84
3. South sheet line of 9,5,4 adjusted north
4. *Printed by Ramshaw* probably deleted
Changes related to revised mapping on 64,65,69,70,83,84:
5. The river network redrawn
6. The road network redrawn, and extended: eg Sutton to Kings Lynn
7. Most placenames moved, added, deleted or altered - not all correctly, since Peterborough was classified a market town, and Yaxley a city! *NEWARK upon TRENT* became *NEWARK* upon *TRENT*
8. Names deleted: DOWNHAM MARKET, BRANDON, *Littleworth Drove*, East Holland, ISLE OF ELY, names of minor topographic features and man-made watercourses
9. *Whittlesea Mere* named (64se)
10. LINCOLNSHIRE, RUTLAND names moved
11. [Nottin]GHAM[shire] deleted (83sw), the name resited
12. "+" symbols positioned more accurately
13. Parkland mapping revised, and several previously anonymous ones named
Changes with the introduction of hydrographic information, presumably on publication of 49,67,68 (?on S-P7):
14. Hydrographic features and associated names added to 17,16,10,11,9,5,4,3,2,1,48,69. Some of these features outside sheet lines, eg *Owers* off Selsea Bill
15. Maplin Sand reshaped (2sw)
16. Coastal lighthouse added (48NE)
Changes associated with revised south-west mapping:
17. 32 sheet line verticals extended, 33 from landscape to portrait, Isles of Scilly sheet lines added
18. *Wolf Rock* added (33sw)
19. Somerset, Devon and Cornwall coast reshaped
20. South west river systems developed
21. Many roads added, especially on 19,21,22,25,26,30,31
22. Names of principal and minor headlands and bays added on 20,27,28,29,30,33,32,31,24,23,22; therefore Minehead, Watchet (20sw), *S^T IVES* (33ne) moved
23. Placenames revised: Bream to Brean (20ne), Tor Mohams to Tor Moham (22sw), *DOCK* to *DEVONPORT* (24nw); *DULVERTON* (21nw), Otterton (22ne), Alternan, Sth Petherwin, Lawannick (25nw), Welcombe, Fremington (26nw), Poughill (26sw) added; Buckland St Mary (21se), Lawhitton (25nw) moved
24. *Longships Lighthouse* named (33sw)
Changes on 47,48 (Essex-Suffolk border area)
25. Placenames revised: Barley (47nw), Glemsford (47ne), Nedging (48NW) deleted; Buckland, Bartlow (47nw), Boyton, Sutton (48NE) added; Tharfield replaced by Kelshall, Hinkeston by Hadstock (47nw), Castle Camps by Helion Bumpstead, Stoke by Sturmer (47ne); Duxworth to Duxford, Melborne to Melbourne, Linton to *LINTON* (47nw), Keddington to Redington! (47ne). Others moved or affected by names on adjacent sheets
26. County boundary revised south of Ickleton (47nw)
27. Suffolk-Cambridgeshire county boundary added (47ne)
28. Some roads revised, redrawn or added, eg Ipswich
29. Rivers in Linton-Haverfield area revised
Other changes
30. E.Grinsted to E^T. *GRINSTEAD* (6sw)
31. General addition of county names
32. Some roads re-engraved, eg Wareham-Bere Regis (16nw)
33. A section of Fosse Way north of Bath deleted (35se)
34. Underwater rocks added (9sw,16nw)

Railways (+ Italic names)
   NB: Some railways in this list are unopened, and are included according to Parliamentary Act alignments
LONDON BRIDGE-GREENWICH 8.2.36-24.12.38
   CORBETTS LANE JN (SOUTHWARK PARK)-WEST CROYDON 5.6.39
      NORWOOD-BRIGHTON+   12.7.41-21.9.41
      NORWOOD-DOVER+   26.5.42-7.2.44
NB: Alignment from L&C as in 1837 Act. The unbuilt section to Edenbridge occasionally reappears later
      BRIGHTON-SHOREHAM   12.5.40
LONDON NINE ELMS-SOUTHAMPTON+   21.5.38-11.5.40
PADDINGTON-BRISTOL+   4.6.38-31.5.41
   SWINDON-CHELTENHAM   4.11.40-12.5.45
      GLOUCESTER STN BRS   4.11.40-8.7.44
      CHELTENHAM ST JAMES BR   23.10.47
   THINGLEY JCN (CHIPPENHAM)-TROWBRIDGE   5.9.48
      BRADFORD JCN-BRADFORD-ON-AVON=   2.2.57
BRISTOL-EXETER+   14.6.41-1.5.44
   BRANCH AT NAILSEA - ERROR?
   WESTON JCN-WESTON BR   14.6.41
   BRANCH AT BLEADON - ERROR?
   BRANCH TO BURNHAM - ERROR
   TIVERTON JCN-TIVERTON BR   12.6.48
EUSTON-BIRMINGHAM+   20.7.37-17.9.38
BIRMINGHAM-MADELEY (ONE MILE SOUTH)   4.7.37-1.39
LONDON BISHOPSGATE-NORWICH-YARMOUTH+   20.6.39-7.3.43
NB: The unused alignment from Colchester as in the EC 1836 Act: the Norwich-Yarmouth part was never deleted
LONDON ISLINGTON-CAMBRIDGE+   15.9.40-30.7.45
NB: Alignment north from Islington as in the N&E 1836 Act: the unused site at Cambridge was never deleted

Lbl OSD Serial No 510 (Portfolio 28)
Lit: Hodson (1989) 91

State S-1 (?1841) (?first complete state)

Dating evidence: the agents were appointed on 7.8.1840. Grattan & Gilbert were bankrupt by autumn 1841

Dimensions (derived from later flat copies):
   Paper: 1064mm W-E by 712mm S-N
   Plate: 960mm W-E by 662mm S-N
   Neat line: 926mm W-E (no northern neat line present)

Introduction of new standard features
A.5   Scale of ten Statute Miles to an Inch [scale-bar 10+50 miles]: bottom right, 25mm anl
A.6   *Price four Shillings*: bottom right, 2mm anl
A.7   PLATES of the one-inch map listed, with prices, sheet numbers and names: top left, 35mm rnl
A.8   *Agents for the Sale of Ordnance Survey Maps*: below list of plates. First wording: *West End* J. Arrowsmith, *10, Soho Square. / City* Grattan & Gilbert *51, Paternoster Row*

Changes
1. Title altered to: **Index to the Ordnance Survey, of England and Wales**
2. A piano key border added on west, south and east sides, c.12mm beyond the original neat line
3. Quarter sheet lines added generally except 1
4. Farnham to Petersfield road added
5. Longstock "+" moved from left to below (12sw)
6. Liss to Lyss (11ne)
7. DOWNHAM MARKET, (65ne) BRANDON (65se) reinstated
   NB: Garston mispelled Gratston

Additions to railways (+ Italic names)
London Bridge-Greenwich
   Corbetts Lane Jcn (Southwark Park)-West Croydon
      Norwood-Brighton+
         REDHILL.....Dover+   26.5.42-7.2.44
NB: Alignment from Redhill as built, as in SE 1839 Act
Paddington-Bristol+
NB: DIDCOT-GOOSEY REALIGNED. The original alignment was not properly deleted, and the name "Great Western Railway" always followed this
   DIDCOT-OXFORD OLD STN   12.6.44
      Swindon-Cheltenham BIRMINGHAM+   24.6.40-17.8.41
Bristol-Exeter+
   Weston Jcn-Weston RE-ENGRAVED FURTHER NORTH
   Branch to Burnham DELETED
PENARTH-MERTHYR   8.10.40-12.4.41
NB: Routes from Penarth and Cardiff were in the TV 1836 Act: the Penarth branch was not built
   PONTYPRIDD-TREHAFOD   10.6.41
   PONTYPRIDD-NELSON   1841
Euston-Birmingham+
   CHEDDINGTON-AYLESBURY HIGH STREET BR   10.6.39
Birmingham-Madeley EARLESTOWN+   25.7.31-4.7.37=
   CREWE-BIRKENHEAD+   23.9.40-1.10.40
   CREWE-[MANCHESTER]+   4.7.40-10.8.42

LIVERPOOL-[MANCHESTER]   16.9.30-15.8.36=
   LIVERPOOL CROWN STREET BR   16.9.30=
   WIDNES-ST HELENS=   2.1.32-21.2.33=
      BROAD OAK BR   21.2.33=
DERBY-NOTTINGHAM   4.6.39
   TRENT-RUGBY+   5.5.40-12.8.40
DERBY-STECHFORD+   12.8.39
NB: Routes to L&B at Stechford and Hampton in the B&D Jcn 1836 Act: Stechford route from Whitacre not built
DERBY-[LEEDS]+   11.5.40-1.7.40
   ROTHERHAM-SHEFFIELD   31.10.38=
London Bishopsgate-Norwich-Yarmouth+
   ROMFORD-SHELL HAVEN
NB: Thames Haven Dock & Ry Co line authorised 1836 but not built. The alignment was never deleted from the map

Lbl Maps 148 e27; SOrm (TEW 1837)
Documentary source: PRO WO 47/1865 7829-30

State S-2 (?1843)

Introduction of new standard feature
C.9   Italic
   aa   Railway station: (Ammanford, 41se)

Changes
1. *City* agent altered to Letts & Son, *8, Cornhill*
2. Queenborough (3nw), Wily (14sw) added
3. Lyndhurst, Minstead, Brokenhurst added (15se)
4. *ISLE OF ELY* reinstated (65sw)
5. *Old* and *New Bedford River* names reinstated (65sw)
6. Names of further estuarial rivers and principal bays added: *RIVER MERSEY* (80NW), *RIVER DEE* (79NE), *PORTH DINLLEYN* (75NW), *SEVERN RIVER* (35), *BUDE BAY* (29se)
7. General addition of river and canal names
8. Most railways re-entered, (not Corbetts Lane Jcn-West Croydon, Swindon-Cheltenham, Thingley Jcn-Trowbridge, Colchester-Norwich, Norwich-Yarmouth)

Additions to railways (+ Italic names)
CANTERBURY-WHITSTABLE   3.5.30-19.3.32=
London Nine Elms-Southampton+
   EASTLEIGH-GOSPORT   29.11.41
Paddington-Bristol+
   Swindon-Birmingham Camp Hill+=
      CHELTENHAM ST JAMES BR LACKING
CARDIFF......Merthyr   8.10.40-12.4.41
NB: Faint traces of the Penarth branch remain
LLANELLY-CROSS HANDS COLLIERY   1.6.39-6.5.41
   PANTYFFYNON-BRYNAMMAN   6.5.41-6.42
Liverpool-[Manchester]+   NAMED
Derby......HAMPTON-IN-ARDEN+   12.8.39=
NB: Faint traces of the Stechford alignment remain
   WHITACRE-BIRMINGHAM   10.2.42

SOos

I.1 (S-3)

State S-3 (?1845)

Dating evidence: Longman appointed agent in 1845

Change
1. A further *City* agent added: Messrs Longman, Brown & C{o}. *Paternoster Row*

PC (RT 1845)

State S-4 (?1847)

Changes
1. "London" heading added to list of agents
2. Letts entry: Letts Son & Steere, *8, Royal Exchange*
   NB: This address first appeared in 1845 directories
3. Hordwell to Hordle (16ne)

Og A/2/1 (RT 1847)

State S-5 (?1849)

This issue was apparently current for eight years. No railway marked was opened later than 1849, it is recorded on 1854 paper, and a possible 1856 EPD is known.

Introduction of new standard features
A.9   List of fifteen *Country Agents*: west of 39
A.10 *Printed from an Electrotype*: bc, 1mm anl

Changes
1. Price reduced to *Price Two Shillings*
2. Prices reduced in list of plates (on 1.7.1848)
3. Sheet 1 noted as quartered in the list of plates
4. *West End* agent T.W.Saunders, *6, Charing Cross* added
   NB: This agency was probably awarded in late 1846
5. Middle Level Main Drain added, unnamed (65nw)

Additions to railways (+ Italic names)
London Bridge-Greenwich
  Corbetts Lane Jcn-West Croydon EPSOM STN 10.5.47
    Norwood-Brighton+
      Redhill-Dover+
        TONBRIDGE-TUNBRIDGE WELLS 20.9.45-25.11.46
        PADDOCK WOOD-MAIDSTONE 29.5.44
        ASHFORD-RAMSGATE 6.2.46-13.4.46
          MINSTER-DEAL 1.7.47
          RAMSGATE-MARGATE 1.12.46
        THREE BRIDGES-HORSHAM OLD STN 14.2.48
        KEYMER JCN (WIVELSFIELD)-LEWES 1.10.47
        BRIGHTON-ST LEONARDS 8.6.46-7.11.46
          SOUTHERHAM JCN (LEWES)-NEWHAVEN 8.12.47
          POLEGATE-EASTBOURNE BR 14.5.49
        Brighton-Shoreham FAREHAM 24.11.45-1.10.48
          COSHAM-PORTSMOUTH 15.3.47-14.6.47
London Nine Elms-Southampton+
  CLAPHAM JCN-RICHMOND 27.7.46
  WEYBRIDGE-CHERTSEY 14.2.48

WOKING-GUILDFORD 5.5.45
EASTLEIGH-SALISBURY MILFORD 27.1.47
NORTHAM J (SOUTHAMPTON)-DORCHESTER STN 1.6.47-29.7.47
  HAMWORTHY JCN-HAMWORTHY BR 1.6.47
Paddington-Bristol+
READING-HUNGERFORD 21.12.47
  SOUTHCOTE JCN (READING)-BASINGSTOKE 1.11.48
Thingley Jcn (Chippenham)-Trowbridge WESTBURY 5.9.48
BRISTOL Standish-Birmingham Camp Hill+ 8.7.44
  MANGOTSFIELD-KEYNSHAM TRAMWAY 17.7.32=
    SOUNDWELL COLLIERY BR
    WESTERLEIGH MINERAL BR
    OXWICK FARM MINERAL BR
Bristol-Exeter+ PLYMOUTH 30.5.46-4.4.49
  YATTON-CLEVEDON BR 28.7.47
  ALLER JCN (NEWTON ABBOT)-TORQUAY 18.12.48
Euston-Birmingham+
  LEIGHTON BUZZARD-DUNSTABLE 29.5.48
  BLETCHLEY-BEDFORD 17.11.46
    BEDFORD GOODS BR
  BLISWORTH-PETERBOROUGH 13.5.45-2.6.45
  COVENTRY-LEAMINGTON 9.12.44
LIVERPOOL STREET.....50+ *Norwich* 1.6.46-3.12.49
NB: Alignment from Colchester as in EU and I&B Acts of 1844, 1845 and 1846. Earlier route to Norwich deleted. The line into Liverpool Street was never drawn
  STRATFORD....Trumpington+ YARMOUTH 1.5.44-15.12.45
NB: Islington alignment to Tottenham deleted. Norwich-Yarmouth alignment as in the 1842 Y&N Act
  BROXBOURNE JCN-HERTFORD 31.10.43
    Great Chesterford 51-NEWMARKET= 3.1.48
    CHESTERTON (CAMBRIDGE)-GODMANCHESTER 17.8.47
      ST IVES-MARCH 1.2.48
    ELY NORTH JCN-65= 10.12.46
      MARCH-WISBECH SOUTH BRINK 3.5.47
        WISBECH-MAGDALEN ROAD 1.2.48
    ELY NORTH JCN-KINGS LYNN 27.10.46-26.10.47
      KINGS LYNN-DEREHAM 27.10.46-11.9.48
    WYMONDHAM-FAKENHAM 7.12.46-20.3.49
    REEDHAM-LOWESTOFT 3.5.47
  STRATFORD-NORTH WOOLWICH 29.4.46-14.6.47
  WITHAM-MALDON 15.8.48
  WITHAM-BRAINTREE OLD STN 15.8.48
  COLCHESTER-HYTHE 1.4.47
  BENTLEY-HADLEIGH BR 21.8.47
  HAUGHLEY-BURY ST EDMUNDS 30.11.46

PC (EPD 22 Dec.1851); Lkg S039752 (JWTM 1851); Ob C17 (26)-(26a) (JWTM 1853) (EPD 2 Sep.1853 at foot of map); PC (JWTM 1854); Lrgs 6 (JW 1854) (EPD 26 Jun.1857?); Lbl Maps 59.c.43

State S-6 (?1857)

Changes
1. The two lists of agents deleted
2. The electrotype note deleted
3. *CREWE* added (73NE)

4. Extra railway names added in italic
5. *Middle Level* [Main] *Drain* named (65nw)

Additions to railways (+ Italic names)
London Bridge-Greenwich
   Corbetts Lane Jcn (Southwark Park)-Epsom Stn
     Norwood-Brighton+
       REDHILL-SHALFORD JCN (GUILDFORD) 4.7.49-20.8.49
       Redhill-Dover+
         Tonbridge-Tunbridge Wells   BOPEEP JCN (ST LEONARDS)   1.9.51-1.2.52
         Ashford-Ramsgate+   NAMED
       Brighton-St Leonards   ASHFORD+ NAMED   13.2.51
       Brighton-Fareham+   NAMED
   NORTH KENT JCN Maidstone-Paddock Wood 23.8.47-18.6.56
London Nine Elms-Southampton+
   Clapham Jcn-Richmond   WINDSOR   22.8.48-1.12.49
     BARNES-BRENTFORD-FELTHAM JCN   22.8.49-1.2.50
   HAMPTON COURT JCN-HAMPTON COURT BR   1.2.49
   Woking-Guildford   GODALMING   15.10.49
     GUILDFORD-ALTON   20.8.49-28.7.52
       ASH JCN-READING   4.7.49-20.8.49
   Northam Jcn (Southampton)-Dorchester Stn+   NAMED
Paddington-Bristol+
   SLOUGH-WINDSOR BR   8.10.49
   Reading-Hungerford+   NAMED
   Didcot-Oxford   SHREWSBURY+   1.6.49-14.11.54
     RADLEY-ABINGDON BR   2.6.56
     WOLVERCOT (OXFORD) - PRIESTFIELD (WOLVERHAMPTON)+
                      5.10.50-30.4.54
       KINGHAM-CHIPPING NORTON   10.8.55
       DROITWICH JCNS-STOKE WORKS   18.2.52
NB: S-E loop at Droitwich disappears on later editions
       DUDLEY-WICHNOR+   1.11.47-1850
   Swindon-Standish
     KEMBLE JCN-CIRENCESTER BR   31.5.41
   Bristol-Birmingham Camp Hill+
     GLOUCESTER LOOP   29.5.54
     GLOUCESTER-HAVERFORDWEST+   18.6.50-2.1.54
       GRANGE COURT - BARRS COURT JCN (HEREFORD)+
                   11.7.53-12.6.55
       BULLO PILL-CHURCHWAY COLLIERY   24.7.54
       LANDORE-SWANSEA   18.6.50
     ASHCHURCH-TEWKESBURY   21.7.40
NB: Engraved in outline and treated as a road
Bristol-Plymouth+   EXETER-PLYMOUTH NAMED
   HIGHBRIDGE-GLASTONBURY+   28.8.54
   DURSTON-YEOVIL+   1.10.53
     COWLEY BRIDGE JCN (EXETER)-BIDEFORD+   1.8.48-2.11.55
NEWPORT-BLAENAVON   1.7.52-2.10.54
NEWPORT-NANTYGLO   12.12.50 or 4.8.52-5.55
   ABERBEEG-EBBW VALE   19.4.52 or 5.55
Cardiff-Merthyr
   ABERCYNON-ABERDARE-NEATH+   24.9.51-6.8.56
NB: Incorrectly transferred from TV to VN route
     GELLY TARW JCN (HIRWAIN)-MERTHYR   2.11.53
Euston-Birmingham+   BUSHBURY (WOLVERHAMPTON) 2.52-1.6.54
   CAMDEN JCN-? *Stratford*   26.9.50-15.2.51

BLETCHLEY-BANBURY   1.5.50
   VERNEY JCN-OXFORD+   1.10.50-20.5.51
     YARNTON CURVE   1.4.54
Blisworth PETERBOROUGH-65 Ely North Jcn+   10.12.46=
RUGBY-LUFFENHAM   1.5.50-2.6.51
TRENT VALLEY JCN (RUGBY)-STAFFORD+   15.9.47
Coventry-Leamington   RUGBY   1.3.51
COVENTRY-NUNEATON   2.9.50
Birmingham-Earlestown+
   STAFFORD-WELLINGTON+   1.6.49
   NORTON BRIDGE-STOKE-CHEADLE HULME   9.6.45-18.6.49
     STONE-COLWICH   1.5.49
     STOKE JCN-BURTON   7.8.48-11.9.48
       UTTOXETER-NORTH RODE   13.7.49
       MARSTON JCN-WILLINGTON JCN   13.7.49
     HARECASTLE-CREWE   9.10.48
       LAWTON JCN-SANDBACH   21.1.52-12.66
   Crewe-Birkenhead+
     CHESTER-WALTON JCN   18.12.50
NB: Line engraved in outline to Warrington
     CHESTER-PONTYPOOL+   4.11.46-2.1.54
       SALTNEY (CHESTER)-HOLYHEAD+   1.5.48-20.5.51
         MOLD JCN (SALTNEY)-MOLD   14.8.49
           PADESWOOD-COED TALON   11.49
         BANGOR-CAERNARVON   1.3.52-1.7.52
       GOBOWEN-OSWESTRY   23.12.48
       PONTYPOOL-HENGOED   20.8.55-11.1.58
   Crewe-[Manchester]+
     [MANCHESTER]-GARSTON DOCK+   20.7.49-1.5.54
LIVERPOOL EXCHANGE-[LOSTOCK JCN (BOLTON)]   20.11.48
   SANDHILLS (LIVERPOOL)-[SOUTHPORT]   1.10.50
[PENISTONE]-[ULCEBY]+   1.11.48-17.7.49
   WOODHOUSE JCN-BEIGHTON JCN   12.2.49
NB: Causing an uncorrected error to MR line
   CLARBOROUGH JCN-SYKES JCN (SAXILBY)+   7.8.50
Derby-Nottingham [BARNETBY]+ NAMED 4.8.46-18.12.48
   Trent-Rugby+=
     SYSTON-PETERBOROUGH   1.9.46-20.3.48
     KNIGHTON JCN (LEICESTER)-BURTON   27.3.48-7.49
       SWADLINCOTE LOOP   2.10.48-1.9.84
         WOODVILLE GDS BR (?error)
   LONG EATON-MANSFIELD   1.7.47-6.11.51
     KIRKBY-LENTON JCN (NOTTINGHAM)   2.10.48
   ROLLESTON JCN-SOUTHWELL=   1.7.47
Derby-[Leeds]+
NB: Alignment across MS&L incorrectly made stepped
   AMBERGATE-ROWSLEY   4.6.49
   Rotherham-Sheffield=
     WINCOBANK-[ALDAM JCN]=   4.9.54
KINGS CROSS (MAIDEN LANE)-[DONCASTER]+   4.9.49-15.7.52
NB: The line into Kings Cross was never drawn
   HITCHIN-SHELFORD (CAMBRIDGE)   21.10.50-1.4.52
   WERRINGTON J (PETERBORO')-GAINSBORO'+ 17.10.48-9.4.49
     BOSTON-[GRIMSBY]+   1.3.48-2.10.48
     WOODHALL JCN-HORNCASTLE BR   26.9.55
   ESSENDINE-STAMFORD   1.11.56
   LITTLE BYTHAM-EDENHAM   8.12.57
   GRANTHAM-COLWICK JCN (NOTTINGHAM)+   15.7.50-2.8.52

I.1 (S-6)

Liverpool Street-50 NORWICH VICTORIA+   3.12.49
   (Stratford)-Cambridge-Norwich-Yarmouth+
NB: Again showing Islington, not Stratford, alignment
    ANGEL ROAD-ENFIELD   1.3.49
      Chesterton-Godmanchester HUNTINGDON   29.10.51
   FOREST GATE JCN-TILBURY-SOUTHEND   13.4.54-1.3.56
   Witham-Maldon, Witham-Braintree Old Stn LACKING
   MARKS TEY-SUDBURY OLD STN   2.7.49
   Colchester-Hythe   LACKING
   MANNINGTREE-HARWICH   15.8.54
   Haughley BURY-N'M'KET 51-GT CHESTERFORD 3.1.48-1.4.54
    SIX MILE BOTTOM-CAMBRIDGE   9.10.51
   TROWSE UPPER JCN-TROWSE LOWER JCN   8.9.51

Lrgs 4 (JW 1856) (1857 ms date on reverse); Lrgs 5 (JW 185??); PC (JW 1858) (2 copies); Cu Maps 34.03.105; Mp FF912.42 064; PC

State S-7 (?1859)

Additions to railways (+ Italic names)
London Bridge-Greenwich
   Corbetts Lane Jcn (Southwark Park)-Epsom Stn
     WEST CROYDON-WIMBLEDON   22.10.55
   North Kent Jcn-Strood-Paddock Wood
     LEWISHAM JCN-BECKENHAM   1.1.57
WATERLOO London Nine Elms-Southampton+   11.7.48=
   CLAPHAM JCN-CRYSTAL PALACE-SHORTLANDS   1.12.56-3.5.58
     CRYSTAL PALACE-SYDENHAM   27.3.54
   Clapham Jcn-Windsor
     STAINES JCN-WOKINGHAM   4.6.56-9.7.56
   Woking-Godalming HAVANT+ NAMED   28.12.58
   WORTING JCN (BASINGSTOKE)-15 Gillingham 3.7.54-2.5.59
   Eastleigh-Salisbury FISHERTON JCN (SALISBURY) 1.5.57
Paddington-Bristol+
   WEST DRAYTON-UXBRIDGE   8.9.56
   MAIDENHEAD-HIGH WYCOMBE   1.8.54
   TWYFORD-HENLEY-ON-THAMES BR   1.6.57
   Thingley Jcn-Westbury SALISBURY   9.9.51-30.6.56
     WESTBURY-14 Frome   7.10.50=
     HOLT JCN-DEVIZES=   1.7.57
Euston-Bushbury (Wolverhampton)+
   Camden Jcn-7 // N&E-STRATFORD   15.8.54
     STRATFORD LINK TO N&E
   WILLESDEN JCN-KENSINGTON CANAL BASIN   27.5.44=
   WILLESDEN JCN-KEW   15.2.53-1.6.54
   WATFORD JCN-ST ALBANS ABBEY   5.5.58
[MANCHESTER] [Penistone]-[Ulceby]+   11.12.42=
   DINTING-GLOSSOP BR   9.6.45=
Liverpool Street (Bishopsgate)-Norwich Victoria+
   Stratford-Cambridge-Norwich-Yarmouth+
NB: Showing both Islington and Stratford alignments
    LOUGHTON JCN (STRATFORD)-LOUGHTON   22.8.56
   Witham-Maldon REINSERTED ON 1
   FENCHURCH STREET-BARKING   6.7.40-31.3.58

Ob C17 (420) (JW 1858) (EPD Oct.1859); Og A/2/1 (JWTM 1858); RH (JWTM 1858); PC

State S-8 suspected (?1861)

Evidence: Hartlebury-Shrewsbury (1.2.62) was the only additional railway on S-9 to be named in italic. Egyptian lettering was used for other additional railway names, all of which were opened earlier, perhaps on S-8.

Introduction of new standard features
B.2   Topographic feature
   pp Mineral tramway: eg 25sw,36nw
C.6   Egyptian capitals
   b   Railway company name (State S-8 additions only)

Additions to railways (+ Italic names; * Egyptian upper case names added only on this issue). Hypothetical list
London Bridge-Greenwich
   Corbetts Lane Jcn (Southwark Park)-Epsom Stn
     Norwood-Brighton+
       PURLEY-CATERHAM BR   4.8.56
     Redhill-Dover+
       Tonbridge-Bopeep Jcn (St Leonards)* NAMED
       Three Bridges-Horsham PETWORTH*   10.10.59
       THREE BRIDGES-EAST GRINSTEAD   9.7.55
       Keymer Jcn (Wivelsfield)-Lewes
         HAMSEY (LEWES)-UCKFIELD   18.10.58
     Brighton-Ashford+
       POLEGATE-HAILSHAM   14.5.49=
   North Kent Jcn-Strood-Paddock Wood* NAMED
Waterloo-Southampton+
   Clapham Jcn-Shortlands CANTERBURY*   25.1.58-3.12.60
     SITTINGBOURNE-SHEERNESS   19.7.60
   Clapham Jcn-Windsor* NAMED
     RAYNES PARK-LEATHERHEAD   1.2.59
   Woking-Havant+
     Guildford-Alton
       Ash Jcn-Reading* NAMED
   Worting Jcn (Basingstoke)-15 EXETER*   7.5.60-19.7.60
   Northam Jcn-Dorchester Stn+ DORCHESTER JCN 20.1.57
     BROCKENHURST-LYMINGTON BR   12.7.58-19.9.60
     WIMBORNE-BLANDFORD   1.11.60
Paddington-Bristol+
   SOUTHALL-BRENTFORD BR   13.7.59
   Didcot-Birmingham-Shrewsbury+
     Wolvercot (Oxford)-Priestfield (Wolverhampton)+
       HONEYBOURNE-STRATFORD-UPON-AVON=   11.7.59
       TUNNEL JCN (WORCESTER)-MALVERN WELLS   25.7.59-25.5.60
     Dudley-Wichnor+
       WALSALL-RUGELEY   1.2.58-7.11.59
   HATTON-STRATFORD-UPON-AVON OLD STN   10.10.60
   Thingley Jcn (Chippenham)-Salisbury* NAMED
     Westbury-14 WEYMOUTH STN*   1.9.56-20.1.57
     FROME-RADSTOCK   14.11.54
     WITHAM-SHEPTON MALLET=   9.11.58
     MAIDEN NEWTON-BRIDPORT   12.11.57
   BATHAMPTON Bradford-on-Avon-Bradford Jcn   2.2.57
   Bristol-Birmingham Camp Hill+
     COALEY JCN-DURSLEY BR   25.8.56

```
        Gloucester-Haverfordwest+  NEYLAND   15.4.56
            MYRTLE HILL J (CARMARTHEN)-CONWIL  1.3.60-3.9.60
        ASHCHURCH-TEWKESBURY
NB: Second alignment - engraved as a railway
        BARNT GREEN-REDDITCH=   10.9.59
Bristol-Plymouth+  PENZANCE*   11.3.52-11.5.59
    BURNHAM Highbridge-Glastonbury WELLS+=  3.5.58-15.3.59
    Aller Jcn (Newton Abbot)-Torquay  PAIGNTON  2.8.59
    TAVISTOCK JCN (PLYMOUTH)-TAVISTOCK  22.6.59
        LEE MOOR TRAMWAY
    LISKEARD-KILMAR TOR TRAMWAY   28.11.44=
    PAR-TREFFREY SIDING   18.5.47=
        COLCERROW BR   c.1844=
    PENWITHERS JCN-TRURO NEWNHAM   16.4.55
    PORTREATH BR   11.3.52
    CARNBREA BR   11.3.52
    ROSKEAR BR   11.3.52
Newport-Nantyglo
    BASSALEG-RHYMNEY  (ENGRAVED AS A TRAMWAY)
Cardiff-Merthyr
    RADYR-PENARTH   1859
    TAFFS WELL-RHYMNEY   25.2.58
    Pontypridd-Trehafod  TREHERBERT   3.49-7.8.56
        PORTH-FERNDALE   3.49-1856
    DOWLAIS MINERAL TRAMWAYS
SWANSEA-YSTALYFERA   1852-24.12.59
Llanelly-Tirydail  LLANDOVERY*   24.1.57-1.4.58
Euston-Bushbury (Wolverhampton)+
    Camden Jcn-? // N&E-Stratford
        KENTISH TOWN-GOSPEL OAK-WILLESDEN JCN   2.1.60
    Leighton Buzzard-Dunstable  WELWYN*   3.5.58-1.9.60
    Blisworth-Ely North Jcn+
        NORTHAMPTON-MARKET HARBOROUGH*   16.2.59
Birmingham-Earlestown+
    Norton Bridge-Stoke-Macclesfield-Cheadle Hulme
        Stoke Jcn-Burton
            Uttoxeter-North Rode
                ROCESTER-ASHBOURNE   31.5.52
            STOKE JCN-BIDDULPH-CONGLETON   28.8.60
            STOKE JCN-SILVERDALE   6.9.50-31.8.60
    CREWE-SHREWSBURY*   2.9.58
    Crewe-Birkenhead+
        Chester-Walton Jcn  WARRINGTON   11.55
        Chester-Pontypool+
            Saltney (Chester)-Holyhead+
                FORYD JCN-DENBIGH   5.10.58
                LLANDUDNO JCN-LLANDUDNO BR   1.10.58
            WHEATSHEAF JCN (WREXHAM)-GWERSYLLT COLL   7.47=
            Gobowen-Oswestry  LLANIDLOES*   30.4.59-10.6.61
            SHREWSBURY-MINSTERLEY=   14.2.61
            CRAVEN ARMS-KNIGHTON=   1.10.60-6.3.61
                LEOMINSTER-KINGTON   1.56-2.8.57
                PONTYPOOL-MONMOUTH   2.6.56-12.10.57
            Pontypool-Hengoed  QUAKERS YARD   11.1.58
    Crewe-[Manchester]+
        STOCKPORT-WHALEY BRIDGE   9.6.57
        HEATON NORRIS (STOCKPORT)-[GUIDE BRIDGE]   1.8.49=
        [Manchester]-Garston Dock+

        Widnes-St Helens
            ST HELENS NEW STN LOOP   1.2.58
Derby-[Barnetby]+
    Trent-Wigston Jcn (Leicester)+=  HITCHIN*   15.4.57
        MOUNT SORREL (PRIVATE) BR   1861
    Knighton Jcn (Leicester)-Burton
        DESFORD-LEICESTER WEST STREET BR   27.3.48=
Derby-[Leeds]+
    LITTLE EATON-RIPLEY   8.48-9.55
Kings Cross (Maiden Lane)-[Doncaster]+
    Hitchin-Shelford (Cambridge)*  NAMED
    Werrington Jcn (Peterborough)-Gainsborough+
        SPALDING-HOLBEACH   9.8.58
        BOSTON-BARKSTON (GRANTHAM)*   16.6.57-13.4.59
    ESSENDINE-BOURNE   16.5.60
Liverpool Street (Bishopsgate)-Norwich Victoria+
    Stratford-Cambridge-Norwich-Yarmouth+
        Broxbourne Jcn-Hertford  WELWYN*   1.3.58
        Ely North Jcn-Kings Lynn
            Kings Lynn-Dereham*  NAMED
        Wymondham-Fakenham  WELLS   1.12.57
    Witham-Maldon  REINSERTED
    Witham-Braintree Old Stn  REINSERTED
    Marks Tey-Sudbury Old Stn*  NAMED
        CHAPPEL-HALSTEAD   16.4.60
    COLCHESTER-HYTHE STN  (REINSERTED)   2.7.49=
    IPSWICH-YARMOUTH SOUTH TOWN*   20.11.54-1.6.59
        WICKHAM MARKET-FRAMLINGHAM BR   1.6.59
        SNAPE BR   1.6.59
        SAXMUNDHAM-ALDEBURGH BR   1.6.59-12.4.60
        BECCLES-OULTON BROAD JCN (LOWESTOFT)   1.6.59
            KIRKLEY GDS BR   1.6.59
    TIVETSHALL-BUNGAY   1.12.55-2.11.60

Not found.

State S-9 (?1863)

NB: Some or all State S-8 details may apply here

Additions to railways (+ Italic names, * Egyptian names)
London Bridge-Greenwich
    Corbetts Lane Jcn (Southwark Park)-Epsom Stn
        Norwood-Brighton+
            Three Bridges-Petworth*
                ITCHINGFIELD JCN (WEST HORSHAM) - SHOREHAM
                                        1.7.61-16.9.61
LONDON BATTERSEA PIER-HERNE HILL   25.8.62
    STEWARTS LANE GDS BR
    BRIXTON-LOUGHBOROUGH JCN   1.5.63
    HERNE HILL-ELEPHANT & CASTLE   6.10.62
Waterloo-Southampton+
    Clapham Jcn-Canterbury*  DOVER PRIORY   22.7.61
        SWANLEY-SEVENOAKS   2.6.62
        FAVERSHAM-HERNE BAY   1.8.60-13.7.61
    Worting Jcn (Basingstoke)-Exeter*
        EXMOUTH JCN (EXETER)-EXMOUTH   1.5.61
COWES-NEWPORT   16.6.62
```

I.1 (S-9)

Paddington-Bristol+
  Maidenhead-High Wycombe THAME 1.8.62
  Didcot-Birmingham-Shrewsbury+
    Wolvercot (Oxford)-Priestfield (Wolverhampton)+
      YARNTON-WITNEY 14.11.61
      KINGHAM-BOURTON 1.3.62
      Tunnel Jcn (Worcester)-Malvern Wells SHELWICK JCN (HEREFORD) 13.9.61
      HARTLEBURY-SHREWSBURY+ 1.2.62
        BUILDWAS-MUCH WENLOCK 1.2.62
     MADELEY JCN-LIGHTMOOR 1.6.54
     KETLEY JCN-LIGHTMOOR 1.5.57-1.11.64
  Bristol-Birmingham Camp Hill+
    Gloucester-Neyland+
      LLANTRISANT-PENYGRAIG 2.8.60-12.62
       CASTELLA BR (NEVER OPENED)
       COED ELY-GELLYRHAIDD COLLIERY 8.1.62
      PORTHCAWL TRAMWAY
      JOHNSTON-MILFORD HAVEN BR 7.9.63
Bristol-Penzance+*
  Burnham-Glastonbury+= CASTLE CARY 3.2.62
    Glastonbury WELLS-SHEPTON MALLET Witham= 1.3.62
  NORTON FITZWARREN-WATCHET 31.3.62
Euston-Bushbury (Wolverhampton)+
  Willesden Jcn-Kensington CLAPHAM JCNS 2.3.63-6.7.65
  WATFORD JCN-RICKMANSWORTH 1.10.62
  Bletchley-Bedford* CAMBRIDGE 23.6.57-1.8.62
  Trent Valley Jcn (Rugby)-Stafford+
    NUNEATON-HINCKLEY 1.1.62
Birmingham-Earlestown+
  ASTON-SUTTON COLDFIELD 2.6.62
  Stafford-Wellington+
    HADLEY-COALPORT 17.6.61
  Crewe-Birkenhead+
    Chester-Pontypool+
      Saltney (Chester)-Holyhead+
        Foryd Jcn-Denbigh RUTHIN 1.3.62
        Bangor-Caernarvon PENYGROES 2.9.67
NB: Nantlle Ty open since 1828: it carried passengers
      RUABON-LLANGOLLEN 1.12.61
      Shrewsbury CRUCKMEOLE JCN-BUTTINGTON 27.1.62
      WOOFFERTON-TENBURY 1.8.61
  Crewe-[Manchester]+
    [Manchester]-Garston Dock+
      TIMPERLEY-KNUTSFORD 12.5.62
[Manchester]-[Ulceby]+
  [HYDE JCN]-COMPSTALL (MARPLE) 1.3.58-5.8.62
Derby-[Leeds]+
  Ambergate-Rowsley HASSOP 1.8.62
  CLAY CROSS-ALFRETON= 1.11.61
Kings Cross (Maiden Lane)-[Doncaster]+
  Werrington Jcn (Peterborough)-Gainsborough+
    Spalding-Holbeach SUTTON BRIDGE 3.7.62
Liverpool Street (Bishopsgate)-Norwich Victoria+
  Stratford-Cambridge-Norwich-Yarmouth+
    Ely North Jcn-Kings Lynn HUNSTANTON 3.10.62
  Marks Tey-Sudbury Old Stn*
    Chappel-Halstead YELDHAM 26.5.62

Cu Maps 34.01.112 (EPD Mar.1863); PC (TJH) (EPD Apr. 1863); Dtc (TJH 1860) (EPD Jul.1863)

State S-10 (?1865)

Introduction of new standard feature
B.2   Topographic feature
    qq Underground railway: Paddington-Farringdon

Additions to railways (+ Italic names, * Egyptian names)
London Bridge-Greenwich
  DEPTFORD WHARF BR 2.7.49=
  Corbetts Lane Jcn (Southwark Park)-Epsom Stn
    Norwood-Brighton+
      Three Bridges-Petworth*
        HARDHAM JCN (PULBOROUGH)-FORD 3.8.63
      Brighton-Fareham+
        FORD-LITTLEHAMPTON BR 17.8.63
London Battersea Pier-Herne Hill CRYSTAL PALACE 1.7.63
NB: Incorrect alignment: should go to Penge Jcn
Waterloo-Southampton+
  Clapham Jcn-Crystal Palace-Strood-Dover Priory*
    BALHAM-WINDMILL BRIDGE JCN (EAST CROYDON) 1.12.62
    Faversham-Herne Bay RAMSGATE HARBOUR 5.10.63
  Clapham Jcn-Windsor*
    TWICKENHAM-KINGSTON 1.7.63
    Guildford-Alton
      BR AT ALDERSHOT - ERROR?
  Worting Jcn (Basingstoke)-Exeter*
    CHARD JCN-CHARD 8.5.63
  Eastleigh-Fisherton Jcn (Salisbury)
    KIMBRIDGE JCN-ANDOVER TOWN (NOT CONNECTED) 6.3.65
  Eastleigh-Gosport
    FORTON JCN (GOSPORT)-STOKES BAY BR 6.4.63
  Northam Jcn (Southampton)-Dorchester Jcn+
    REDBRIDGE-ROMSEY (NOT CONNECTED) 6.3.65
    RINGWOOD-CHRISTCHURCH 13.11.62
    Wimborne BLANDFORD-CASTLE CARY Burnham+ 3.2.62-31.8.63

Paddington-Bristol+
  PADDINGTON-FARRINGDON 10.1.63
  Maidenhead-Thame
    PRINCES RISBOROUGH-AYLESBURY 1.10.63
  Reading HUNGERFORD-DEVIZES Holt Jcn+ 11.11.62
  Didcot-Birmingham-Shrewsbury+
    Wolvercot (Oxford)-Priestfield (Wolverhampton)+
      STOURBRIDGE-CRADLEY 1.4.63
      Dudley-Wichnor+
        WEDNESBURY-TIPTON JCN 14.9.63
        WEDNESBURY-JAMES BRIDGE JCN 14.9.63
  BRISTOL-NEW PASSAGE PIER 8.9.63
  Bristol-Birmingham Camp Hill+
    Gloucester-Neyland+
      PYLE-PORTHCAWL= 10.8.61
NB: Converted from ty to ry - still ty to north
    CHELTENHAM ST JAMES BR REINSERTED
Bristol-Penzance+*
  TRURO-FALMOUTH 24.8.63

Cardiff-Merthyr
  Taffs Well-Rhymney
    PANT-BRECON≡   1.5.63
  Abercynon-Aberdare-Neath+   SWANSEA   15.7.63
Euston-Bushbury (Wolverhampton)+
  Camden Jcn-? // N&E-Stratford
    VICTORIA PARK-POPLAR   26.9.50-1.1.52=
Birmingham-Earlestown+
  Crewe-Shrewsbury*
    NANTWICH-MARKET DRAYTON   20.10.63
    WHITCHURCH-ELLESMERE   20.4.63
  Crewe-Birkenhead+
    Chester-Warrington
      HELSBY-HOOTON   1.7.63
    Chester-Pontypool+
      Saltney (Chester)-Holyhead+
        LLANDUDNO-LLANRWST   17.6.63
      Gobowen-Llanidloes*
        ABERMULE-KERRY BR   2.3.63
        MOAT LANE JCN-MACHYNLLETH // DOVEY ESTUARY-
                BORTH   3.1.63-24.10.63
        // ABERDOVEY QUAY-LLWYNGWRIL   24.10.63
      HEREFORD-EARDISLEY   24.10.62-30.6.63
      ABERGAVENNY JCN-BRYNMAWR   29.9.62
  Crewe-[Manchester]+
    Stockport-Whaley Bridge   BUXTON≡   15.6.63
  [Manchester]-Garston Dock+
    Timperley-Knutsford   NORTHWICH   1.1.63
[Manchester]-[Ulceby]+
  [Hyde Jcn]-Compstall (Marple)
    WOODLEY-STOCKPORT   12.1.63
Derby-[Barnetby]+
  Long Eaton-Mansfield
    PYE BRIDGE   Alfreton-Clay Cross   10.53-1.11.61
Derby-[Leeds]+
  Ambergate-Hassop   BUXTON   1.6.63
Liverpool Street (Bishopsgate)-Norwich Victoria+
  BOW JCN-BOW JCN (Fenchurch St link)   2.4.49-13.4.54=
  Stratford-Cambridge-Norwich-Yarmouth+
    Broxbourne Jcn-Welwyn*
      ST MARGARETS-BUNTINGFORD BR   3.7.63
    Marks Tey-Sudbury Old Stn*
      Chappel-Yeldham   HAVERHILL   10.5.63
    Colchester-Hythe   WIVENHOE   8.5.63
    Tivetshall-Bungay   BECCLES   2.3.63

Lmd (TJH 1865) (EPD Jun.1866); Cjc (TJH 1865); PC (TJH ?1865); Lmd (1868 military overprint (qv)); PC (TJH)

State S-11 (?1869)

Changes
1. Title deleted (moved to middle sheet, State M-P14)
2. Scale-bar deleted bottom right (reinstated on S-12)
3. Price deleted bottom right
4. List of plates deleted top left
5. Geology symbols deleted below 24

Additions to railways (+ Italic names, * Egyptian names)
London Bridge-Greenwich
  Corbetts Lane Jcn-Epsom Stn
    Norwood-Brighton+
      Three Bridges-Petworth*   PETERSFIELD   1.9.64-
                17.12.66
        WEST HORSHAM-PEASMARSH JCN (G'LDF'D) 2.10.65
      Three Bridges-East Grinstead   GROVE JCN (TUN-
                BRIDGE WELLS)   1.10.66
    SUTTON-EPSOM DOWNS   22.5.65
  North Kent Jcn-Strood-Paddock Wood*
    ST JOHNS-TONBRIDGE   1.7.65-3.2.68
      HITHER GREEN-DARTFORD   1.9.66
  Lewisham Jcn-Beckenham   ADDISCOMBE ROAD   1.4.64
VICTORIA   Battersea Pier - Sydenham Hill   PENGE JCN
                1.10.60-1.7.63
NB: Crystal Palace alignment error remains
    BRIXTON-SOUTH BERMONDSEY   13.8.66
    PECKHAM RYE-CRYSTAL PALACE BR   1.8.65
Waterloo-Southampton+
  Clapham-Windsor*
    Twickenham-Kingston≡
      STRAWBERRY HILL-SHEPPERTON BR   1.11.64
  Raynes Park-Leatherhead   HORSHAM   4.3.67-1.5.67
  WALTON-ON-THAMES-HAMPTON COURT-TEDDINGTON
NB: Error: no railway here: deleted on NS-S-2
  Weybridge-Chertsey   VIRGINIA WATER   1.10.66
  Woking-Havant+
    Guildford-Alton   WINCHESTER JCN   2.10.65
  Worting Jcn (Basingstoke)-Exeter*
    Chard J-Chard   CREECH JN (TAUNTON)   11.9.66-26.11.66
  ST DENYS (SOUTHAMPTON)-NETLEY   5.3.66
  Northam Jcn (Southampton)-Dorchester Jcn+
    Redbridge-Romsey   CONNECTED BOTH ENDS
    WEST MOORS-ALDERBURY JCN (SALISBURY)   20.12.66
RYDE-VENTNOR   23.8.64-10.9.66
Paddington-Bristol+
  Maidenhead-Thame   KENNINGTON JCN (OXFORD)   24.10.64
    Princes Risborough-Aylesbury   VERNEY JCN   23.9.68
  Reading-Holt Jcn+
    SAVERNAKE-MARLBOROUGH   14.4.64
  CHOLSEY & MOULSFORD-WALLINGFORD BR   2.7.66
  Didcot-Birmingham-Shrewsbury+
    NORTH WALES JCN (SHREWSBURY)-NANTMAWR   13.8.66
      KINNERLEY JCN-CRIGGION BR   13.8.66
  UFFINGTON-FARINGDON BR   1.6.64
  CHIPPENHAM-CALNE BR   29.10.63
  Thingley Jcn (Chippenham)-Salisbury*
    Westbury-Weymouth Stn*   PORTLAND   16.10.65
  Bristol-New Passage Pier   SEVERN TUNNEL JCN   1.9.86
  Bristol-Birmingham Camp Hill+
    STONEHOUSE JCN-35 *Nailsworth*   1.2.67
  Gloucester-Neyland+
    AWRE JCN-NEW FANCY COLLIERY   25.5.68-1.70
    Llantrisant-Penygraig
      MWYNDY JCN (LLANTRISANT)-BROFISCIN   8.1.62
        MAESARAUL J (L'TRISANT)-TREFOREST J 12.63
        GELLYNOG COLLIERY BR

LLANTRISANT-COWBRIDGE   18.9.65
BRIDGEND-NANTYFFYLLON   10.8.61
   TONDU Pyle-Porthcawl   10.8.61
   TONDU-NANTYMOEL   1.8.65
  Myrtle Hill Jcn (Carmarthen)-Conwil ABERYSTWYTH
                             28.3.64-12.8.67
    PENCADER JCN-LLANDYSSIL   3.6.64
  WHITLAND-PEMBROKE DOCK   30.7.63-4.9.68
ASHCHURCH-ALCESTER   1.7.64-16.6.66
 Ashchurch-Tewkesbury MALVERN JCN   1.7.62-16.5.64
HOTWELLS (BRISTOL)-AVONMOUTH   6.3.65
Bristol-Penzance+*
  BEDMINSTER JCN (BRISTOL)-PORTISHEAD   18.4.67
  Norton Fitzwarren-Watchet
    WATCHET-GUPWORTHY   28.9.59-9.64
  Cowley Bridge Jcn-Bideford+
    COLEFORD JN (YEOFORD)-SAMPFORD COURTENAY   1.11.65-
                                       8.1.67
  NEWTON ABBOT-MORETONHAMPSTEAD   4.7.66
  Aller Jcn-Paignton KINGSWEAR   14.3.61-16.8.64
  Tavistock Jcn (Plymouth)-Tavistock LAUNCESTON   1.7.65
Newport-Nantyglo
  BASSALEG-RHYMNEY CONVERTED FROM TY TO RY   1864
    MACHEN-PENRHOS   1865
  RISCA-SIRHOWY   11.55-19.6.65
NB: Alignment converted from a road into a railway
Cardiff-Merthyr
  Taffs Well-Rhymney
    BARGOED Pant-Brecon NEATH   2.10.64-1.9.68
Swansea-Ystalyfera BRYNAMMAN   1.1.64-1869
Llanelly LLANDOVERY-KNIGHTON Craven Arms* 1862-8.10.68
  PONTARDULAIS-SWANSEA   1.66
    GOWERTON-PENCLAWDD   14.12.67
    PENCLAWDD JCN   1899
    SWANSEA SOUTH DOCK BR   1.66
  LLANDILO JCN-ABERGWILI JCN   11.64
Euston-Bushbury (Wolverhampton)+
  Camden Jcn 7-N&E Stratford   26.9.50=
    Kentish Town-Gospel Oak-Willesden Jcn
      GOSPEL OAK-TOTTENHAM   21.7.68-23.1.1916
    DALSTON JCN-BROAD STREET BR   1.11.65
  Watford Jcn-St Albans Abbey
    NAPSBURY LINK (NEVER IN PUBLIC USE)
  WOLVERTON-NEWPORT PAGNELL BR   2.9.67
  Blisworth-Ely North Jcn+
    WANSFORD-STAMFORD   9.8.67
    WHITEMOOR JCN (MARCH)-SPALDING   1.4.67
  BLISWORTH-TOWCESTER   1.5.66
  Coventry-Hinckley WIGSTON JCN (LEICESTER)   1.1.64
Birmingham-Earlestown+
  STAFFORD-BROMSHALL JCN (UTTOXETER)   23.12.67
  Crewe-Shrewsbury*
    Nantwich-Market Drayton WELLINGTON   16.10.67
    Whitchurch-Ellesmere OSWESTRY   27.7.64
  Crewe-Birkenhead+
    Chester-Pontypool+
      Saltney (Chester)-Holyhead+
        Mold Jcn (Saltney)-Mold DENBIGH JCN   12.9.69

        Foryd Jcn-Ruthin CORWEN   6.10.64-1.9.65
      Bangor-Penygroes AFONWEN   2.9.67
      GAERWEN-AMLWCH   16.12.64-3.6.67
    WREXHAM-CONNAHS QUAY   1.1.66
    Ruabon-Llangollen LLANDRILLO   8.5.65-16.7.66
    Gobowen-Llanidloes* TALYLLYN JCN   1.9.64
      LLANYMYNECH-LLANFYLLIN   10.4.63
      Moat Lane Jcn MACHYNLLETH - DOVEY, BORTH -
                    ABERYSTWYTH   1.7.63-5.64
        CEMMES ROAD-DINAS MAWDDWY   1.10.67
        DOVEY JCN Aberdovey Quay-Llwyngwril PEN-
                MAENPOOL=   3.7.65-14.8.67
        BARMOUTH JCN-PWLLHELI   3.6.67-10.10.67
      LLANIDLOES-LLANGURIG   1864
    STRETFORD BRIDGE JCN-BISHOPS CASTLE   24.10.65
    Woofferton-Tenbury BEWDLEY   13.8.64
    Hereford-Eardisley THREE COCKS J 11.7.64-1.9.64
    Abergavenny Jcn-Brynmawr NANTYBWCH   1.3.64
  HOOTON-PARKGATE   1.10.66
  BIRKENHEAD-HOYLAKE=   2.7.66
  Crewe-[Manchester]+
    SANDBACH-NORTHWICH   11.11.67
  [Manchester]-Garston+ LIVERPOOL CENTRAL   1.6.64
    SPEKE JCN-BOOTLE   15.2.64-15.10.66
[Manchester]-[Ulceby]+
  [Hyde Jcn]-Compstall (Marple) BLACKWELL MILL JCNS
                  (BUXTON)   1.7.65-1.10.66
    Woodley-Stockport BROADHEATH JCN   1.12.65-1.12.66
      NORTHENDEN Stockport-Buxton   1.8.66
Derby-[Barnetby]+
  Trent-Bedford+* ST PANCRAS   9.9.67
    KETTERING JCN-HUNTINGDON   21.2.66
Derby-Hampton-in-Arden+
  LONDON ROAD JCN (DERBY)-SPONDON JCN   27.6.67
  PEARTREE (DERBY)-MELBOURNE   1.9.68
NB: Chellaston Jcn shown
  WHITACRE-NUNEATON   1.11.64
Derby-[Leeds]+
  DUFFIELD-WIRKSWORTH BR   1.10.67
Kings Cross (Maiden Lane)-[Doncaster]+
  FINSBURY PARK-EDGWARE   22.8.67
  HATFIELD-ST ALBANS ABBEY   16.10.65-1.11.66
NB: Not connected at Hatfield or St Albans
  HOLME-RAMSEY BR   22.7.63
  PETERBOROUGH-SUTTON BRIDGE   1.6.66
  Werrington Jcn (Peterborough)-Gainsborough+ [BLACK
                CARR JCN (DONCASTER)]   1.7.67
    Spalding-Sutton Bridge SOUTH LYNN   1.11.64
    Boston-Barkston* (Grantham)
      HONINGTON-SINCIL JCN (LINCOLN)   15.4.67
    Boston-[Grimsby]+
      FIRSBY-SPILSBY BR   1.5.68
  Essendine-Bourne SPALDING   1.8.66
Liverpool Street (Bishopsgate)-Norwich Victoria+
  Stratford-Cambridge-Norwich-Yarmouth+
    Loughton Jcn (Stratford)-Loughton ONGAR   24.4.65
    AUDLEY END-BARTLOW   23.11.65-22.10.66
    ELY JCN-SUTTON   16.4.66

Ely North Jcn-Kings Lynn-Hunstanton
    HEACHAM-WELLS 17.8.66
Forest Gate Jcn-Tilbury-Southend-on-Sea
    MUCKING-THAMESHAVEN BR 7.6.55=
Witham-Braintree BISHOPS STORTFORD 22.2.69
Marks Tey-Sudbury* SHELFORD (CAMBRIDGE) 1.6.65-9.8.65
    Chappel-Haverhill HAVERHILL JCN 1.6.65
    LONG MELFORD-BURY ST EDMUNDS 9.8.65
Colchester-Wivenhoe BRIGHTLINGSEA 18.4.66
    WIVENHOE-WALTON-ON-THE-NAZE 8.1.66-17.5.67
MELLIS-EYE BR 2.4.67

Lmd (1869 military overprint (qv))

State S-12 (?1873)

With the publication of the first true Anglo-Scottish middle sheet, the south sheet was radically revised and re-entered. Some features were not re-entered, such as marsh, forest, heath, geology. These faded away.

Changes
1. Many sheet lines re-entered, but not quarters on 45, which by this stage had become invisible. See S-15
2. Quarter sheet lines of 1 engraved
3. There are faint traces of the correct northern sheet line of 13 to a common intersection with 7, as well as the original incorrect line
4. Scale-bar reinstated in new position, br, 9mm anl
5. New roads added: eg at Bridgend, Holsworthy
6. Balcombe Tunnel re-engraved as double dotted line
7. Waterlining deleted

Additions to railways (+*Italic names, * Egyptian names)
London Bridge-Greenwich CHARLTON 1.1.73-1.2.78
  Corbetts Lane Jcn-Epsom Stn EPSOM JCN 8.8.59=
    Norwood-Brighton+
      Keymer Jcn (Wivelsfield)-Lewes
        Hamsey (Lewes)-Uckfield GROOMBRIDGE 3.8.68
      Brighton-Ashford+
        LONDON ROAD-BRIGHTON KEMP TOWN BR 2.8.69
        STONE CROSS JCN-WILLINGDON JCN 2.8.71
      Brighton-Fareham+
        BARNHAM JCN-BOGNOR BR 1.6.64
Victoria-Penge Jcn
  Brixton-South Bermondsey
    PECKHAM RYE-MITCHAM-SUTTON 1.10.68
NB: Disconnected in error at Tulse Hill
      STREATHAM JCN-WIMBLEDON 1.10.68
        TOOTING JCN-MERTON PARK 1.10.68
    Peckham Rye-Crystal Palace Br=
      NUNHEAD-BLACKHEATH HILL BR 18.9.71
  Herne Hill-Elephant FARRINGDON 1.6.64-20.2.66
  HERNE HILL-TULSE HILL 1.1.69
Waterloo-Southampton+
  Clapham Jcn-Crystal Palace-Strood-Dover Priory*
    Swanley-Sevenoaks TUBS HILL 1.8.69
    MALDEN Kingston-Twickenham 1.1.69

Woking-Havant+
  Guildford-Winchester Jcn
    FARNHAM JCN-PIRBRIGHT JCN 2.5.70
Worting Jcn (Basingstoke)-Exeter*
  SEATON JCN-SEATON BR 16.3.68
Eastleigh-Fisherton Jcn (Salisbury)
  Kimbridge Jcn-Andover Town CONNECTED AT KIMBRIDGE
Eastleigh-Gosport
  BOTLEY-BISHOPS WALTHAM BR 6.6.63
Northam Jcn (Southampton)-Dorchester Jcn+
  Ringwood-Christchurch BOURNEMOUTH EAST 14.3.70
Paddington-Bristol+
  Paddington-Farringdon
    PRAED STREET JCN-VICTORIA 1.10.68-24.12.68
      EARLS COURT MD-WL CONNECTIONS 12.4.69-1.2.72
  WESTBOURNE PARK-HAMMERSMITH 13.6.64
    LATIMER ROAD-UXBRIDGE ROAD LINK 1.7.64
  Maidenhead-Kennington Jcn (Oxford)
    PRINCES RISBOROUGH-WATLINGTON BR 15.8.72
  Didcot-Birmingham-Shrewsbury+
    Wolvercot (Oxford)-Priestfield (Wolverhampton)+
      Yarnton-Witney FAIRFORD BR 15.1.73
      Hartlebury-Shrewsbury+
        Buildwas-Much Wenlock MARSH FARM JN (CRAVEN ARMS) 5.12.64-16.12.67
      Stourbridge-Cradley HANDSWORTH JN 1.1.66-1.4.67
    FENNY COMPTON-KINETON= 1.6.71
  Bristol-Camp Hill+ BIRMINGHAM CURZON STREET 17.8.41=
    Mangotsfield-Keynsham BATH 4.8.69
    MANGOTSFIELD JCN MADE A TRIANGLE
    Ashchurch ALCESTER-REDDITCH Barnt Green 4.5.68
Bristol-Penzance+*
  YATTON Wells-Witham 3.8.69-5.4.70
  NORTON FITZWARREN-WIVELISCOMBE 8.6.71
  Cowley Bridge Jcn-Bideford+ TORRINGTON 18.7.72
    Coleford Jcn (Yeoford)-Sampford Courtenay OKEHAMPTON 3.10.71
  TOTNES-BUCKFASTLEIGH 1.5.72
Euston-Bushbury (Wolverhampton)+
  Willesden Jcn-Clapham Jcns
    KENSINGTON-GUNNERSBURY 1.1.69
  Willesden Jcn-Kew
    SOUTH ACTON-HAMMERSMITH & CHISWICK BR 1.5.57=
    ACTON-RICHMOND 1.1.69
  Blisworth-Towcester COCKLEY BRAKE JCN (BANBURY) 31.8.71-1.6.72
  CRANE STREET JCN (WOLVERHAMPTON)-WALSALL 1.11.72
Birmingham-Earlestown+
  Norton Bridge-Stoke-Macclesfield-Cheadle Hulme
    Stoke Jcn-Biddulph-Congleton
      MILTON JCN-CHEDDLETON 1.11.67
    Stoke Jcn-Silverdale MARKET DRAYTON 1.2.70
      SILVERDALE-ALSAGER 24.7.70
      KEELE LOOP 1.10.81
      CHESTERTON BR
NB: There is an uncorrected error at the junction
      CHATTERLEY-JAMAGE BR 24.7.70
        TALK O' TH' HILL BR

I.1 (S-12)

```
                ETRURIA-HANLEY   20.12.61
    Crewe-Shrewsbury*
        WHITCHURCH-TATTENHALL JCN (WAVERTON)   1.10.72
    Crewe-Birkenhead+
        Chester-Pontypool+
            Saltney (Chester)-Holyhead+
                Mold Jcn (Saltney)-Denbigh Jcn
                    MOLD-FFRITH   16.3.69-7.1.72
                Llandudno-Llanrwst  BETWS-Y-COED   18.11.67
                Bangor-Afonwen
                    CAERNARVON-LLANBERIS BR   1.7.69
                Ruabon   LLANDRILLO - PENMAENPOOL   Dovey Jcn
                                                    1.4.68-1.8.69
                    Barmouth Jcn-Pwllheli
                        PORTMADOC-BLAENAU FFESTINIOG   20.4.36=
    Crewe-[Manchester]+
        [Manchester]-Garston-Liverpool Central
            Timperley-Northwich   HELSBY   1.9.69-1.7.71
                NORTHWICH SALT BR-80SE   17.12.67
NB: Junction placed west of Sandbach line in error
                HARTFORD-WINNINGTON & ANDERTON GDS   1.6.70
                CUDDINGTON-WINSFORD & OVER BR   1.6.70
    WEAVER JCN-DITTON   1.2.69
[Manchester]-[Ulceby]+
    [Hyde Jcn]-Blackwell Mill Jcns (Buxton)
        MARPLE-MACCLESFIELD   2.8.69-3.4.71
        NEW MILLS-HAYFIELD BR   1.3.68
    WOODBURN JCN (SHEFFIELD)-ROTHERHAM   1.8.64-1.8.68
        TINSLEY Wincobank-[Aldam Jcn]   1.8.64
Derby-[Barnetby]+
    Trent-St Pancras+*
        WELLINGBOROUGH MR-L&NW CURVE   1.1.59=
        OAKLEY JCN (BEDFORD)-NORTHAMPTON ST JOHNS   10.6.72
        BRENT JCN-ACTON WELLS   1.10.68
        Long Eaton-Mansfield   SHIREOAKS JCNS   18.1.75
            MANSFIELD   Southwell-Rolleston Jcn   3.4.71
Derby-Hampton-in-Arden+
    Peartree (Derby)-Melbourne   WORTHINGTON   1.9.69
        CHELLASTON JCN-CASTLE DONINGTON JN (TRENT)   6.12.69
Derby-[Leeds]+
    TAPTON JCN (CHESTERFIELD) Sheffield-Rotherham   1.2.70
Kings Cross (Maiden Lane)-[Doncaster]+
    Finsbury Park-Edgware
        FINCHLEY CENTRAL-HIGH BARNET BR   1.4.72
    WOOD GREEN-ENFIELD   1.4.71
    Essendine-Spalding
        BOURNE-SLEAFORD   10.71
Liverpool Street (Bishopsgate)-Norwich Victoria+
    Stratford-Cambridge-Norwich-Yarmouth+
        ROUDHAM JCN (THETFORD)-WATTON   26.1.69

PC
```

State S-13 (?1874)

Change
1. *BOURNEMOUTH* added (16ne)

```
Additions to railways (+ Italic names, * Egyptian names)
Waterloo-Southampton+
    Clapham Jcn-Crystal Palace-Strood-Dover Priory*
        Swanley-Sevenoaks Tubs Hill
            OTFORD-MAIDSTONE   1.6.74
    Northam Jcn (Southampton)-Dorchester Jcn+
        Wimborne-Burnham+
            EVERCREECH JCN-BATH JCN   20.7.74
Cowes-Newport
    NEWPORT-SANDOWN   1.2.75-6.10.75
    NEWPORT-SMALLBROOK JCN   20.12.75
Paddington-Bristol+
    Maidenhead-Kennington Jcn (Oxford)
        BOURNE END-MARLOW BR   28.6.73
    Didcot-Birmingham-Shrewsbury+
        Wolvercot (Oxford)-Priestfield (Wolverhampton)+
            Tunnel Jcn (Worcester)-Shelwick Jcn (Hereford)
                BRANSFORD RD JN (WORCESTER)-YEARSETT   2.5.74
        Hatton  STRATFORD STNS LINK  Honeybourne   24.7.61=
    Thingley Jcn (Chippenham)-Salisbury*
        Westbury-Portland*
            Frome-Radstock   BRISTOL JCN   3.9.73
Bristol-Penzance+*
    Norton Fitzwarren-Wiveliscombe   BARNSTAPLE   1.11.73
    Cowley Bridge Jcn (Exeter)-Torrington+
        BARNSTAPLE-ILFRACOMBE   20.7.74
        Coleford Jcn-Okehampton   LYDFORD   12.10.74
Cardiff-Merthyr
    Taffs Well-Rhymney
        Bargoed-Brecon-Neath
            MAES-Y-MARCHOG BR
Swansea-Brynamman
    YNYSYGEINON JCN-CAPEL COELBREN   26.7.69
Euston-Bushbury (Wolverhampton)+
    Blisworth-Cockley Brake Jcn (Banbury)
        TOWCESTER  Fenny Compton-Kineton  STRATFORD   1.7.73
    Trent Valley Jcn (Rugby)-Stafford+
        NUNEATON-OVERSEAL & MOIRA   1.8.73
            HIGHAM-HINCKLEY LINK   (NEVER OPENED)
            SHACKERSTONE-COALVILLE JCN   1.8.73
Birmingham-Earlestown+
    Norton Bridge-Stoke-Macclesfield-Cheadle Hulme
        Etruria-Hanley  KIDSGROVE   1.11.73-15.1.75
            HANLEY MINERAL BRS
                NEWFIELDS BR
            LONGPORT-TUNSTALL   1.6.75
Crewe-Birkenhead+
    Chester-Pontypool+
        Saltney (Chester)-Holyhead+
            Bangor-Afonwen
                PENYGROES-NANTLLE BR   1.8.72
            Gobowen-Talyllyn Jcn*
                Moat Lane Jcn-Aberystwyth
                    CAERSWS-VAN BR   14.8.71
Crewe-[Manchester]+
    [Manchester]-Garston Dock-Liverpool+
        Timperley-Helsby
            MOULDSWORTH-CHESTER   2.11.74
```

[Manchester]-[Ulceby]+
   [Hyde Jcn]-Blackwell Mill Jcns (Buxton)
      APETHORNE JCN-GODLEY   1.2.66
        Woodley-Broadheath Jcn CRESSINGTON   1.3.73
          GLAZEBROOK-[MANCHESTER]   2.9.73
            ROMILEY-[ASHBURYS]   17.5.75
NB: Terminates in error at L&NW
      REDDISH JCN-BRINNINGTON JCN (STOCKPORT)   2.8.75
Derby-Hampton-in-Arden+
   Peartree (Derby)-Worthington ASHBY   1.1.74
Kings Cross (Maiden Lane)-[Doncaster]+
   Werrington Jcn (Peterboro')-Gainsborough-[Doncaster]+
      Boston-[Grimsby]+
        FIRSBY-SKEGNESS BR   24.10.71-28.7.73
Liverpool Street (Bishopsgate)-Norwich Victoria+
   BETHNAL GREEN-CHINGFORD   26.4.70-17.11.73
     TWO LINKS WITH N&E   26.4.70-1.8.72

ABn (HC) (EPD Apr.187?6); LVg

State S-14 (?1876)

Additions to railways (+ Italic names, * Egyptian names)
London Bridge-Charlton
   Corbetts Lane Jcn (Southwark Park)-Epsom Jcn
      Norwood-Brighton+
         Redhill-Dover+
            SANDLING JCN-SANDGATE BR   9.10.74
Waterloo-Southampton+
   Worting Jcn (Basingstoke)-Exeter*
      SIDMOUTH JCN-SIDMOUTH BR   6.7.74
   Northam Jcn (Southampton)-Dorchester Jcn+
      BROADSTONE JCN-BOURNEMOUTH CENTRAL≡ 2.12.72-6.3.88
      POOLE BR   (ERROR)
Paddington-Bristol+
   Bristol-Birmingham+
      Gloucester-Neyland+
         Llantrisant-Penygraig
            Coed Ely-Gellyrhaidd Coll BLACKMILL 1.9.75
                HENDREFORGAN-GILFACH GOCH   10.65
            WHITLAND-CRYMMYCH ARMS   24.3.73-10.74
              GLOGUE QUARRIES BR   24.3.73
Bristol-Penzance+*
   Norton Fitzwarren-Watchet MINEHEAD   16.7.74
Cardiff-Merthyr
   MERTHYR-PONTSTICILL JCN   1.8.68
Birmingham-Earlestown+
   Crewe-Birkenhead+
      Chester-Pontypool+
         PONTYPOOL-MAINDEE JN (NEWPORT) 17.9.74-21.12.74
            CWMBRAN COLLIERY BR
            LLANTARNAM-CWMBRAN JCNS   4.78
Kings Cross (Maiden Lane)-[Doncaster]+
   Finsbury Park-Edgware
      HIGHGATE-ALEXANDRA PALACE BR   24.5.73

Ob C17 (420) (TJHK) (EPD 1876); Lpro WO 78/4813 (THSL) (EPD 1876); PC (APSL) (EPD Dec.1881); Lpro WO 78/935 (1) (APSLB) (EPD Oct.1883); Lpro WO 78/935 (2) (used by OS 31.3.1884); Lsa; RH; PC

State S-15 (?1885) (Sheet 3 of a three-sheet set)

Introduction of new standard features
A.13   Sheet 3 inserted 4mm above price, bottom right
B.2   Topographic features
   p   Large town: Dublin
   r   Medium town: eg Wicklow
   t   Village: eg Wissant
C.1   Roman capitals
   d   Country: France
   e   Marine name: eg English Channel

Changes
1. Scilly Isles sheet cutting line deleted
2. *Price 2/-* reinstated, above scale-bar, bottom right
3. *BRIGHTHELMSTON* to *BRIGHTON* (9se), St Dye to St Day, *Piran Bay* to *Perran Bay* (31nw), *DODMAN* to *DODMAN P^r*, *Carac dû* to *Carrack dû* (31ne), *Girrick R* to *Carracks R* (33nw), Aberafon to Aberavan (37ne), CAERMARTHEN(SHIRE) to CARMARTHEN(SHIRE) (41sw)
4. *Varne Ridge* deleted, *THE DOWNS*, *STRAIT OF DOVER*, ENGLISH CHANNEL, S^T GEORGE'S CHANNEL added
5. Railway revision reverts to State S-13
6. Coastal outline of France and Ireland added

Lpro MPHH 419 (EPD ?Dec.?1884) (in use 25.11.1885); Lgh Old Series set (EPD Feb.1885); Cu Atlas 1.08.1

State S-16

Changes
1. Dublin changed from hatched circle to built-up area
2. *Part of Oxon* deleted (7nw)
3. 45 quarter sheet lines re-entered

EXg

State NS-S-1 (?1885)

One change at least must be assumed:
1. New Series sheet lines, numbers replace Old Series

Not found.
Documentary source: PRO OS 1/144 70B (see pp.5,73)

State NS-S-2 (OSPR 10/1890)

Introduction of new standard features
A.14   ARRR: below 332, 26mm bnl
A.15   *Railways inserted to 1890*: bottom left, 20mm bnl

New dimensions: (from Ob and NS-S-3 Lpro OS 5/49)
   Paper: 1142mm W-E by 771mm S-N
   Plate: 1082mm W-E by 730mm S-N
   Neat line: 986mm W-E (no northern neat line present)

I.1 (NS-S-2)

Changes
1. Plate extended westwards: new dimensions as above. Map and plate not square due to border realignment
2. Restyled border of five lines (one thick), surrounding 3-sheet map (therefore W,S,E sides only)
3. Scale-bar removed to below 353-355, 12mm anl
4. *Price 2/-* removed to above "ARRR", 21mm bnl
5. Sheet 3 removed to top right, 21mm rnl
6. Scilly Islands added
7. French outline mapping redrawn and extended to include *LE HAVRE, CHERBOURG* and Guernsey; placenames revised and placename dots all hatched
8. Ireland coastline redrawn and extended to *WATERFORD*; *ARKLOW, Cahore Point* added; *KINGSTON* to *KINGSTOWN*
9. Fathom lines deleted
10. Princetown added (338)

Railways inserted to 1889 (+ Italic, * Egyptian names)
NB: New railways are additional to those on State S-14
London Bridge-Charlton
    Corbetts Lane Jcn (Southwark Park)-Epsom Jcn
        Norwood-Brighton+
            SOUTH CROYDON-CULVER JCN (LEWES) 1.8.82-10.3.84
                HURST GREEN JCN (OXTED)-ASHURST JCN (GROOM-
                          BRIDGE) 2.1.88-1.10.88
            HORSTED KEYNES-HAYWARDS HEATH 3.9.83
        Reigate-Dover+
            Ashford-Ramsgate+
                CANTERBURY-CHERITON JCN 4.7.87-1.7.89
                Minster-Deal KEARNSEY JCN (DOVER) 17.6.81
        Three Bridges-Petersfield*
            MIDHURST-CHICHESTER 11.7.81
        Brighton-Ashford+
            Polegate-Hailsham REDGATE MILL JCN (ERIDGE)
                          5.4.80-1.9.80
            APPLEDORE-DUNGENESS 7.12.81
                LYDD-NEW ROMNEY BR 19.6.84
        Brighton-Fareham+
            DYKE JCN (HOVE)-THE DYKE BR 1.9.87
            HAVANT-HAYLING ISLAND BR 19.1.65-16.7.67
            Cosham-Portsmouth
                FRATTON-EAST SOUTHSEA BR 1.7.85
    North Kent Jcn-Strood-Paddock Wood*
        St Johns-Tonbridge
            GROVE PARK-BROMLEY NORTH BR 1.1.78
            DUNTON GREEN-WESTERHAM BR 7.7.81
        Lewisham Jcn-Addiscombe Road
            ELMERS END-HAYES BR 29.5.82
            WOODSIDE-SELSDON ROAD 10.8.85
        HOO JCN-PORT VICTORIA 1.4.82-11.9.82
Waterloo-Southampton+
    Clapham Jcn-Crystal Palace-Strood-Dover Priory*
        Swanley-Sevenoaks Tubs Hill
            Otford-Maidstone ASHFORD 1.7.84
        FAWKHAM JCN-GRAVESEND WEST STREET PIER 10.5.86
    Raynes Park-Horsham
        LEATHERHEAD JCN-GUILDFORD 2.2.85
        HAMPTON COURT JCN-EFFINGHAM JCN 2.2.85

Woking-Havant+
    Guildford-Winchester Jcn
        Farnham-Pirbright
            ASH VALE-ASCOT 18.3.78-2.6.79
    Worting Jcn (Basingstoke)-Exeter*
        ANDOVER-WOLFHALL JCN (SAVERNAKE) 1.5.82-5.2.83
    Eastleigh-Fisherton Jcn (Salisbury)
        Kimbridge Jcn-Andover Town
            FULLERTON-HURSTBOURNE 1.6.85
    St Denys (Southampton)-Netley FAREHAM 2.9.89
    Northam Jcn (Southampton)-Dorchester Jcn+
        BROCKENHURST-CHRISTCHURCH, BOURNEMOUTH E-CENTRAL
                Broadstone Jcn 20.7.85-6.3.88
        BRANKSOME-BOURNEMOUTH WEST BR 15.6.74
Cowes-NEWPORT-Sandown 1.6.79
    NEWPORT-FRESHWATER 10.9.88
Ryde-Ventnor
    BRADING-BEMBRIDGE 23.8.64-27.5.82
Paddington-Bristol+
    Paddington-Farringdon
        Praed St-Victoria TOWER HILL 24.12.68-6.10.84
        Earls Court MD-WL Connections
            EARLS COURT-WIMBLEDON 1.3.80-3.6.89
            WEST KENSINGTON-HAMMERSMITH 9.9.74
    West Drayton-Uxbridge
        WEST DRAYTON-STAINES 9.8.84-2.11.85
    Reading-Holt Jcn+
        ENBORNE JCN (NEWBURY)-WINCHESTER 4.5.85
        Savernake-Marlborough CIRENCESTER 27.7.81-1.11.83
    DIDCOT-NEWBURY 13.4.82
    Didcot-Birmingham-Shrewsbury+
        Wolvercot (Oxford)-Priestfield (Wolverhampton)+
            Kingham-Chipping Norton KINGS NORTON 6.4.87
            Kingham-Bourton CHELTENHAM 1.6.81
            STOURBRIDGE JCN-STOURBRIDGE TOWN BR 1.10.79
            Stourbridge-Handsworth Jcn
                LANGLEY GREEN-OLDBURY BR 7.11.84
        SWAN VILLAGE-GREAT BRIDGE 1.9.66=
    SWINDON-HIGHWORTH BR 9.5.83
    DAUNTSEY-MALMESBURY BR 18.12.77
    Thingley Jcn (Chippenham)-Salisbury*
        Westbury-Portland*
            Maiden Newton-Bridport WEST BAY 31.3.81
            UPWEY JCN-ABBOTSBURY BR 9.11.85
    Bristol-Birmingham+
        BERKELEY ROAD-LYDNEY 2.8.75-20.10.79
        Stonehouse Jcn-35 NAILSWORTH 1.2.67
        Gloucester-Neyland+
            OVER JCN (GLOUCESTER)-LEDBURY 27.7.85
            CHEPSTOW-REDBROOK 1.11.76
            Llantrisant-Penygraig
                LLANTRISANT COMMON BR 1863=
NB: Both ends point south in error
            Bridgend-Nantyffyllon CYMMER 1.7.78
                CRYTHAN PFM-CYMMER-NORTH RHONDDA COLL 6.61=
            Whitland-Crymmych Arms CARDIGAN 1.9.86
        NORTHFIELD JCN-OLD HILL-DUDLEY 1.3.78-10.9.83
        KINGS NORTON-BIRMINGHAM SUBURBAN LOOP 3.4.76

Bristol-Penzance+
  Norton Fitzwarren-Barnstaple+
    MOREBATH JCN-STOKE CANON 1.8.84-1.5.85
    BARNSTAPLE GW-L&SW LINK 1.6.87
  TIVERTON JCN-HEMYOCK BR 29.5.76
  Cowley Bridge Jcn (Exeter)-Torrington+
    Coleford Jcn (Yeoford)-Lydford
      MELDON JCN (OKEHAMPTON)-HOLSWORTHY 21.1.79
        HALWILL JCN-LAUNCESTON 21.7.86
  Newton Abbot-Moretonhampstead
    HEATHFIELD-TEIGN HOUSE CROSSING 9.10.82
  Totnes-Buckfastleigh ASHBURTON BR 1.5.72
  Tavistock Jcn (Plymouth)-Launceston
    YELVERTON-PRINCETOWN BR 11.8.83
  Kilmar Tor-Liskeard LOOE 27.12.60=
  BODMIN ROAD-BOSCARNE JCN (BODMIN) 27.5.87-3.9.88
  FOWEY Par-Treffrey Siding NEWQUAY 1.6.74
  GWINEAR ROAD-HELSTON BR 9.5.87
  ST ERTH-ST IVES BR 1.6.77
BODMIN-WADEBRIDGE 4.7.34=
Cardiff-Merthyr
  Radyr-Penarth
    PENARTH-BARRY 8.2.89-20.12.89
    PENARTH-CADOXTON VIA COAST 1878-20.12.88
  PONTYPRIDD-PENRHOS 7.7.84
  DOWLAIS MINERAL TRAMWAYS DELETED
PORT TALBOT DOCK-CYMMER-BLAENGWYNFY 2.11.85-2.6.90
BAKER STREET-CHESHAM 13.4.68-8.7.89
Euston-Bushbury (Wolverhampton)+
  Willesden Jcn-Clapham Jcns
    Kensington-Gunnersbury
      TURNHAM GREEN-HOUNSLOW 1.5.83
      ACTON TOWN-EALING BROADWAY 1.7.79
  Blisworth-Ely North Jcn+
    Northampton-Market Harborough*
      KINGSTHORPE JCN (NORTHAMPTON)-RUGBY 1.8.81
  Blisworth-Cockley Brake Jcn (Banbury)
    Towcester-Stratford-upon-Avon BROOM 2.6.79
  WEEDON-DAVENTRY 1.3.88
  Rugby-Luffenham
    SEATON-YARWELL JCN (WANSFORD) 21.7.79
    BERKSWELL-KENILWORTH 2.3.84
    STECHFORD-ASTON 7.9.80
    HARBORNE JCN-HARBORNE BR 10.8.74
  Crane St Jcn (W'hampton)-Walsall WATER ORTON 19.5.79
    CASTLE BROMWICH CURVE 19.5.79
Birmingham-Earlestown+
  Aston-Sutton Coldfield LICHFIELD 1.9.84
  PERRY BAR-HARBORNE JCN 3.10.87-1.3.88
  Crewe-Birkenhead+
    Chester-Warrington
      FRODSHAM JCN-HALTON JCN (RUNCORN) 1.5.73
    Chester-Pontypool+
      Saltney (Chester)-Holyhead+
        Saltney-Denbigh
          Padeswood-Coed Talon
            Mold-Ffrith WHEATSHEAF 7.47=
NB: Connected to WM&CQ line in error

        Llandudno-Betws BLAENAU 28.2.79-1.4.81
        BANGOR-BETHESDA BR 1.7.84
      Bangor-Afonwen
        DINAS-BEDGELLERT 21.5.77-14.5.81
          TRYFAN JCN-BRYNGWYN 5.77
    Ruabon-Aberdovey Quay
      BALA-BLAENAU FFESTINIOG 1.11.82-10.9.83
      TOWYN-ABERGYNOLWYN 1.10.66
    Gobowen-Talyllyn Jcn*
      Moat Lane Jcn-Aberystwyth
        MACHYNLLETH-ABERLLEFENI 4.59=
    Hereford-Three Cocks Jcn
      EARDISLEY-TITLEY (NOT CONNECTED) 3.8.74
      PONTRILAS-HAY 1.9.81-21.4.89
    Pontypool-Quakers Yard
      LLANCAIACH-DOWLAIS 10.1.76
    Hooton PARKGATE-HOYLAKE B'head 1.4.78-19.4.86
      BIRKENHEAD-NEW BRIGHTON 2.1.88-30.3.88
      BIRKENHEAD WOODSIDE BR 1.4.78
      BIRKENHEAD DOCKS BR
  Crewe-[Manchester]+
  [Manchester]-Garston-Liverpool+
    Widnes-St Helens
      ST HELENS NEW STN BR LACKING
Liverpool-[Manchester]+
  HUYTON-ST HELENS 16.12.71
[Manchester]-[Ulceby]+
  [Hyde Jcn]-Blackwell Mill Jcns (Buxton)
    Woodley-Cressington
      GLAZEBROOK-(WIGAN) 16.10.79-1.4.84
      WARRINGTON LOOP 13.8.83
      WIDNES LOOP 3.4.77-1.7.79
      HALEWOOD-DERBY 1.12.79
      HUNTS CROSS CURVE (HALEWOOD) 1.12.79
  Woodburn J (Shef'ld)-Rotherham [MEXB'GH] 8.63-13.3.71
Derby-[Barnetby]+
  Trent-St Pancras+*
    GLENDON JCN (KETTERING)-MANTON 1.12.79
NB: Connected to L&NW in error
    HARPENDEN-HEMEL HEMPSTEAD 16.7.77
  Long Eaton-Shireoaks Jcns
    Pye Bridge-Clay Cross
      TIBSHELF JCN-PLEASLEY EAST JCN 1.5.66-3.8.82
    Kirkby-Lenton Jcn (Nottingham)
      BASFORD-BENNERLEY J (ILKESTON) 13.10.77-3.12.77
NB: Partly duplicating the GN line
    RADFORD-TROWELL 6.1.75
NB: Pointing south at Trowell in error
    NOTTINGHAM-MELTON MOWBRAY JCN 1.11.79
Derby-Hampton-in-Arden+
  STENSON JCN (WILLINGTON)-CHELLASTON JCN 3.11.73
Derby-[Leeds]+
  CRICH JCN (AMBERGATE)-CODNOR PARK 1.6.74-1.2.75
NB: Connected to GN in error
  GRASSMOOR AND PILSLEY COLLIERIES LOOP 5.7.69-1.3.76
  MONKWOOD MINERAL BR
    NESFIELD MINERAL BR 1.4.70
  STAVELEY-ELMTON & CRESSWELL 1.8.66-1.6.75

I.1 (NS-S-2)

Kings Cross (Maiden Lane)-[Doncaster]+
   FINSBURY PARK-CANONBURY JCN   14.12.74
   Werrington Jcn (Peterboro')-Gainsborough-[Doncaster]+
     SPALDING-LINCOLN   6.3.82-1.8.82
NB: Connected to MR in error
     SLEAFORD LOOP   6.3.82
    Boston-Barkston* ALLINGTON JCN (SEDGEBROOK)   1875
    Boston-[Grimsby]+
      WILLOUGHBY-SUTTON   4.10.86
      LOUTH-MABLETHORPE=   17.10.77
     BARDNEY-LOUTH   9.11.74-26.6.76
   Grantham-Nottingham+ EGGINTON   23.8.75-28.1.78
NB: Alignment at Derby broken by MR in error
    AWSWORTH-PINXTON   1.8.76-18.12.76
   NEWARK-WELHAM JCN (MARKET HARBOROUGH)   1.4.78-1.11.79
    BOTTESFORD JCNS   30.6.79-1.3.80
    HARBY-SAXONDALE JCN   30.6.79
    HALLATON JCN-DRAYTON JCN   2.7.83
Liverpool Street (Bishopsgate)-Norwich Victoria+
   Bethnal Green-Copper Mill Jcn
    Copper Mill-Tottenham error DELETED
    HACKNEY DOWNS-EDMONTON   27.5.72-22.7.72
     SEVEN SISTERS-PALACE GATES   1.1.78-7.10.78
   Stratford-Cambridge-Norwich-Yarmouth+
    Chesterton (Cambridge)-Huntingdon
     St Ives-March
      SOMERSHAM-RAMSEY BR   16.9.89
    Ely Jcn-Sutton NEEDINGWORTH JCN (ST IVES)   10.5.78
    ELY JCN-SOHAM   1.9.79
NB: Line opened that day to Warren Hill Jcn (Newmarket)
    Ely North Jcn-Kings Lynn-Hunstanton
     DENVER-STOKE FERRY BR   1.8.82
     KINGS LYNN-NORWICH CITY   16.8.79-2.12.82
      MELTON CONSTABLE-CROMER BCH   1.10.84-16.6.87
      MELTON CONSTABLE-YARMOUTH BCH   7.8.77-5.4.83
       YARMOUTH NORTH QUAY BR   15.5.82
    Roudham Jcn (Thetford)-Watton SWAFFHAM   15.11.75
    WHITLINGHAM-CROMER   20.10.74-26.3.77
     WROXHAM-COUNTY SCHOOL   8.7.79-1.5.82
    BRUNDALL-ACLE-BREYDON JN (YARMOUTH)   12.3.83-1.6.83
   Forest Gate Jcn-Tilbury-Southend SHOEBURYNESS   1.2.84
    BARKING-PITSEA   1.5.85-1.6.88
   SHENFIELD-SOUTHEND-ON-SEA VICTORIA   19.11.88-1.10.89
    WICKFORD-SOUTHMINSTER   1.6.89
     WOODHAM FERRERS-MALDON   1.10.89
   Colchester-Brightlingsea
    Wivenhoe-Walton-on-the-Naze
     THORPE-LE-SOKEN-CLACTON BR   4.7.82
   Ipswich-Yarmouth South Town*
    SOUTHWOLD RY   24.7.79
   Haughley-Great Chesterford+
    BURY ST EDMUNDS-THETFORD JCNS   15.11.75-1.3.76
    FORCETT-WYMONDHAM   2.5.81
Fenchurch Street-Barking
   STEPNEY-BLACKWALL   6.7.40=

Dtc, En, Lbl, Ob C16 (554) (EPD Oct.1890); Mg (MK) (EPD Mar.1891). Dtc, Mg, Ob: flat sheets with excess paper

State NS-S-3 (?1892)

Changes
1. S^T GEORGE'S CHANNEL moved to off 177-117
2. IRISH [Channel] added (off 117-below S1)
3. Sheet 3 moved, top right, 4mm lnl

Lpro OS 5/49 (flat sheet with excess paper); SOrm

State NS-S-4 (?1893)

Change
1. Severn Tunnel shown dotted

Railways inserted to ?1893 (+ Italic, * Egyptian names)
London Bridge-Charlton
   Corbetts Lane Jcn (Southwark Park)-Epsom Jcn
    Norwood-Brighton+
     Redhill-Dover+
      PADDOCK WOOD-HAWKHURST BR   1.10.92-4.9.93
     Brighton-Ashford+
      Southerham Jcn-Newhaven SEAFORD   1.6.64=
Victoria-Penge Jcn
   Brixton-South Bermondsey
    Peckham Rye-Nunhead SHORTLANDS   1.7.92
Waterloo-Southampton+
   Northam Jcn (Southampton)-Dorchester Jcn+
    Wimborne-Burnham+
     EDINGTON JCN-BRIDGWATER   21.7.90
    WORGRET JCN (WAREHAM)-SWANAGE BR   20.5.85
Paddington-Bristol+
   Maidenhead-Kennington Jcn (Oxford)
    Princes Risborough-Verney Jcn
     QUAINTON ROAD-BRILL   1.4.71-11.71=
   Reading-Holt Jcn+
    (Enborne Jn) Newbury-Winchester SHAWFORD JN   4.9.91
    Savernake-Cirencester ANDOVERSFORD   16.3.91
   Didcot-Birmingham-Shrewsbury+
    Wolvercot (Oxford)-Priestfield (Wolverhampton)+
     MORETON-IN-MARSH-SHIPSTON-ON-STOUR BR   1.7.89
     KIDDERMINSTER-BEWDLEY   1.6.78=
     KIDLINGTON-BLENHEIM & WOODSTOCK BR   19.5.90
    Hatton-Honeybourne
     BEARLEY-ALCESTER   4.9.76=
    Ketley Jcn-Lightmoor BUILDWAS   1.11.64=
   Swindon-Standish
    KEMBLE JCN-TETBURY BR   2.12.89
   Thingley Jcn (Chippenham)-Salisbury*
    Westbury-Portland*
     Frome-Bristol Jcn
      HALLATROW-CAMERTON   1.3.82
   Bristol-Birmingham+
    YATE-THORNBURY BR   8.68-2.9.72=
    Stonehouse Jcn-Nailsworth
     WOODCHESTER-STROUD BR   16.11.85
   Gloucester-Neyland+
    Grange Court-Barrs Court Jcn (Hereford)+
     ROTHERWAS JCN-RED HILL JCN   16.7.66=

```
                LYDNEY-BILSON JCN (CINDERFORD)  19.4.69-15.9.73
                  COLEFORD JCN-COLEFORD-WYESHAM JCN (MONMOUTH)
                                                    19.7.75-1.9.83
                    SERRIDGE JCN-LYDBROOK  26.8.74
                  Chepstow-Redbrook MONMOUTH  1.11.76
                    MONMOUTH-COLEFORD  1.9.83
                    CLYNDERWEN-ROSEBUSH  19.9.76=
Bristol-Penzance+
    WORLE JCN-UPHILL JCN (WESTON LOOP)  1.3.84
    Cowley Bridge Jcn (Exeter)-Torrington+
        Coleford Jn-Lydford DEVONPORT JN (PLYMOUTH) 1.6.90
            Meldon-Holsworthy
                Halwill-Launceston TRESMEER  28.7.92
    Aller Jcn (Newton Abbot)-Kingswear
        CHURSTON-BRIXHAM BR  28.2.68=
Newport-Nantyglo
    Bassaleg-Rhymney
        Machen-Penrhos
            CAERPHILLY-HEATH JCN  1.4.71=
NB: The line opened that day through to Cardiff
            TIR PHIL-NEW TREDEGAR
Cardiff-Merthyr
    Pontypridd  TREHERBERT-BLAENGWYNFI  Port Talbot Dock
                                              2.7.90-14.7.90
Swansea-Brynamman
    UPPER BANK (SWANSEA)-GLAIS LOOP  2.12.71-1.3.75=
Euston-Bushbury (Wolverhampton)+
    HARROW-STANMORE BR  18.12.90
    ROADE-NORTHAMPTON  1.8.81
    Trent Valley Jcn (Rugby)-Stafford+
        Nuneaton-Overseal & Moira
            Shackerstone-Coalville Jcn
                HUGGLESCOTE JN-LOUGHBOROUGH DERBY RD 16.4.83
Birmingham-Earlestown+
    Crewe-Birkenhead+
        Chester-Pontypool+
            Wrexham-Connahs Quay
                BUCKLEY-HAWARDEN BRIDGE-CHESTER  3.8.89
            Leominster-Kington NEW RADNOR  25.9.75=
            LEOMINSTER-STEENS BRIDGE  1.3.84
            Hereford-Three Cocks Jcn
                Eardisley-Titley TITLEY JCN  3.8.74=
            Pontypool-Monmouth ROSS  1.8.73=
[Manchester]-[Ulceby]+
    [Hyde Jcn]-Blackwell Mill Jcns (Buxton)
        Woodley-Cressington
            Glazebrook-[Manchester Central]
                [Throstle Nest Jcn]-HEATON MERSEY JCN 1.1.80
        Woodhouse Jcn-Beighton Jcn  CHESTERFIELD  1.12.91-
                                                     -24.10.92
    Derby-[Barnetby]+
        Trent-St Pancras+*
            Knighton Jcn (Leicester)-Burton-on-Trent
                Swadlincote loop REFORMED AS A RING (ERROR)
                    Woodville unidentified Gds Br DELETED
                Glendon Jcn (Mkt Harborough)-Manton CORRECTED
            Long Eaton-Shireoaks Jcns
                HEANOR JCN-BUTTERLEY  2.9.89-17.11.90
```

```
Derby-[Leeds]+
    Staveley-Elmton & Cresswell
        SEYMOUR JCN (STAVELEY)-PLEASLEY  31.8.66-1.9.90
Kings Cross (Maiden Lane)-[Doncaster]+
    Werrington Jcn (Peterboro')-Gainsborough-[Doncaster]+
        Spalding-South Lynn  BAWSEY JCN  1.11.85
        Spalding-Lincoln  PYEWIPE JCN (LINCOLN)  1.8.82
NB: Error at Lincoln corrected
        Boston-[Grimsby]+
            Willoughby SUTTON-MABLETHORPE Louth  14.7.88
        Grantham-Nottingham-Egginton (wrong at Derby)+
            NOTTINGHAM-DAYBROOK  2.12.89
NB: Connected to MR in error
            LEEN VALLEY JCN-ANNESLEY  7.81-27.10.81
            ILKESTON-HEANOR  7.6.86-1.7.91
        Newark-Welham Jcn (Market Harborough)
            MAREFIELD JCN-LEICESTER BELGRAVE ROAD  15.5.82
Liverpool Street (Bishopsgate)-Norwich Victoria+
    Bethnal Green-Copper Mill Jcn
        Hackney Downs-Edmonton CHESHUNT  1.10.91
    Stratford-Cambridge-Norwich-Yarmouth+
        BARNWELL (CAMBRIDGE)-MILDENHALL  2.6.84
        Ely Jcn-Soham WARREN HILL JCN (NEWMARKET) 1.9.79=
        Ely North Jcn-Kings Lynn-Hunstanton
            Kings Lynn-Norwich City
                Melton Constable-Yarmouth Beach
                    Yarmouth North Quay Br  REDRAWN LARGER
    Forest Gate Jcn-Tilbury-Shoeburyness
        WEST THURROCK JCN (GRAYS)-UPMINSTER  1.7.92
    Ipswich-Yarmouth South Town*
        WESTERFIELD-FELIXSTOWE  1.5.77=

Lpro MR 934 (EPD May 1905); Mg (EPD Aug.1911)
NB: defective copies, railway revision date uncertain
```

I.2. Middle sheet, ?1824

State M-P1 suspected (?1824)

Evidence: none, but 83,84,85,86 were published 1 March 1824. There may be a thinly mapped state comparable to 64,65,69,70 on S-P6 (?1830), and ?S-P5
Not found.

State M-P2 (?1830)

One-inch sheets completed: 83,84,85,86 (1824)

Dimensions: unknown, since the paper of the only known copy is cut to within 5mm of the sheet lines

   NB: Standard features used conform with those on S-P1,S-P2,S-P3,S-P4,S-P6 (qv)

County names used: LINCOLN[shire], [Nottin]GHAM[shire]

Lrgs (trimmed), attached to Lrgs 2 (S-P6) (see p.4)

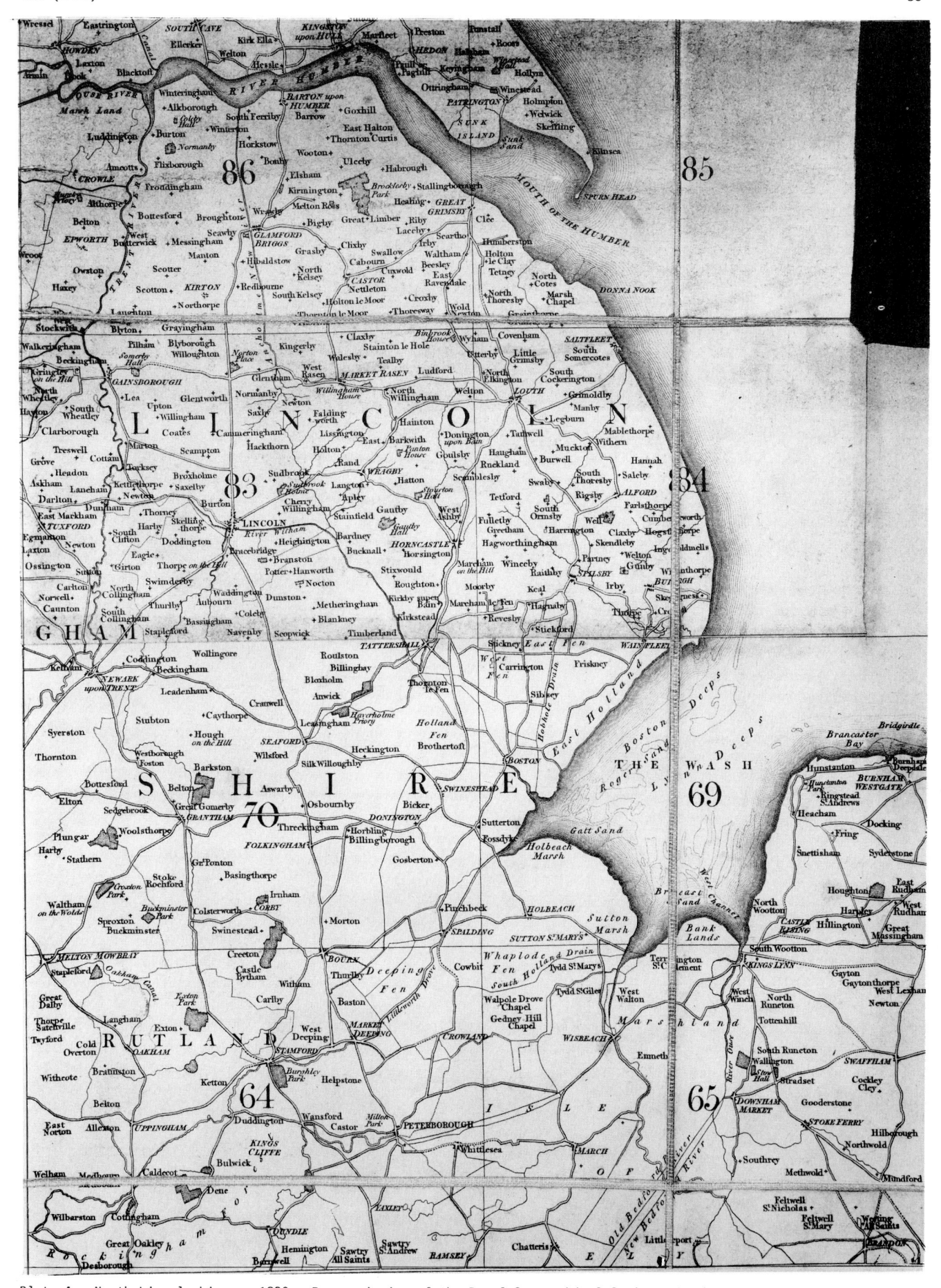

Plate 4. North Lincolnshire, c.1830. By permission of the Royal Geographical Society, London.

State M-P3 (?1839)

Dating evidence: 87 (1840) absent
NB: The map is completely re-engraved, 85,86 here, with 83,84 removed to an enlarged south sheet State S-P8

Visible constructional lines (no neat line visible):
    Delamere Meridian: from top to bottom
    Preston to Hull line: to the western edge, and to the east from 30mm outside 85. Traces visible on 86

One-inch sheets re-engraved: 85,86 (1824)
Sheet lines added: 87-90 (unnumbered, unquartered)
    NB: No title, border, price, scale-bar

Sample points of comparison against M-P2
1. 83,84 no longer part of this sheet
2. Hullshire boundary added
3. River systems more developed: eg the old course of Ancholme, River Eau, some Axholme drains added
4. Inland river names, (Market Weighton) *Canal* lacking
5. Many roads lacking, added, realigned or replaced
6. Some parkland deleted or redrawn, and names lacking
7. Sand and mudbanks added or redrawn
8. *SUNK ISLAND*, *Sunk Sand* (85nw), *Marsh Land* (85nw), names lacking, *Old Warp*, *Paull Sand* names added
9. Many placenames lacking, added, moved or replaced
10. *BARTON* upon *HUMBER* to *BURTON* upon *HUMBER* (86ne)
11. Easington added in error at Kilnsea
12. Kilnsea located in error at Easington
13. *Goole* added (as a hamlet)

Lbl OSD Serial No 510 (Portfolio 28)
Lit: Hodson (1989) 91

State M-P4 suspected (?1844)

Evidence: none, but perhaps showing 87-90 (1840-44)?
Not found.

State M-P5 suspected (?1847)

Evidence: none, but perhaps showing 91SW (1847)?
Not found.

State M-P6 (?1852)

Dating evidence: 91NW,91SE published 12.1852. No 91NW

Dimensions: (derived from M-P8 and M-P12 in Dtc):
    Paper: 1024mm W-E by 695mm S-N
    Plate: 977mm W-E by 674mm S-N
Distances from one-inch sheet lines to edge of plate:
    21mm below 90, 200mm west of 100, 197mm east of 85, 230mm north of 110

One-inch sheets added: 87-90 (1840-44), 91SE (1852)
    Partially filled: 91SW (1847), with land area completed, and 92SW (1857), with the road under Ribchester and the county boundary above Hurst Green present.
    Sheets numbered: 85-90
    Quarter sheet lines for 87-99, 101-110 now included. 100 is shown as a portrait full sheet. In addition extensions down to the Preston to Hull line of the western cutting lines of 99 and 107, together with the extension west of the 91SW,91NW cutting line to supply the south border of 100, form the framework of a complete unnumbered (?unintended) sheet comprising four quarters in the Irish Sea (some information derived from M-P8).

There is a skeleton outline of northern England, comprising the east coastline to Berwick with Holy Island and the Farne Islands, the west coastline to the Scotland portion on 107, principal rivers and Ullswater (without name), county and national boundaries, and certain location points marked by names and crosses. Most of these are market towns, as well as some parish names and high points such as lighthouses or churches (presumably triangulation points): four are in Scotland.

NB: For standard features used above the Preston to Hull line, see the list given with State M-P7

Changes
1. Kilnsea and Easington reversal corrected (85nw)
2. Delamere Meridian deleted below Preston-Hull line
3. Waterlining added to 90 (only on 85,86,90)
4. River names added, rewritten or reinstated on 85,86

Railways (+ Italic names)
[LIVERPOOL]-89SW *Lostock Jcn (Bolton)* 20.11.48=
    [LIVERPOOL]-SOUTHPORT 24.7.48-22.8.51=
    WALTON JCN (LIVERPOOL)-HOUGHTON 1.6.46-2.4.49=
        ORMSKIRK-SKELMERSDALE 1.3.58
        FARINGTON L&Y-L&NW CURVE 1.6.46=
        BAMBER BRIDGE-PRESTON 2.9.50=
[LIVERPOOL]-MANCHESTER+ 16.9.30=
    PARKSIDE JCN-91SE *Lancaster* 3.9.32-16.6.40=
        PRESTON-LONGRIDGE BR 2.5.40=
        PRESTON-FLEETWOOD 15.7.40=
            KIRKHAM-LYTHAM 16.2.46=
    KENYON JCN-BOLTON+ 1.8.28-1.31=
    SALFORD-EUXTON JCN 29.5.38-22.6.43=
MANCHESTER OLDHAM ROAD - GOOSE HILL JCN (NORMANTON)+
                                      4.7.39-1.3.41=
        HEYWOOD JCN-HEYWOOD 15.4.41=
    Dewsbury Jcn (Mirfield) MORLEY-87 18.9.48=
    HORBURY JCN-BARNSLEY= 1.1.50=
        SILKSTONE BR 15.1.50=
    WAKEFIELD-GOOLE 1.4.48=
        PONTEFRACT-METHLEY 12.9.49=
MANCHESTER-[DINTING] // 87-ULCEBY 17.11.41-17.7.49=
    ARDWICK JCN (MANCHESTER)-[CREWE] 4.7.40=
    WRAWBY JCN (BARNETBY) [Lincoln-Derby] 1.11.48=
[DERBY]-LEEDS HUNSLET+ 1.7.40=
LEEDS MARSH LANE // HULL MANOR HOUSE ST= 22.9.34-2.7.40=
NB: Not on 93 at Leeds and Selby
    DAIRYCOATES JCN (HULL)-86 *Bridlington* 6.10.46=

I.2 (M-P6)

```
[KINGS CROSS]-DONCASTER-KNOTTINGLEY   6.6.48-4.9.49=
    [WERRINGTON JCN-BOSTON]-NEW HOLLAND   1.3.48=
        GRIMSBY DOCK BR   1.8.53
        NEW HOLLAND-BARTON-UPON-HUMBER BR   1.3.49=
    DONCASTER-SWINTON   10.11.49=
York 87-ALTOFTS JCN (NORMANTON)   11.5.40-1.7.40=
    SHERBURN JCN-GASCOIGNE WOOD JCN   29.5.39=
    MILFORD JCN-GASCOIGNE WOOD JCN   11.5.40=
    BURTON SALMON-KNOTTINGLEY   3.3.50=
```

Lpro T1/5968B 19661 (JWTM 1854) (trimmed 8mm west of 90)

State M-P7 (?1854)

One-inch sheets added: 91NW,91NE,92-96,97SE (1852-61)
    From partial to complete: 91SW (1847), 92SW (1857)
    Sheets numbered: 85-96

Standard features present on 91-99, 101-110
B.1   Boundaries
    a   County boundary
B.2   Topographic features
    a   River
    b   Lake: eg Malham Tarn (92NW)
    c   Canal
    g   Turnpike
    j   Minor road
    l   Built-up area
    m   Buildings blocked together in towns
    n   Anglican church, usually with name of parish
    ff   Parkland: stipple infill, no paling boundary
    ll   Railway
    ss   Railway in tunnel: eg Thackley Tunnel (92se)
B.4   Waterlining: around coast, sandbanks, lakes
B.5   Hydrographic features
    a   Low water line
    b   Sandbank, mudbank: eg Morecambe Bay
C.3   Roman
    a   Parish (with "+")
    g   Market town (Northern England skeleton mapping)
C.4   Roman sloping capitals
    a   Market town
    b   Principal bay, estuarial river
    d   Principal headland: eg Flamborough Head (95SE)
    g   City, county town
C.9   Italic
    b   River

Additions to railways (+ Italic names)
PIEL-91 *Foxfield*   3.6.46=
    ROOSE-BARROW-IN-FURNESS   12.8.46=
[Liverpool]-89SW *Lostock Jcn (Bolton)*
    Walton Jcn (Liverpool)-Houghton
        Blackburn 92-CHATBURN   22.6.50=
[Liverpool]-Manchester+
    Parkside Jcn-91SE 91 *Oxenholme*   16.6.40-22.9.46=
Manchester Oldham Road-Goose Hill Jcn (Normanton)+
    Dewsbury Jcn Morley-87 LEEDS CENTRAL   18.9.48=

```
[Derby]-Leeds Hunslet+ BRADFORD   1.7.46-7.9.46=
    LEEDS: ENGINE SHED JCN-WELLINGTON BR   1.7.46=
    HOLBECK JCN (LEEDS)-93SW // RIPLEY-96   *Stockton*
                                             1.9.48-2.6.52
        MELMERBY-THIRSK BR   5.1.48=
    SHIPLEY-SKIPTON-COLNE-92 *Burnley*   16.3.47-2.10.48=
        SKIPTON JCN // 92 *Ingleton*   1.8.49=
            CLAPHAM-MORECAMBE   12.6.48-1.6.50=
                LANCASTER: GREEN AYRE-CASTLE   19.12.49=
Leeds Marsh Lane-Hull Manor House Street  COMPLETED
    BARLBY JCN (SELBY)-MARKET WEIGHTON   1.8.48=
    Dairycoates Jcn (Hull)-86 SEAMER  6.10.46-20.10.47=
YORK-96 *Darlington*   4.1.41=
    YORK 87-Altofts Jcn (Normanton)   29.5.39=
    CHURCH FENTON-SPOFFORTH   10.8.47=
    YORK-SCARBOROUGH   8.7.45=
        BOOTHAM JCN (YORK)-MARKET WEIGHTON   4.10.47=
        RILLINGTON-95 *Whitby*   8.6.35-8.7.45=
    POPPLETON JCN (YORK)-KIRK HAMMERTON   30.10.48=
    SESSAY WOOD JCN (PILMOOR) // DRIFFIELD   1.6.53
    PILMOOR-BOROUGHBRIDGE   17.6.47=
    NORTHALLERTON-BEDALE   6.3.48-1.2.55
    ERYHOLME-96 *Richmond*   10.9.46=
```

Lrgs 3 (JW 1854); Lrgs 8 (JW 185?4) (EPD 8 Sep.1856);
Lrgs 7 (JW 1856) (EPD 23 J??)

State M-P8 (?1857)

Details of new mapping:
    To Lancashire-Yorkshire border across 98,97,102,103,
104, together with some of County Durham on sheet 103,
north to Staindrop and Heighington, then south of the
Clarence and east of the Stockton & Darlington Railways.
    Partially filled is the Furness district of Lanca-
shire, with some roads and a beck missing, and some
names missing or in draft alphabets.
    Sheets numbered: still 85-99,103,104

Introduction of new standard features
C.3   Roman
    b   Minor island: eg Isle of Walney (91NW)
C.9   Italic
    d   Minor lake: eg Coniston Water (98NW)
    bb   Minor range of hills, peak: Whernside (97SW)

Change
1. Waterlining extended to top of 91, and to Redcar

Additions to railways (+ Italic names)
Piel-91 BROUGHTON-IN-FURNESS   3.6.46-2.48=
    BR AT KIRKBY - ERROR?
[Liverpool]-Manchester+
    Parkside Jcn-91 LANCASHIRE BORDER   22.9.46-17.12.46=
[Derby]-Leeds-Bradford+
    Holbeck Jcn-93SW // Ripley-96 STOCKTON JCN   2.6.52=
    Shipley-Skipton-Colne-92 *Burnley*=
        Skipton Jcn // 92 INGLETON   1.8.49=

York-96 AYCLIFFE (CLARENCE RY X)   4.1.41-15.4.44=
   York-Scarborough
      Rillington-95 WHITBY   8.6.35=
   Eryholme-96 RICHMOND BR   10.9.46=
SIMPASTURE JN-DARLINGTON-REDCAR OLD STN≡ 28.9.25-4.6.46=
   SIMPASTURE JN (SHILDON)-PORT CLARENCE 8.33-29.10.33=
      NORTON-STOCKTON NORTH SHORE BR   8.33=
   HOPETOWN JCN (DARLINGTON)-*BARNARD CASTLE*   8.7.56
   ALLENS CURVE (YARM)
   STOCKTON QUAYSIDE BR   28.9.25=
   MIDDLESBROUGH DOCKS BR   1.2.42=
   MIDDLESBROUGH-GUISBOROUGH   11.11.53=
   TEESSIDE-ESTON   6.1.51=
   UPLEATHAM MINERAL TRAMWAY (private)

Ob C17 (420) (JW 1857) (EPD 24 Sep.1857) (flat sheet: for dimensions see M-P3); Cu Maps 34.03.105; PC

State M-P9 (?1858)

Details of new mapping:
   Mostly complete in County Durham south and west of a line from Blackhall Mill to Brancepeth. There is a dotted line in 102NE, apparently along the watershed, south of which road mapping is incomplete, as is the Barnard Castle area. The north east of the county has fragmentary coverage, including pockets at Whickham, Washington to South Shields, and Houghton le Spring. The Kendal area of Westmorland through Tebay to Asby is mapped, though road and rail connections into Lancashire are not complete, and some placenames are missing. There is no new mapping in the Furness district of Lancashire.
   Sheets numbered: still 86-99,103,104

Introduction of new standard features
B.2   Topographic feature
   b   Additional use: Reservoir (98NE)
   ll   Additional use: Private mineral line
C.4   Roman sloping capitals
   h   Principal lake: eg Ullswater (102SW)
C.9   Italic
   cc   Industrial location: Collieries (103NE)
   dd   Reservoir: (Killington) Reservoir (98NE)

Change
1. *Winder Mere* (98NE) to *WINDERMERE* (98NW)

Additions to railways (+ Italic names)
[Liverpool]-Manchester+
   Parkside Jcn-Lancashire border // *TEBAY*+   17.12.46=
      OXENHOLME-WINDERMERE BR+   22.9.46-21.4.47=
York-Aycliffe // WASHINGTON   24.8.38-15.4.44=
   LEASINGTHORNE COLLIERY BR   1.36-28.6.41=
   Leamside BRANCEPETH-BISHOP AUCKLAND   19.8.56
   PENSHAW-*SUNDERLAND*   20.12.52=
STANHOPE // SOUTH SHIELDS   15.5.34-10.9.34=
   WEATHERHILL-ROOKHOPE (private)   1846=
   WASKERLEY Simpasture Jcn-Redcar   28.9.25-16.5.45=

   WEAR VALLEY JCN-BISHOPLEY QUARRY   3.8.47=
   SHILDON TUNNEL JCN-COCKFIELD FELL   28.9.25-13.9.56
   SHILDON-BRUSSELTON INCLINE-WEST AUCKLAND   28.9.25=
   Simpasture Jcn (Shildon)-Port Clarence
      STILLINGTON-FERRYHILL   16.1.34=
      BILLINGHAM // WEST HARTLEPOOL   12.11.40=
CONSETT IRONWORKS BR
MEDOMSLEY COLLIERY BR   1834=
ANNFIELD-*TANFIELD MOOR COLLIERY*≡   1835=
   Pelton HOLMSIDE COLLY-BURNHOPE COLLY   1826-1845=
WASHINGTON-PELAW   1.9.49=
SEAHAM-NORTH HETTON COLLIERY (Londonderry) 1.3.25-?=
   SEAHAM-*NEW HESLEDEN* (South Hetton Colly Ry) 5.8.33=
SUNDERLAND // HARTLEPOOL VIA MURTON   1.1.35-5.7.36=
   MURTON-*BELMONT*   13.10.36=
   CASTLE EDEN // WHITE LEA COLLIERY   3.37-11.46=
      EAST HETTON COLLIERY BR   18.3.39?=
      PAGE BANK COLLIERY BR   1855=
SUNDERLAND-ELEMORE COLLIERY (Hetton Ry)   18.11.22-?=
   SUNDERLAND-*BOURNMOOR* (Lambton Colly Ry) 1814-1835=
   HOUGHTON COLLIERY BR
   LAMBTON LOW SPOUTS BR
JARROW-SPRINGWELL BANK FOOT (Bowes)   17.1.26=
BROCKLEY WHINS // *RIVER DERWENT*   1.3.37-30.8.39=
   CLEADON LANE JCN-SOUTH SHIELDS   19.6.39=
   EAST BOLDON JCN-HEDWORTH LANE JCN   9.9.39=
   BROCKLEY WHINS JCN-HARTON JCN   30.8.39=

PC (JW 1858); Mp volume FF912.42 064

State M-P10 (?1859)

Details of new mapping:
   Westmorland: no increase, though the mapping is revised (see below), and the connections into Lancashire are complete. County Durham: the Barnard Castle and St John's Chapel areas are complete (and the dotted line deleted), and complete south of the line from Brancepeth through Durham and Houghton to Sunderland, except Sunderland and Seaham themselves. There is no new mapping in the Furness district of Lancashire.
   Sheets numbered: still 86-99,103,104 (not 105)

Introduction of new standard feature
C.9   Italic
   t   Railway company name

Changes
1. St John's Chapel to *ST JOHN'S CHAPEL* (102NE)
Kendal area (98) mapping revised, which entailed:
2. Kirkby Lonsdale to *KIRKBY LONSDALE*
3. Barbon, Casterton deleted
4. Old Town, Row, Beethwaite Green, Milnthorpe, Hutton Roof deleted, but leaving "+"
5. Beetham, *BURTON*, Heversham added
6. Kendal to Kirkby Lonsdale roads revised
7. *Levens Hall* deleted
8. *R.Rawthey* deleted

I.2 (M-P10)

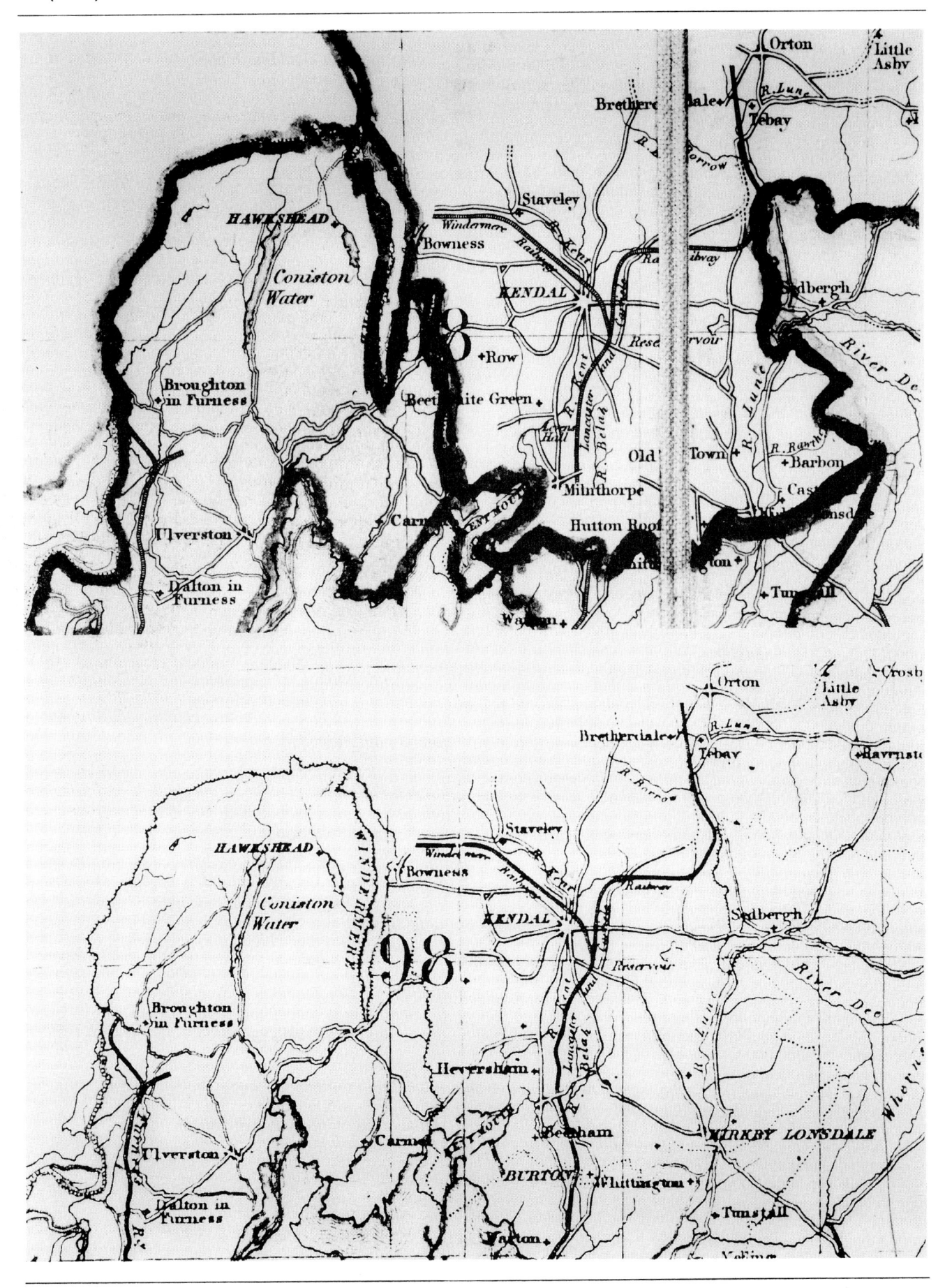

Plate 5. The Kendal area on States M-P9, c.1858 and M-P10, c.1859. From the collections of Peter Clark and the author.

Additions to railways (+ Italic names)
Piel-Broughton-in-Furness+≡
   Millwood Jcn ARNSIDE-LANCASHIRE BORDER 10.8.57
Liverpool-89SW LOSTOCK JCN (BOLTON)+ NAMED 20.11.48=
  [Liverpool]-Southport+ NAMED
    Walton Jcn (Liverpool) HOUGHTON-BURNLEY Skipton Jn+
                         NAMED 19.6.48-18.9.48=
      DAISY FIELD JCN (BLACKBURN) 92-Chatburn 22.6.50=
    WIGAN-SOUTHPORT+ 9.4.55
  [Liverpool]-Manchester+
    Parkside Jcn-*Tebay*+ COMPLETED 17.12.46=
    PATRICROFT-MOLYNEUX JCN 2.2.50=
  Salford-Euxton Jcn+
    CLIFTON JCN-ACCRINGTON+   28.9.46-17.8.48=
      STUBBINS JCN-WATERFOOT   28.9.46-27.3.48=
    BOLTON-BLACKBURN+ 3.8.47-12.6.48=
Manchester Oldham Road-Goose Hill Jcn (Normanton)+
  MILES PLATTING-STALYBRIDGE 13.4.46-5.10.46=
    PARK-ARDWICK JCN 20.11.48=
  MIDDLETON JCN-OLDHAM 3.42-1.11.47=
  Heywood Jcn-Heywood BOLTON 1.5.48-20.11.48=
  TODMORDEN-GANNOW JCN (BURNLEY) 12.9.49=
  GREETLAND-HALIFAX-LOW MOOR 7.8.50=
  MIRFIELD-BRADFORD EXCHANGE 18.7.48-9.5.50=
  Dewsbury Jcn DEWSBURY Morley-Leeds Central 18.9.48=
  INGS ROAD JCN (WAKEFIELD)-MILL LANE JCN (BRADFORD)
                                1.8.54-5.10.57
    ARDSLEY-LAISTERDYKE 20.8.56-10.10.57
  Wakefield-Goole+ NAMED
Manchester [DINTING]-87 Ulceby+ NAMED 8.8.44-23.12.45=
  GUIDE BRIDGE-HEATON LODGE JCN (MIRFIELD)+ 23.12.45-
                                        1.8.49=
    SPRINGWOOD JCN (HUDDERSFIELD)-PENISTONE+ 1.7.50=
      BROCKHOLES-HOLMFIRTH BR 1.7.50=
  PENISTONE JCN-BARNSLEY 15.5.54-10.55
[Derby]-Leeds-Bradford+
  Holbeck Jcn 93SW-RIPLEY Stockton Jcn+ NAMED 1.9.48=
    PICTON-KILDALE+ 3.3.57-6.4.58
      BATTERSBY-ROSEDALE BANK FOOT 6.4.58
  Shipley-Ingleton+ COMPLETED AND NAMED 1.8.49=
    Clapham-Morecambe+ NAMED
Leeds Marsh Lane-Hull Manor House Street+ NAMED
  Barlby Jcn (Selby)-Market Weighton+ NAMED
  Dairycoates Jcn (Hull)-Seamer+ NAMED
  SOUTHCOATES (HULL)-WITHERNSEA+ 27.6.54
[Kings Cross]-Doncaster-Knottingley
  Doncaster SWINTON-BARNSLEY Horbury Jcn+ 1.2.50-1.7.51
    ELSECAR BR 1.2.50=
    ALDAM JCN-[WINCOBANK] 4.9.54
      MINERAL BR
    ALDAM JCN-WEST SILKSTONE 6.50-4.52=
  DONCASTER-THORNE+ 1.7.51-11.12.55
    THORNE WATERSIDE BR 1.56
York // Washington+ MORE COMPLETE - NAMED 15.4.44=
  York-Altofts Jcn (Normanton)+ NAMED
    Church Fenton-Spofforth HARROGATE 20.7.48=
  York-Scarborough+ NAMED
    Bootham Jcn (York)-Market Weighton+ NAMED

    Rillington-Whitby+ NAMED
  Poppleton Jcn (York)-Kirk Hammerton STARBECK 1.10.51=
  Sessay Wood Jcn (Pilmoor)-Driffield+ COMPLETED AND
                                      NAMED 1.6.53=
FERRYHILL-COXHOE BR 16.1.34=
LEAMSIDE-DURHAM GILESGATE BR 15.4.44=
LEAMSIDE Brancepeth-Bishop Auckland 19.8.56
Stanhope // South Shields
  Weatherhill-Rookhope MIDDLEHOPE (private) 1855
  Waskerley-Darlington-Redcar Old Stn+ NAMED
    Wear Vy Jn-Bishopley FROSTERLEY+ NAMED 3.8.47=
    Shildon Tunnel Jn-Cockfield HAGGERLEASES 2.10.30=
    Simpasture Jcn (Shildon)-Port Clarence
      Billingham-W Hartlepool+ COMPD, NAMED 12.11.40=
    Hopetown Jcn (Darlington)-BARNARD CASTLE 8.7.56
Seaham-North Hetton Colliery (Londonderry)
  Seaham-New Hesleden HASWELL COLLIERY 2.7.35=
    SOUTH HETTON COLLIERY-SOUTH HETTON JCN 6.10.36=
    TUTHILL BR
Sunderland // Hartlepool+ MORE COMPLETE - NAMED
  Murton-*Belmont* HOUGHALL COLLIERY 6.11.37-17.2.42=
  LUDWORTH COLLIERY BR 1.1.35=
    THORNLEY COLLIERY BR 1.1.35=
  Castle Eden-White Lea Colliery COMPLETED 3.37-11.46=
Sunderland-Elemore Colliery (Hetton Ry)
  Sunderland-*Bournmoor*
    BOURNMOOR-FRANKLAND WOOD COLLIERY
    RAINTON CROSSING-ALEXANDRINA PIT
      BROOMSIDE COLLIERY BR
    RESOLUTION PIT-SHERBURN HOUSE COLLIERY
      LITTLETOWN COLLIERY BR

RH (JW 1859); PC

State M-P11 (?1862)

Details of new mapping:
    98,99 but for north of St Bridget Beckermet; Westmorland, except the Asby to Orton road; Cumberland on 102, except for Penrith and some detail south west to north west of Penrith; the remainder of County Durham; Northumberland to a line approximately from Willyshaw Rigg through Whitfield, Hexham, Morpeth to the coast at Druridge Bay.
    Berwick area sheets completed (England only): 110NW
    Berwick area sheets partially filled (England only): 108NE,110NE,110SW,110SE
    Sheets numbered: 85-99,102-105,110

Introduction of new standard feature
C.1   Roman capitals
    a   County

Changes
1. Northern England outline deleted
2. Waterlining extended from 91 and Redcar to all mapped areas except 110
3. Urswick added (98SW)

I.2 (M-P11)   41

4. *DALTON IN FURNESS, BROUGHTON IN FURNESS, ULVERSTON*
   (98SW) from Roman to Roman sloping capitals
5. Carmel to *CARTMEL*
6. *R.Lune* deleted making way for WESTMORLAND (98NE)

Additions to railways (+ Italic names)
Piel FOXFIELD-*BRAYSTONES*+=   21.7.49-1.11.50=
   MILLWOOD JCN // Arnside-Lancs border 3.6.46-10.8.57=
[Liverpool]-Manchester+
   Parkside Jcn-*Tebay*+ // 102 *Carlisle* 17.12.46=
NB: This gap at Penrith is forgotten until NS-M-2
York-Washington+ COMPLETED   15.4.44=
   Penshaw-SUNDERLAND   20.12.52=
Stanhope-South Shields COMPLETED   10.9.34=
   PELTON Holmside Colliery-Burnhope Colliery   1826=
Seaham-North Hetton Colliery (Londonderry)
   Seaham-Haswell Colliery (S Hetton Coll. Co)
      SEAHAM-RYHOPE (Londonderry)   17.1.54=
SUNDERLAND-Hartlepool+ COMPLETED   9.8.36=
Sunderland-Elemore Colliery
   Sunderland-Bournmoor COCKEN PIT
MONKWEARMOUTH (SUNDERLAND) *Brockley Whins-River Derwent*
                  HEXHAM COMPLETED   26.11.34-30.8.39=
   CLEADON LANE JCN-South Shields   19.6.39=
   EAST BOLDON JCN-Hedworth Lane Jcn   9.9.39=
   Brockley Whins Jcn-Harton Jcn
   REDHEUGH Tanfield Moor-Annfield   26.11.39-11.11.40=
   BLAYDON-NEWCASTLE-TYNEMOUTH   12.5.39-1.1.51=
      HEATON-*CHEVINGTON* // 110-TWEEDMOUTH 1.3.47-1.7.47=
         KILLINGWORTH COLLIERY-105NE *Willington Quay*
                                    (private)   1806-c.1830=
      Tw'dm'th 110NW-BORDER *Sprouston* 27.7.49-1.6.51=
JARROW-*Springwell Bank Foot* DIPTON   17.1.26-20.9.54=
NORTHUMBERLAND DOCK-BLYTH   1.6.40-1846=
   NORTHUMBERLAND DOCK-WEST BRUNTON (private) -1826=
   NORTHUMBERLAND DOCK-WEST CRAMLINGTON (private) -1838=
   HARTLEY-SEATON SLUICE   1846=
   NEWSHAM-MORPETH   12.6.50-1.10.57=

Cu Maps 34.01.111, Dtc (JW 1860) (both EPD Jul.1862); PC

State M-P12 (?1864)

One-inch sheets added:
   From partial to complete: 99NW,102NW,102SW,106SE,
110NE,110SW,110SE (England coverage only)
   Partially filled: 101,105NE,106NW,106NE,106SW,107,
108,109 (England coverage only)
   Sheets numbered: 85-99,101-110

Additions to railways (+ Italic names)
Piel-*Braystones* WORKINGTON-MARYPORT-CARLISLE+ (ROUGH
                           ALIGNMENT)   15.7.40-30.9.52=
   Millwood Jcn-Lancs border COMPLETED   10.8.57=
   Foxfield-Broughton // 98NW-CONISTON   18.6.59=
   CORKICKLE-CLEATOR MOOR   11.1.56=
      MOOR ROW-EGREMONT=   19.1.56=
   WORKINGTON-COCKERMOUTH (ROUGH ALIGNMENT)   28.4.47=

[Liverpool]-Lostock Jcn (Bolton)+
   Walton Jcn (Liverpool)-Skipton Jcn+
      Ormskirk SKELMERSDALE-ST HELENS [Widnes]   1.2.58=
         ST HELENS NEW STN LOOP   1.2.58=
[Liverpool]-Manchester+
   Parkside Jcn-102+ GRETNA   17.12.46-10.9.47=
      LOW GILL-98NE *Ingleton*=   24.8.61
      GRETNA-LONGTOWN (ROUGH ALIGNMENT)   1.11.61
   Salford-Euxton Jcn+
      Clifton Jcn-Accrington+
         Stubbins Jcn-Waterfoot BACUP   1.10.52=
Manchester-Ulceby+ (parts on south sheet)
   Ardwick Jcn (Manchester)-[Crewe]
      [HEATON NORRIS]-GUIDE BRIDGE   1.8.49=
      GUIDE BRIDGE-ASHTON-OLDHAM-GREENFIELD   5.7.56-1.7.62
      HYDE-[BLACKWELL MILL JCNS (BUXTON)]   1.3.58=
[Derby]-Leeds-Bradford+
   Holbeck Jcn (Leeds)-Stockton Jcn+
      RIPLEY JCN-PATELEY BRIDGE   1.5.62
      Picton-Kildale+ CASTLETON   1.4.61
[Kings Cross]-Doncaster-Knottingley
   Doncaster-Thorne+ KEADBY   10.9.59=
York-Washington+
   York-Altofts Jcn (Normanton)+
      Church Fenton-Harrogate BILTON JCN   1.8.62
         HARROGATE-STARBECK   1.8.62
   Leamside-Bishop Auckland
      RELLY MILL JCN (DURHAM)-BLACKHILL   1.9.62
Stanhope-South Shields
   Waskerley-Darlington-Redcar+ SALTBURN   19.8.61
      Simpasture Jcn (Shildon)-Port Clarence
         Billingham - West Hartlepool+ HARTLEPOOL
                                    12.11.40-7.12.40=
            WEST HARTLEPOOL-HARTLEPOOL VIA DOCKS 1.6.47=
      Darlington-Barnard Castle TEBAY   26.3.61-4.7.61
         KIRKBY STEPHEN-CLIFTON   8.4.62
      Middlesbrough-Guisborough
         NUNTHORPE-BATTERSBY   1.6.64
      UPLEATHAM MINERAL TRAMWAY - NEW ROUTE
Monkwearmouth-Hexham CARLISLE (M&C)   28.6.36-18.6.38=
   Blaydon-Newcastle-Tynemouth
      Heaton *CHEVINGTON*-110+ Tweedmouth 29.3.47-1.7.47=
         TWEEDMOUTH 110NW-Border *Sprouston* 27.7.49=
      HEXHAM JCN-BORDER *Riccarton Jcn*   5.4.58-24.6.62
      HALTWHISTLE-ALSTON   3.51-17.11.52=

Lrgs 9 (TJH)

State M-P13 (?1866) (?first complete English state)

One-inch sheets from partial to complete: 101,105NE,
106NW,106NE,106SW,107,108,109 (England coverage only)
   Sheets numbered: still 85-99,101-110. NB: no 100
   NB: Publication of one-inch sheets completed by 1869

Changes
1. Withernsea (85nw), Ingleton (97SW) added
2. Waterlining added on 110 (no more on west coast)

3. Sheet lines re-entered. But the N-S quarter sheet line in 89 was not touched, and instead the sheet line dividing 91 and 92 was extended in error, with the Delamere Meridian, to the bottom of sheet.

Additions to railways (+ Italic names)
Piel-Workington-Maryport-Carlisle+ (PROPER ALIGNMENT)
    Millwood Jcn-Lancs border CARNFORTH   10.8.57=
    Foxfield BROUGHTON-98NW Coniston Br   18.6.59=
    Workington-Cockermouth BASSENTHWAITE LAKE   26.10.64
[Liverpool]-Manchester+
    Parkside Jcn-Gretna+
        Preston-Fleetwood+
            Kirkham-Lytham BLACKPOOL   6.4.63
            POULTON-BLACKPOOL NORTH BR   29.4.46=
Manchester Oldham Road-Goose Hill Jcn (Normanton)+
    Middleton Jcn-Oldham ROCHDALE   2.11.63
        ROYTON JCN (OLDHAM)-ROYTON BR   21.3.64
    DEWSBURY J (MIRFIELD) Dewsbury-Leeds Central 18.9.48=
    Ings Road Jcn (Wakefield)-Mill Lane Jcn (Bradford)=
        WRENTHORPE JCN (WAKEFIELD)-FLUSHDYKE   7.4.62
Manchester-Ulceby+ (parts on south sheet)
    Guide Bridge-Heaton Lodge Jcn (Mirfield)+
        DELPH JCN (SADDLEWORTH)-DELPH BR   1.9.51
[Derby]-Leeds-Bradford+
    Shipley INGLETON-98NE Low Gill+   24.8.61=
Leeds Marsh Lane-Hull Manor House Street+
    WILMINGTON (HULL)-HORNSEA+   28.3.64
York-Washington+
    Northallerton-Bedale LEYBURN   24.11.55
    Leamside-Bishop Auckland
        DEARNESS VALLEY JCN (DURHAM)-CROOK   19.10.58
Stanhope-South Shields
    Waskerley-Darlington-Saltburn+
        Wear Vy Jn-Frosterley+ STANHOPE 30.4.62-22.10.62
        Shildon Tunnel Jcn-Haggerleases
            SPRING GARDENS JCN-BARNARD CASTLE   1.8.63
Monkwearmouth-Carlisle PORT CARLISLE   9.3.37-22.5.54=
    Blaydon-Newcastle-Tynemouth
        Heaton-Tweedmouth+
            AMBLE JCN (CHEVINGTON)-AMBLE BR   5.9.49=
    BRAMPTON-LAMBLEY   17.11.52
    PORT CARLISLE JCN-BORDER Newcastleton 12.10.61-1.3.62
    DRUMBURGH-SILLOTH   4.9.56=

Lmd (TJH 1865) (EPD Apr.1866) (TDWO stamp 9.6.1866);
Lmd (TJH) (with 1868 military overprint (qv)); Cjc; PC

State M-P14 (?1869)

Introduction of new standard features
A.1   Title: **Index to the Ordnance Survey, of England and Wales, and Part of Scotland**
A.5   Scale of ten Statute Miles to an Inch [Scale-bar 10+50 miles]: 31mm below title

Changes
1. Scotland coastal outline drawn to north and west
2. Scotland main line railways added in English format
3. *STALEY BRIDGE* to *STALY-BRIDGE* (88SW)

Additions to railways (+ Italic names)
Piel-Workington-Maryport-Carlisle+
    Millwood Jcn-Carnforth WENNINGTON   10.4.67
    CORKICKLE-Cleator Moor MARRON JCNS   11.62-15.1.66
    Workington-Bassenthwaite Lake PENRITH 26.10.64
        BRIGHAM-BULGILL   12.4.67
    ASPATRIA-MEALGATE-AIKBANK JCN   2.4.66
[Liverpool]-Manchester+
    Parkside Jcn-Gretna+ BEATTOCK   10.9.47=
        GRETNA-CARRONBRIDGE=   23.8.48-28.10.50=
            DUMFRIES-MAXWELLTOWN   7.11.59=
    Kenyon Jcn-Bolton+
        PENNINGTON-TYLDESLEY   1.9.64
    ECCLES JCN-WIGAN   1.9.64
NB: Drawn on too southerly an alignment - not deleted
Manchester Oldham Road-Goose Hill Jcn (Normanton)+
    THORNHILL-HECKMONDWIKE   10.5.69
Manchester-Ulceby+ (parts on south sheet)
    Guide Bridge-Heaton Lodge Jcn (Mirfield)+
        Springwood Jcn (Huddersfield)-Penistone+
            MELTHAM JCN (HUDDERSFIELD)-MELTHAM BR   8.8.68
        DEIGHTON-KIRKBURTON BR   7.10.67
[Derby]-Leeds-Bradford+
    Holbeck Jcn (Leeds)-Stockton Jcn+
        ARTHINGTON-ILKLEY   1.2.65
        Picton-Castleton+ GROSMONT   2.10.65
    APPERLEY JCN-BURLEY   1.8.65
    Shipley-Low Gill+
        KEIGHLEY-OXENHOPE   15.4.67
[Kings Cross]-Doncaster-Knottingley
    [Werrington Jcn-Gainsborough]
        [Boston]-New Holland
            Grimsby Dock Br CLEETHORPES   6.4.63
    Doncaster-Keadby+
        DONCASTER-THORNE (REVISED ROUTE)   10.9.66-1.12.66
    DONCASTER Wakefield-Mill Lane Jcn (Bradford) 1.2.66
    ADWICK JCN-STAINFORTH   1.11.66
    Wrenthorpe Jcn-Flushdyke BATLEY   2.4.64-15.12.64
Stanhope-South Shields
    Waskerley-Darlington-Saltburn+
        Darlington-Tebay
            TEES VY J (BARNARD CASTLE)-MIDDLETON BR 13.5.68
Monkwearmouth (Sunderland)-Port Carlisle
    Blaydon-Newcastle-Tynemouth
        MANORS-GOSFORTH-WHITLEY BAY   1.5.63-27.6.64
NB: The line has no cross hatching
        Heaton-Tweedmouth+ EDINBURGH   22.6.46-20.7.50=
            Tweedmouth-Border ST BOSWELLS   27.7.49-1.6.51=
            MONKTONHALL-EDINBURGH ST LEONARDS   7.7.47=
            PORTOBELLO JCN Border-Port Carlisle Jcn (Car-
                  lisle)*   21.6.47-24.6.62=
NB: A section is marked in tunnel from the Edinburgh-shire border to Galashiels: this is blank on State M-P15
    Hexham Jcn-Border RICCARTON JCN   24.6.62=
    HEXHAM JCN-ALLENDALE BR   19.8.67-13.1.68

I.2 (M-P14)

Northumberland Dock-Blyth
   Hartley-Seaton Sluice
      DAIRY HOUSE-TYNEMOUTH   31.10.60=
   Newsham-Morpeth
      BEDLINGTON-NEWBIGGIN (ERROR - DIRECT)   1.3.72

Lmd (1869 military issue (qv))

State M-P15 (?1873) (?first true Anglo-Scottish state)

Plate size increased. Dimensions:
   Paper: c.1092mm W-E by 718mm S-N
   Plate: 1072mm W-E by 689mm S-N, extended W and N
Distances from one-inch sheet lines to edge of plate:
   20mm below 90; 22mm above S58; 38mm above S67
One-inch sheets added:
   100 (1873). *NB*: In Scottish style
   Scotland mainland sheets: they are thus complete as far west as the S1-62 column. S12 is also complete
   Partially filled (mainland areas only): S20,28,36,44,51,52,61
   Sheets numbered: 85-110, S1-18,20-26,28-34,36-41,44-49,52-57,61-67. *NB*: Not S51,57A

Sheet lines: a full grid of Scotland sheet lines to the S58-67 row, including blank and unnumbered boxes for S19,27,35,42,43,50,58,59,60. There is an unused box west of S50, below S58. West from S12 (west side) sheet lines are elongated southwards.
   NB: BEN AVON, the name cut in half on top border

Standard features present in Scotland and Isle of Man
B.1   Boundaries
  a County boundaries, including detached parts
  b National border
B.2   Topographic features
  a River
  b Inland loch
  c Canal: eg Monkland (S31)
  e Ship canal: Caledonian (S62)
  g Turnpike
  j Minor road
  l Built-up area: Glasgow (S30)
  m Building, blocked together in towns
  n Parish church, usually with name of parish
  nn Railway
B.5   Hydrographic features
  a Low water line
  e Lighthouse: Inchcape or Bell Rock (S49)
C.1   Roman capitals
  a County, large detached part of county
  b City, county town, including Berwick upon Tweed (final form) (110NE)
  c Large island: eg Isle of Man (100), Rum (S60)
  e Marine name: eg Atlantic Ocean
C.3   Roman
  a Parish (with "+")
  b Minor island: eg Ailsa Craig (S7)
  e Offshore lighthouse: Inchcape or Bell Rock (S49)
  p "Church" (with "+"): unnamed parish (S37), mostly in Aberdeenshire, Banffshire and Nairnshire
C.4   Roman sloping capitals
  a Market town
  b Principal bay, channel, loch, estuarial river
  d Principal headland: eg Mull of Galloway (S1)
C.6   Egyptian capitals
  b Railway company name
  c Important ranges of hills: eg Paps (S28), Grampian Mountains (S46-65)
C.7   Egyptian
  a Peak: Ben Nevis (S53)
C.8   Egyptian sloping capitals
  a Ship canal: Caledonian Canal (S62), Glen Roy! (S63), Canal (S83)
C.9   Italic
  b River
  d Inland lochs (all): eg Loch Tummel (S55)
  e Minor headland: eg Cock of Arran (S21)
  f Small island, rock: eg Scares (S2)
  g Hamlet: eg Fortwilliam (S53)
  h Isolated building: Gentleman's seat, some in parkland: eg Balmoral Castle (S65)
  l Detached part of county: eg Perth "(Det)" (S39)
  m Minor bay, channel: eg Crinan Loch (S36)
  q Important valley and other lowland topographic feature: eg Forest of Athole (S64)
  bb Minor range of hills, peak: eg Snaefell (100), Cheviot Hills (108), Lammermuir Hills (S33), Ben Lomond (S38)
C.10   Gothic
  a Antiquity: Blair Castle (S55), Fyvie Castle (S86), Dunrobin Castle (S103), John O'Groats House (S117)

Changes
1. Plate dimensions increased (see above)
2. Preston-Hull line and 99,100 extension to it deleted
3. Waterlining deleted
4. *BERWICK UPON TWEED* to BERWICK UPON TWEED (110NE)
5. Ancroft relocated above "+" (110NE)
6. State M-P14 Scottish railways rescribed

Additions to railways (+ Italic names, * Egyptian names)
Piel-Workington-Maryport-Carlisle+
   SELLAFIELD Egremont-Moor Row   2.8.69
[Liverpool]-Lostock Jcn (Bolton)+
   Walton Jcn (Liverpool)-Skipton Jcn+
      EARBY-BARNOLDSWICK BR   8.2.71
   HINDLEY-BLACK ROD   15.7.68
[Liverpool]-Manchester+
   Parkside Jcn-Beattock+ GLASGOW*   15.2.48-1.6.49=
      Preston-Fleetwood+
         BURN NAZE-FLEETWOOD DEVIATION   13.1.51=
      GARSTANG-PILLING   5.12.70
      HEST BANK-MORECAMBE   8.8.64
      LOCKERBIE-DUMFRIES   1.9.63

SYMINGTON JCN-PEEBLES   5.11.60-1.2.64
CARSTAIRS JCNS-EDINBURGH LOTHIAN ROAD*   15.2.48=
    CAMPS BR
CLEGHORN-LANARK   5.1.55=
    LANARK JCNS-HAPPENDON   1.4.64
Motherwell ROSS-COALBURN   1.12.56=
    DALSERF-COTCASTLE   1.9.62-1.9.64
        OVERWOOD QUARRY BR
MOTHERWELL-GLASGOW BUCHANAN STREET 1.10.26-8.10.57
NB: The functions of CR and NB routes through Coatbridge
are combined. This leads to several E-W errors
        MOSSEND-MIDCALDER JCN   25.1.34-1.1.69
        CLELAND-MORNINGSIDE   1.11.64
            DRUMBOWIE JCN-LANRIDGE COLLIERY
                LANGBYRES JCN-OMOA (NOT JOINED)
                LANRIDGE JCN-DUNSYSTON COLLIERY
                LINRIGG COLLIERY BR
                DEWSHILL COLLIERY BR
            MINERAL BR (AFTER CROSSING WM&C)
            Woodmuir Jcn NB-LIMEFIELD JCN MINERAL LOOP
NB: Error - lacking west of NB Line
        COATBRIDGE-BO'NESS*   18.8.28-17.3.51
            COATBRIDGE-BATHGATE*   11.8.62
                WESTCRAIGS-SHOTTS-BLACKHALL
        GARTSHERRIE-DUMGOYNE   1.10.26-5.11.66
NB: Line to Gartness Jcn never shown
            GARNQUEEN CURVE   7.8.48=
            GARTCOSH-ARBROATH*   4.12.38-2.66
                GREENHILL: LOWER JCN-UPPER JCN   1.3.48=
                LARBERT JCN-DENNY STONEYWOOD   1.4.58=
                    INGLISTON BR
                ALLOA JCN (LARBERT)-SOUTH ALLOA   2.9.50=
                STIRLING-TAYPORT*   20.9.47-17.5.48=
                    ALLOA-TILLICOULTRY   3.6.51=
                    WHITEMYRE JCN (DUNFERMLINE)-KELTY*   5.58=
                        LILLIEHILL-STEELEND
                        GASK BR
                        ST LETHANS BR
                        LASSODIE BR
                        CHARLESTOWN TRAMWAY
                        HALBEATH TRAMWAY
                    LUMPHINNANS JCN-LADYBANK*   6.7.57-20.6.60
                        KINROSS-RUMBLING BRIDGE   1.5.63
                        THORNTON JCN-BURNTISLAND   20.9.47=
                        THORNTON JCN-ANSTRUTHER*   3.7.54-1.9.63
                        MARKINCH-LESLIE BR   1857=
                    LADYBANK-HILTON J(PERTH) 20.9.47-25.7.48=
                    MILTON JCN (LEUCHARS)-ST ANDREWS   1.7.52=
                DUNBLANE-CALLANDER   1.7.58=
                GLENEAGLES-CRIEFF=   16.3.56=
                PERTH JCN-DUNDEE-ARBROATH*   6.10.38-1.3.49=
                    DUNDEE-NEWTYLE-MEIGLE   16.12.31=
                ALMOND VALLEY JCN (PERTH)-METHVEN   1.1.58=
                STANLEY JCN (PERTH)-[FORRES]*   7.4.56-9.9.63
                    BALLINLUIG-ABERFELDY BR   3.7.65
                COUPAR ANGUS-BLAIRGOWRIE BR   1.8.55=
                ALYTH JCN (MEIGLE)-ALYTH BR   12.8.61
                KIRRIEMUIR JCN-KIRRIEMUIR BR   12.8.61

FORFAR-BROUGHTY FERRY   12.8.70
    GUTHRIE-[ABERDEEN]*   1.2.48-1.4.50=
        GLASTERLAW-FRIOCKHEIM   1.2.48=
        BRIDGE OF DUN-BRECHIN   1.2.48=
        DUBTON JCN-MONTROSE   1.2.48=
            BROOMFIELD J (MONTROSE)-BERVIE 1.11.65
        [ABERDEEN]-BALLATER   8.9.53-17.10.66
NEWTON-HAMILTON WEST   17.9.49=
    HAMILTON-STRATHAVON   6.8.60-1.12.62
        QUARTER BR   6.8.60
RUTHERGLEN-WHIFFLET   20.9.65
    TANNOCHSIDE COLLIERY BR (DRUMPELLER RY)
    LANGLOAN-COATBRIDGE   20.9.65
ECCLES-TYLDESLEY REDRAWN ON CORRECT ALIGNMENT
Salford-Euxton Jcn+
    CHORLEY-CHERRY TREE (BLACKBURN)   1.11.69
Manchester Oldham Road-Goose Hill Jcn (Normanton)+
ROCHDALE-FACIT   5.10.70
Leeds Marsh Lane-Hull Manor House Street+
    MICKLEFIELD-CHURCH FENTON   1.4.69
[Kings Cross]-Doncaster-Knottingley
    [Werrington Jcn - Gainsborough]   BLACK CARR JCN
                        (DONCASTER)   1.7.67
    Doncaster-Keadby+ WRAWBY JCN (BARNETBY)   1.5.66
        Doncaster-Thorne STADDLETHORPE   2.8.69
    SHAFTHOLME JCN-CHALONERS WHIN JCN (YORK)   2.1.71
        JOAN CROFT JCN-APPLEHURST JCN   1.7.77
York-Washington
    Sessay Wood Jcn (Pilmoor)-Driffield+
    GILLING-HELMSLEY   9.10.71
NB: Station sited incorrectly - never altered
    TURSDALE J (FERRYHILL)-RELLY MILL J (DURHAM) 1.10.71
    Leamside-Bishop Auckland
        Newton Hall Jcn (Durham) 103NW-GATESHEAD   2.3.68
        Relly Mill Jcn (Durham)-Blackhill SCOTSWOOD BRIDGE
                                        JCN   18.6.67
Stanhope-South Shields
    HOWNES GILL JCN-BLACKHILL   1.10.68
Monkwearmouth (Sunderland)-Port Carlisle
Blaydon-Newcastle-Tynemouth
    Heaton-Edinburgh GLASGOW QUEEN STREET+*   21.2.42=
        ALNMOUTH-ALNWICK   19.8.50=
        Tweedmouth-St Boswells
            ROXBURGH-JEDBURGH BR   17.7.56=
        RESTON-RAVENSWOOD JCN (ST BOSWELLS)*
                                13.8.49-20.10.65=
        DREM-NORTH BERWICK BR   13.8.49-17.6.50=
        LONGNIDDRY-HADDINGTON BR   22.6.46=
        Portobello Jcn // Port Carlisle Jcn (Carlisle)*
NB: Gap north of Galashiels instead of M-P14 tunnel
            HARDENGREEN JCN-SMEATON
            HARDENGREEN JCN-PEEBLES*   4.7.55=
                LEADBURN-CARSTAIRS*   4.6.64-1.3.67=
            GALASHIELS-SELKIRK BR   5.4.56=
        PORTOBELLO-LEITH   1859=
        EDINBURGH-GRANTON   31.8.42-17.5.47=
        RATHO-BATHGATE-BLACKSTONE   12.11.49-2.7.55=
            CAMPS BR

I.2 (M-P15)

BATHGATE-GARRIONGILL JCN*   8.5.43-1.5.50=
ADDIEWELL JCN (BENTS)-ADDIEWELL
POLMONT-LARBERT JCN   1.10.50=
FALKIRK-GRANGEMOUTH   1860=
COWLAIRS-HELENSBURGH   15.7.50-28.5.58=
DALREOCH-BALLOCH PIER   15.7.50=
BALLOCH-STIRLING*   18.3.56-26.5.56=
COWLAIRS-SIGHTHILL   28.5.58=
Hexham Jcn-Riccarton Jcn
REEDSMOUTH-BARMOOR JCN (MORPETH)   7.6.64-1.5.65
SCOTS GAP-ROTHBURY BR   1.11.70
Drumburgh-Silloth
KIRKBRIDE JCN-KIRTLEBRIDGE   13.9.69-8.3.70
ABBEYHOLME-BRAYTON   13.9.69
Northumberland Dock-Blyth
Newsham-Morpeth
Bedlington 109-NEWBIGGIN (CORRECT ALIGNMENT)
GLASGOW BRIDGE ST-AYR-GIRVAN HARBOUR   5.8.39-24.5.60
GLASGOW-NEILSTON-STEWARTON   27.9.48-27-3-71
BUSBY JN (POLLOKSHAWS)-EAST KILBRIDE 1.1.66-1.9.68
KENNISHEAD-SPIERSBRIDGE BR   27.9.48=
IBROX-GOVAN BR   1.5.68
PAISLEY-RENFREW   3.4.37=
PAISLEY-GREENOCK   29.3.41=
PORT GLASGOW-WEMYSS BAY   13.5.65
ELDERSLIE-CART   30.8.69
JOHNSTONE-GREENOCK   25.4.64-30.8.69
NB: Connected in error to the Wemyss Bay line
BARKIP RY
DALRY JCN Carronbridge-Gretna Jcn*   4.4.43-28.10.50=
HURLFORD (KILMARNOCK)-NEWMILNS   9.8.48-20.5.50=
AUCHINLECK-MUIRKIRK   9.8.48=
GILMILNSCROFT BR
GASWATER BR   9.8.48=
Dumfries-Maxwelltown PORTPATRICK*   7.11.59-28.8.62
CASTLE DOUGLAS-KIRKCUDBRIGHT BR   17.2.64
KILWINNING-ARDROSSAN   1.7.40=
DUBBS JCN-DOURA
NB: Joined in error to Perceton Br
IRVINE-CROSSHOUSE   22.5.48=
PERCETON BR   26.6.48=
BARASSIE-KILMARNOCK   1.3.47=
FAIRLIE COLLIERY BR   26.2.49=
AYR HARBOUR BR   28.2.48=
AYR-MAUCHLINE   1.9.70
ANNBANK-CRONBERRY   11.6.72
DALRYMPLE JCN-DALMELLINGTON   15.5.56=
HOLEHOUSE-BELSTON JCN   31.3.73
PEEL-DOUGLAS*   1.7.73

Lpro WO 78/4813 (TJHK) (EPD Feb.1875)

State M-P16 (?1876)

Changes
1. *BARROW IN FURNESS* added (91NW)
2. *BURTON* upon *HUMBER* to *BARTON* upon *HUMBER* (86ne)
3. S51 sheet number added

Additions to railways (+ Italic names, * Egyptian names)
Piel-Workington-Maryport-Carlisle+
Roose-Barrow-in-Furness HAWCOAT QUARRY 29.4.63-1864
CHANNEL PIER (BARROW-IN-FURNESS) BR   1870
[Liverpool]-Manchester+
Parkside Jcn-Carlisle-Glasgow+*
Motherwell-Glasgow Buchanan Street
Gartcosh-Arbroath*
Dunblane-Callander TYNDRUM   1.6.70-1.4.77
Almond Valley Jcn-Methven Jn CRIEFF   21.5.66
Monkwearmouth (Sunderland)-Port Carlisle
Blaydon-Newcastle-Tynemouth
Heaton-Edinburgh-Glasgow Queen Street+*
Portobello Jcn-Port Carlisle Jcn (Carlisle)*
NB: Gap north of Galashiels CLOSED
Hardengreen Jcn-Peebles* KILNKNOWE JCN
(GALASHIELS)   1.10.64-18.6.66

ABn (TJHK) (EPD Mar.1877?6); Lpro WO 78/4813 (TJHK) (EPD Oct.1876) (2 copies); PC (TJHK) (EPD 1876); RH (TJHK) (EPD May 1877); PC

State M-P17 (?1878)

Changes
1. Delamere Meridian deleted
2. Sheet line errors in 89 corrected

Ob C17 (420) (TJHK) (EPD Jan.1879); Lsa (TJHK) (EPD May 1879); PC

State M-1 (?1881) (?first complete state)

One-inch sheets added: S19,27,35,42,45,50,58,59,60
From partial to complete: S20,28,36,44,51,52,61
Sheets numbered: 85-110, S1-57,58-67. *NB*: Not S57A
NB: The addition of the Western Isles causes some coastal alterations, especially choice and positioning of placenames

Introduction of new standard feature
C.9   Italic
ee County to which island is attached

PC (TJHK) (EPD Aug.1881)

State M-2 (?1883)

Introduction of new standard feature
A.11   *Note_For Sheets 91 to 110, refer to Index to the One Inch Map of England & Wales, New Series*:
22mm below scale-bar

Lpro WO 78/935 (defective: for OS contouring, 31.3.1884)

State M-3 (?1885) (Sheet 2 of a three-sheet set)

Introduction of new standard features

A.6   *Price 2/-*: below index sheet number
A.13   Sheet 2: below the New Series note
B.2   Topographic features
   p   Large town: Dublin
   r   Medium town: eg Drogheda
C.1   Roman capitals
   d   Country: Ireland
   e   Marine name: eg Irish Sea

Changes
1. Title altered to: **Index to the Ordnance Survey of England and Wales and Scotland**
2. Seamer, Bishopton, Redmarshall, Elton (103SE), Upleatham (104SW) names added at "+"
3. Sheet lines west of S50 and below S58 deleted
4. *Fortwilliam* to *Fort William* (S53)
5. Chapelhill (S47), *Ballater* (S65) added
6. *SOLWAY FIRTH* lettering enlarged
7. Offshore features at *Scares* added (S2)
8. S57A sheet number added
9. 89 vertical quarter sheet line missing
10. Coastal outline of Ireland added
11. *LIVERPOOL BAY* added

Lsa (EPD Jun.1887); Lgh Old Series set; Cu Atlas 1.08.1, 1.08.11 defective)

State NS-M-1 (?1884)

Dating evidence: reference A.11 on State M-2 (known copy dated 3.1884)

Some changes must be assumed:
1. Sheet lines and numbers below Preston-Hull line changed from Old Series to New Series
2. Other England sheets renumbered
3. A New Series title would seem probable

Not found.
Documentary source: PRO OS 1/144 70B (see pp.5,73)

State NS-M-2 (OSPR 10/1890)

Dimensions: (from NS-M-3 Lpro OS 5/48)
   Paper: 1116mm W-E by 763mm S-N
   Plate: 1074mm W-E by 679mm S-N
   Neat line: 986mm W-E (horizontal neat lines lacking)

Introduction of new standard features
A.14   ARRR: bottom centre, below price, 5mm bnl
A.15   *Railways inserted to 1890*: bottom left, 2mm bnl
A.16   *N.B. N°s 36,45,46,56,57 (Isle of Man) are combined in one sheet*: 3mm below 100

Changes
1. Paper, plate and map dimensions altered
2. Restyled border of five lines (one thick), surrounding 3-sheet map (therefore W,E sides only)

3. Title altered to: **Index to the New Series One Inch Map of the Ordnance Survey of England and Wales, and Scotland**
4. *Price 2/-* removed to below 84, 2mm bnl
5. Sheet 2 removed to top right hand corner, 21mm rnl
6. Unnecessary sheet lines deleted: 90NW,SW, 95NE, 99NW,SW, NW,NE,SE boxes below 99, 104NW,NE, 109NE
7. S58 width cut off at 39mm by the middle line of the border: it breaks the neat line
8. Ireland outline mapping developed: Dublin, Belfast, Londonderry made built-up areas; river at Coleraine, islands named
9. *Callton Mor* to *Poltalloch* (S36), Cocquet I^d to Coquet I^d (10), Burgh on the Sands to Burgh by Sands (17), *STALY-BRIDGE* to *STALYBRIDGE* (85), Rooss to Roos (81), Wortley to Bolsterstone (87)
10. Beeford (64), Foss (S55) added to extant "+"
11. *Garliestown* (S4), *Grangemouth* (S31), *Methil*, East Wemyss (S40) added
12. "+" deleted in road triangle west of Skidby (72)
13. "+" deleted by the railway west of Gateforth (78)
14. *BEN AVON* name deleted from top (S65)

Railways inserted to 1890 (+ Italic, * Egyptian names)
Piel-Workington-Maryport-Carlisle+
   Roose-Barrow-Hawcoat Quarry THWAITE FLAT JCN 2.6.82
   Millwood Jcn-Carnforth-Wennington
      PLUMPTON JCN (ULVERSTON)-CONISHEAD BR 27.6.83
      PLUMPTON JCN-WINDERMERE LAKESIDE BR 18.3.69-1.6.69
      ARNSIDE-HINCASTER JCN 26.6.76
   RAVENGLASS-BOOT 24.5.75
   Corkickle-Marron Jcns
      CLEATOR MOOR JCN-SIDDICK JCN 4.8.79
         DISTINGTON-KELTON FELL 1.77-1.5.82
         CALVA JCN (WORKINGTON)-LINEFOOT 24.3.87
[Liverpool]-Lostock Jcn (Bolton)+
   Walton Jcn (Liverpool)-Skipton Jcn+
      Daisy Field Jcn (Blackburn)-Chatburn HELLIFIELD
                                               2.6.79-1.6.80
      GREAT HARWOOD JCN (BLACKBURN)-ROSE GROVE JN (BURN-
                                           LEY) 1.1.75-1.6.77
   FAZAKERLEY JCN-NORTH MERSEY GDS 27.8.66=
Wigan-Southport+
   SOUTHPORT-PRESTON 19.2.78-16.9.82
[Liverpool]-Manchester+
   [HUYTON]-ST HELENS-HAIGH JCN 1.11.69-5.6.82
      INCE MOSS JCN-WIGAN CURVE 1.11.69
   Parkside Jcn-Carlisle-Glasgow+*
      BOARS HEAD-ADLINGTON 1.11.69
      Cleghorn-Lanark
         Lanark Jcns HAPPENDON-MUIRKIRK 1.11.73
      MOTHERWELL Ross-Coalburn 1.12.56=
      Motherwell-Glasgow Buchanan Street
         Mossend-Midcalder Jcn
            CARFIN JCN-LAW JCN 1.6.80
         Cleland-Morningside
            OMOA-LANGBYRES JCN
         COATBRIDGE-GLASGOW COLLEGE 19.12.70-1.2.71

```
                SHETTLESTONE-HAMILTON   1.11.77
         Gartcosh-Arbroath*
            GREENHILL-BONNYBRIDGE BR   2.8.86
            Stirling-Tayport*   WORMIT   13.5.79
               FORDELL RY
               THORNTON JCN-METHIL HARBOUR  1.8.81-5.5.87
               THORNTON: SOUTH JCN-WEST JCN
                Milton Jcn-St Andrews  BOARHILLS   1.6.87
               LEUCHARS JCN-DUNDEE   1.6.78
            Dunblane-Tyndrum  OBAN*   1.7.80
               KILLIN JCN-LOCH TAY   13.3.86
            Perth Jcn-Dundee-Arbroath*
               ELLIOT JCN-CARMYLLIE BR   19.6.65
               ST VIGEANS JCN (ARBROATH) - KINNABER JCN
                                              1.10.80-1.3.81
               MONTROSE-BROOMFIELD JCN   1.3.81
         Newton-Hamilton West  ROSS   1.6.76-18.9.76
         Rutherglen-Whifflet  AIRDRIE   19.4.86
            AIRDRIE-LANRIDGE COLLIERY   1.9.87-2.7.88
         DALMARNOCK JCN-GERMISTON JCN   24.6.61-2.8.86
   Kenyon Jcn-Bolton+
      PENNINGTON-BICKERSHAW JCN   9.2.85
   Eccles-Wigan
      WORSLEY JCN-BOLTON   1.7.70-16.11.74
   Salford-Euxton Jcn+
      WINDSOR BRIDGE JCN (SALFORD)-HINDLEY   1.10.88
      Clifton Jcn-Accrington+
         BURY-HOLCOMBE BROOK BR   6.11.82
Manchester Oldham Road-Goose Hill Jcn (Normanton)+
   MANCHESTER VICTORIA-THORPES BRIDGE JCN LOOP   4.11.77
      CHEETHAM HILL JCN (MANCHESTER)-BRADLEY FOLD JCN
                                   (BOLTON)   1.8.79-1.12.79
   THORPES BRIDGE JCN-OLDHAM WERNETH   17.5.80
   Rochdale-Facit  BACUP   1.12.81
   SOWERBY BRIDGE-RISHWORTH BR   1.10.78-1.3.81
   Greetland-Halifax-Low Moor
      HALIFAX-QUEENSBURY   17.8.74-14.10.78
   DEWSBURY JN (THORNHILL)-DEWSBURY MARKET PL BR  27.8.66
   Wakefield-Goole+
      Pontefract-Methley  LOFTHOUSE JCNS   6.65
Manchester-Ulceby+ (parts on south sheet)
   LONDON ROAD JCN (MANCHESTER)-[Timperley]   20.7.49=
   Guide Bridge-Heaton Lodge Jcn (Mirfield)+
      STALYBRIDGE-DIGGLE LOOP   1.12.85
      Springwood Jcn (Huddersfield)-Penistone+
         SHEPLEY-CLAYTON WEST BR   1.9.79
   Hyde-[Blackwell Mill Jcns (Buxton)]
      [Woodley-Cressington]
         [GLAZEBROOK]-MANCHESTER CENTRAL   9.7.77-1.7.80
         [GLAZEBROOK]-WIGAN   16.10.79-1.4.84
            WEST LEIGH-PLANK LANE
         [Halewood]-Southport
            HILLHOUSE JCN-HAWKSHEAD STREET JCN (SOUTH-
                                     PORT)   1.10.87-2.9.87
[Derby]-Leeds-Bradford+
   CUDWORTH-HULL ALEXANDRA DOCK+   20.7.85
      MONK BRETTON LINK   20.7.85
      STAIRFOOT LINK   20.7.85
         BEVERLEY ROAD JCN-HULL CANNON STREET   27.7.85
      CUDWORTH-COURT HOUSE JCN (BARNSLEY)   28.6.69
      Holbeck Jcn (Leeds)-Stockton Jcn+
         Arthington-Ilkley  SKIPTON   16.5.88-1.10.88
         MELMERBY-MASHAM BR   9.6.75
         Picton-Grosmont+
            POTTO-WHORLTON BR   3.3.57=
      Apperley Jcn-Burley
         MENSTON JCN-MILNER WOOD JCN (OTLEY)   1.8.65
         SHIPLEY-ESHOLT JCN (GUISELEY)   4.12.76
      Shipley-Low Gill+
         SETTLE JCN-PETTERIL JCN (CARLISLE)   2.8.75
   Leeds Marsh Lane-Hull Manor House Street+
      CROSS GATES-WETHERBY   1.5.76
      GARFORTH-CASTLEFORD   8.4.78
      MINERAL BR TO SOUTH - ERROR?
[Kings Cross]-Doncaster-Knottingley
   Doncaster-Horbury Jcn+
      MEXBOROUGH [Rotherham-Sheffield]   8.63-13.3.71
      STAIRFOOT-NOSTELL   1.8.82
   Doncaster-Wakefield-Mill Lane Jcn (Bradford)
      Ardsley-Laisterdyke  SHIPLEY   8.74-1.11.75
      BRAMLEY-PUDSEY   1877
      ST DUNSTANS (BRADFORD)-THORNTON   11.76-4.78
York-Washington+
   York-Altofts Jcn (Normanton)+
      Burton Salmon-Knottingley  SWINTON   19.5.79
   York-Scarborough+  WHITBY   16.7.85
      BURTON LANE JCN (YORK)-FOSS ISLANDS GDS   1.1.80
      Bootham Jcn (York)-Mkt Weighton+  BEVERLEY   1.5.65
      Rillington-Whitby+
         PICKERING-SEAMER   1.5.82
   Sessay Wood Jcn (Pilmoor)-Driffield+
      Gilling-Helmsley  PICKERING   1.1.74-1.4.75
   Pilmoor-Boroughbridge  KNARESBOROUGH   1.4.75
   Northallerton-Leyburn  HAWES JCN   1.12.77-1.8.78
   Leamside-Bishop Auckland
      NEWTON HALL JCN (DURHAM)  103NW-Gateshead   2.3.68
Stanhope-South Shields
   Waskerley-Darlington-Saltburn+  WHITBY  23.2.65-3.12.83
      BISHOP AUCKLAND-FIELDON BDGE J (W AUCKLAND)  1.2.63
      BISHOP AUCKLAND-BURNHOUSE JN (BYERS GREEN)  1.12.85
      Simpasture Jcn (Shildon)-Port Clarence
         Billingham-Hartlepool+
            W HARTLEPOOL-CEMETERY J (HARTLEPOOL)  28.5.77
      OAK TREE JCN (DINSDALE)-DARLINGTON JCNS   1.1.87
      Middlesbrough-Guisborough  BROTTON   12.62-23.2.65
   SOUTHWICK-MONKWEARMOUTH   1.7.76
   SOUTH SHIELDS-WHITBURN COLLIERY   5.79
Sunderland-Hartlepool+
   WELLFIELD-STOCKTON   1.5.77-1.8.78
Monkwearmouth (Sunderland)-Port Carlisle
   PELAW-TYNE DOCK   1.3.72
   Blaydon-Newcastle-Tynemouth
      Manors-Whitley Bay  TYNEMOUTH VIA COAST   3.7.82
      RIVERSIDE JCN (MANORS)-PERCY MAIN   1.5.79
      Heaton-Edinburgh-Glasgow Queen Street+*
         Alnmouth-Alnwick  COLDSTREAM   2.5.87-5.9.87
```

Portobello Jcn-Port Carlisle Jcn (Carlisle)*
    MILLERHILL-GLENCORSE BR   23.7.74-2.7.77
   Hardengreen Jcn-Kilknowe Jcn (Galashiels)
    ESK VALLEY JCN-POLTON BR   15.4.67
   Hardengreen Jcn-Smeaton INVERESK   1.5.72
    SMEATON-ORMISTON   1.5.72
SAUGHTON-DUNFERMLINE   1.11.77-4.3.90
RATHO-PORT EDGAR   1.3.66-1.10.78
Cowlairs-Helensburgh
   MARYHILL-BONNYWATER JCN   1.6.78-2.7.88
    WESTERTON-MILNGAVIE BR   28.7.63
Newsham-Morpeth
  BEDLINGTON 109-Newbiggin
Glasgow Bridge St-Ayr-Girvan CHALLOCH JCN*   5.10.77
  Glasgow-Stewarton KILMARNOCK   26.6.73
   Busby Jcn-East Kilbride HIGH BLANTYRE   1.3.85
  LUGTON-BEITH   26.6.73
    BARRMILL JCN-ARDROSSAN PIER   4.9.88-3.5.90
     GIFFEN-KILBIRNIE   1.11.89
NB: Partially on the alignment of the Barkip Ry
    KILWINNING-IRVINE BANK STREET   2.6.90
CARDONALD-SHIELDHALL
Paisley-Greenock GOUROCK   1.6.89
Dalry Jcn-Gretna Jcn*
  Hurlford (Kilmarnock)-Newmilns DARVEL   1.6.96
  Dumfries-Portpatrick*
   NEWTON STEWART-GARLIESTON   3.4.75-2.8.75
    MILLISLE-WHITHORN BR   9.7.77
  Kilwinning-Ardrossan LARGS   1.5.78-1.6.85
Peel-Douglas*
  ST JOHNS-RAMSEY*   23.9.79
  DOUGLAS-PORT ERIN*   1.8.74

Dtc; En; Lbl (?all EPD Oct.1890)

State NS-M-3 (?1892)

Changes
1. Sheet 2 moved to top right, 3mm lnl
2. Railway insertion date 189 (*sic*) (but no changes)

Lpro OS 5/48 (a flat sheet with excess paper)

State NS-M-4 (?1893)

Introduction of new standard feature
A.17   Adjacent sheet numbers in margins

Changes
1. [Irish] CHANNEL added (off 117-below S1)
2. *NORTH CHANNEL* name moved (below S19-S3)
3. Reference to *Argyllshire* on Canna, Rum deleted (S60)

Railways inserted to 1893 (+ Italic, * Egyptian names)
Piel-Workington-Maryport-Carlisle+
  Corkickle-Marron Jcns
   CLEATOR MOOR DEVIATION   19.4.66
  PARTON-DISTINGTON   23.10.79

[Liverpool]-Manchester+
 [Huyton]-St Helens-Haigh Jcn
  PARK LANE JCN-PEMBERTON
 Parkside Jcn-Carlisle-Glasgow+*
  BEATTOCK-MOFFAT BR   2.4.83
  Motherwell-Coalburn
   ALTON HEIGHTS JCN-PONEIL JCN   2.4.83
  Motherwell-Glasgow Buchanan Street
   Coatbridge-Glasgow College
    Shettlestone-Hamilton
     PRIESTFIELD COLL BR (INCORRECT ALIGNMENT)
   Gartcosh-Arbroath*
    Stirling-Tayport-Wormit*
     COWDENBEATH-KELTY   2.6.90
    Lumphinnans-Kinross-Ladybank*
     MAWCARSE-BRIDGE OF EARN   4.5.90
    Thornton Jcn-Burntisland INVERKEITHING
      16.4.90
Manchester Oldham Road-Goose Hill Jcn (Normanton)+
  Greetland-Halifax-Low Moor
   Halifax-Queensbury
    HOLMFIELD-HALIFAX ST PAULS   1.8.90-5.9.90
Manchester-Ulceby+ (parts on south sheet)
  Hyde-[Blackwell Mill Jcns (Buxton)]
   [Woodley-Cressington]
    [Glazebrook]-Manchester Central
     THROSTLE NEST JCN-[NEW MILLS] - NOT DRAWN
     [Chorlton]-FAIRFIELD JCN   2.5.92
      HYDE ROAD JCN-GORTON JCN   2.5.92
   [Romiley] L&NW-ASHBURYS   17.5.75
NB: Absence west of L&NW corrected
Leeds Marsh Lane-Hull Manor House Street+
  Barlby Jcn (Selby)-Mkt Weighton+ DRIFFIELD   18.4.90
[Kings Cross]-Doncaster-Knottingley
  Doncaster-Wakefield-Mill Lane Jcn (Bradford)
   Ardsley-Laisterdyke-Shipley
    TINGLEY-BEESTON   1.8.90
    TINGLEY-BATLEY   1887-1.8.90
   St Dunstans-Thornton KEIGHLEY   3.9.82-1.4.84
SUNDERLAND Monkwearmouth-Port Carlisle   4.8.79
  Blaydon-Newcastle-Tynemouth TYNEMOUTH NEW STN   3.7.82
   Heaton-Edinburgh-Glasgow Queen Street+*
    BURNMOUTH-EYEMOUTH BR   13.4.91
    Saughton-Dunfermline (Forth Bridge)
     INVERKEITHING-NORTH QUEENSFERRY   1.11.77
     DUNFERMLINE TOUCH SOUTH-NORTH
    WINCHBURGH JCN-DALMENY   2.6.90
    Cowlairs-Helensburgh
     Dalreoch-Balloch
      Balloch-Stirling*
       BUCHLYVIE-ABERFOYLE BR   1.8.82
  WYLAM-SCOTSWOOD   12.7.75-6.10.76
Northumberland Dock-Blyth
  SEGHILL-DUDLEY COLLIERY (private)   1854=
Newsham-Morpeth
  Bedlington-Newbiggin DIRECT LINE ERROR DELETED
  Bedlington-Newbiggin
   NORTH SEATON COLLIERY BR   1859=

I.2 (NS-M-4)   49

Peel-Douglas*
    ST JOHNS-FOXDALE   17.8.86

Not found.

State NS-M-5 (?1900)

Changes (affecting also NS-N-3: explanation is on p.76):
1. A line 0.1" (1 mile) east of the Delamere Meridian added at intersections with S18-S66, S76-S122
2. E-W constructional ticks added through S68-77, S119-120, the former on NS-M-5 as a line above the map
3. This line is crossed by ticks where extensions of the one-inch lines meet it
    NB: These are square to the border, not the map

Mg (EPD Jan.1909); Lpro MR 934 (defective)

            I.3. North sheet, ?1881

State N-P1 suspected (?1881)

Evidence: "*The North Sheet extends from the north line of the Middle Sheet to the parallel of Duncansby Head in Caithness, and will include the Islands of Orkney and Shetland.*" (OSC 31.12.1881)
Not found.

State N-1 (1884) (first complete state)

    "*In preparation*" (OSC 1.2.1884), and in print by 5.1884. Not in OSPR, but a War Office copy was listed IBWO 382, and used as the base of subsequent military maps (see p.52).
    The only copies known are cut inside the neat line. Paper and plate dimensions are impossible to determine.

One-inch sheets completed: S68-131 (all numbered)
    NB: Flannel (= Flannan) Isles also inset in 104, Sula Sgeir and Rona in 113

    NB: Standard features used here conform to those for Scotland and Isle of Man mapping on the middle sheet. See the list on p.43.
    NB: No title, border, scale-bar, price, sheet number are present (unless removed by trimming)

Introduction of new standard features
A.12   *Parts lettered c belong to Cromartyshire* (S94)
B.2   Topographic features
    ff   Parkland: eg around Guisachan House (S73)
C.3   Roman
    q   "c": detached areas of Cromartyshire, eg S94
    r   Railway terminus: Strome Ferry Terminus (S81)
C.9   Italic
    h   Additional use: Hotel: Altnaharrow Inn (S108)
    ff   Railway junction: Boat of Garten Junction (S74)

Railways (* Egyptian names)
[STANLEY JCN]-FORRES*   3.8.63=
[GUTHRIE]-KITTYBREWSTER (ABERDEEN)   1.4.50-4.11.67=
    FERRYHILL JCN (ABERDEEN)-[BALLATER]* 8.9.53-17.10.66=
ABERDEEN WATERLOO-INVERNESS*   12.9.54-18.8.58=
    DYCE-FRASERBURGH*   18.7.61-24.4.65=
        MAUD JCN-PETERHEAD   18.7.61-3.7.62=
    KINTORE-ALFORD BR*   21.3.59=
    INVERURIE-OLD MELDRUM   1.7.56=
    INVERAMSEY-MACDUFF*   5.9.57-1.7.72=
    GRANGE-BANFF*   2.8.59=
        TILLYNAUGHT-PORTSOY   2.8.59=
    KEITH-BOAT OF GARTEN*   21.2.62-1.6.68=
    ORTON-CRAIGELLACHIE   23.8.58-1.7.63=
    ELGIN-LOSSIEMOUTH   11.8.52=
        ELGIN-ROTHES*   30.12.61=
    ALVES-BURGHEAD PIER   22.12.62=
    KINLOSS-FINDHORN   18.4.60=
INVERNESS-WICK*   11.6.62-28.7.74=
    DINGWALL-STROME FERRY TERMINUS*   5.8.70=
    GEORGEMAS-THURSO BR   28.7.74=

Lgh Old Series set (EPD May 1884); Cu Atlas 1.08.11

State NS-N-1 (OSPR 10/1890)

Dimensions: (from Ob and NS-N-2 Lpro OS 5/47)
    Paper: 1145mm W-E by 751mm S-N
    Plate: 1089mm W-E by 708mm S-N
    Neat line: 986mm W-E (no southern neat line present)

Introduction of new standard features
A.2   Border: five lines (one thick) surrounding 3-sheet map (W,N,E sides only, broken by S131)
A.5   Scale of ten Statute Miles to an Inch [scale-bar: 10+50 miles]: bottom right, 5mm bnl
A.6   *Price 2/-*: bottom centre, 3mm bnl
A.14   ARRR: below price, 8mm bnl
A.15   *Railways inserted to 1890*: under S71,72, 3mm bnl
        NB: Marginalia square with the neat line, not border

Changes
1. The west border cuts through S68,78,88, and the neat line stops at the north sheet lines of S88
2. S131 breaks into and is reduced by the north border

Railways inserted to 1890 (* Egyptian names)
Aberdeen Waterloo-Inverness*
    Grange-Banff*
        GRANGE NORTH JCN-CAIRNIE JCN   3.5.86
        Tillynaught-Portsoy LOSSIE J (ELGIN) 1.4.84-5.4.86
    KEITH JCN-PORTESSIE JCN   1.8.84
Inverness-Wick*
    Dingwall-Strome Ferry Terminus*
        FODDERTY JCN-STRATHPEFFER BR   3.6.85

Dtc, Lbl, Ob C16 (554) (MK), (all EPD Oct.1890)

State NS-N-2 (?1893)

Introduction of new standard feature
A.13   Index sheet number, tr, 22mm anl, 20mm rnl

Changes
1. Railways statement moved to under S68,69, 5mm bnl
2. Scale-bar to bottom centre, above price, 10mm bnl
3. Price below scale-bar, 15mm bnl
4. ARRR moved to bottom right, 1mm bnl
5. County boundaries revised. Some names moved; others including ABERDEENSHIRE, ELGINSHIRE, the administrative subtitles of Orkney & Shetland, deleted. The detached parts of Nairnshire and Cromartyshire and the associated reference system deleted

Railways inserted to 1893 (* Egyptian names)
[Stanley Jcn]-Forres*
   AVIEMORE-CARR BRIDGE   8.7.92

Not found.

State NS-N-3 (?1900)

Changes (see NS-M-5 on p.49, and p.76 for explanation)

Lpro OS 5/47; Mg (EPD Jun.1905)

Supplement 1.   Special issues based on the Index

Index with county meridians overprinted

Presumably on middle and south sheets, for internal use only. Mentioned as lost in an 1879 memo. It was signed by Sir Henry James, so was presumably constructed while he was Director General. (PRO OS 1/144 70B)
Not found.

Eclipse of the Sun on the 15th March 1858

On State S-6, with these changes:
1. Piano key border replaced by a neat line border
2. "Col: James R.E_F.R.S.&c. Superintendent." added between title and scale-bar, 39mm anl
3. *N.B_The Sheets may be obtained from the Ordnance Survey Office Southampton, or through the Agents.* added below the list of plates, 11mm rnl
4. Price deleted
5. List of geology symbols deleted
6. Railways as State S-6, plus Yeovil-Weymouth, on the wrong alignment (perhaps added in haste because it partly lay under the path of the eclipse)
7. With overprint depicting "*Path of the centre of the Moon's shadow on the 15th March 1858*", running from Start Point to the Wash, with times recorded at ten-second intervals while the eclipse is over the sea
8. With additional overprint, of lines showing degrees of obscuration away from the centre crossed by lines giving time intervals. A new note is appended:
   The path of the centre of the moon's shadow, and the limits within which the eclipse will be annular, is shewn on the lines from Start Point to the Wash. / The lines parallel to these have the proportion of the Sun which will be obscured, as seen from the points through which they pass, written on them, and the North and South lines have the time of the commencement of the Eclipse at the points through which they pass, written on them. / From data furnished by Mr Hind in the "Times" of the 4th March 1858.

State 1 (?1858)

Probably incomplete, lacking title and change 8
Mg (JW 1857)

State 2 (1858)

Lpro MPHH 310 (JWTM) (two copies)

Map of the South of England: Shewing the spaces enclosed by the Works of Fortification recommended by the Royal Commission: February 1860 / Zincographed at the Ordnance Survey Office, Southampton, under the direction of Captn A. de C.Scott,R.E., Colonel Henry James,R.E_F.R.S.,&c. Superintendent.   Scale of ten Statute Miles to an Inch [scale-bar 10+50 miles]

On State S-7, with these changes:
1. North to a line from Llanddeinol to Dunwich Bank
2. The base map, including sea, is grey
3. Fortified areas coloured red
4. List of geology symbols deleted

In *Report of the Commissioners appointed to consider the Defences of the United Kingdom* (BPP(HC) 1860 [2682], XXIII, 495).   Another copy: Lpro FO 925/4585

Plan of the Catchment Basins of the Rivers of England and Wales Scale of ten Statute Miles to one inch [scale-bar 10+50 miles]: Zincographed at the Ordnance Survey Office Southampton, under the direction of Captn A de C Scott R.E. Colonel Sir Henry James,RE.,FRS.&c Superintendent February 1861 / To accompany Report on the Salmon Fisheries of England and Wales

Map states: S-7, M-P10, with these changes:
1. In two parts, divided below Aberystwyth to above the Wash, the Trent catchment area on the north sheet
2. Catchment areas are numbered, printed in six colours
3. The title on the south part, the tables on the north
4. The south part has a neat line border

I.Supplement 1

5. On 5.12.1860 the Treasury gave Sir Henry James authority to supply the Royal Commission on Salmon Fisheries with tracings of certain rivers. These named river systems appear in the otherwise blank northern area of State M-P10. The map was made and its areas computed from the plan of the boundaries of the basins traced on the one-inch map

In *Report of the Commissioners appointed to inquire into Salmon Fisheries (England and Wales)* (BPP(HC) 1861 [2768], XXIII, 669-673). Another copy: Lpro OS 5/108
Sources: PRO T1/18741 5/12; T1/6279B 19467; OS 1/144 70B

Geological Map of the South-East of England (Kent, Surrey & Sussex) reduced from the Geological Survey, 1868
The Topography electrotyped by the Ordnance Survey, from the Index Map. (A hand-coloured geology map)

Map shewing the Registration Districts of Kent, Surrey and Sussex
The figures refer to the order of the districts, in respect of their consumption mortality, on the first table of Dr Buchanan's report. (An outline map, with area boundaries and numbers in blue)

Both are maps of the three counties only, with the index lines removed. They use "Brighton" long before the Index. Their railway revision lies between that on States S-10 and S-11, with these additions to S-10: London Victoria extension; Greenwich-Charlton; West Horsham-Peasmarsh Jcn (Guildford); East Grinstead-Grove Jcn (Tunbridge Wells); Uckfield-Groombridge; St Johns-Tonbridge; Hither Green-Dartford

For Dr Buchanan's report in *Public Health: Tenth Report of the Medical Officer of the Privy Council with Appendix, 1867* (BPP(HC) 1867-68 [4004], XXXVI, 737,739)

Supplement 2. Rivers and their Catchment Basins

Ordnance Survey of England and Wales: Rivers and their Catchment Basins / Scale Ten Statute Miles to One Inch / [scale-bar 10+50 miles] / Engraved at the Ordnance Survey Office, Southampton, under the direction of Colonel J. Cameron R.E. Colonel Sir Henry James R.E., F.R.S.,M.R.I.A.&c Superintendent. / The Altitudes given in feet above the Mean Level of the Sea, and those indicated thus (+110) refer to Marks made on Buildings, Walls &c / Price 21/- Mounted and Colored (*sic*)

At Sir Henry James's behest, the Royal Commission on Water Supply wrote on 11.4.1867 to the Secretary of the Treasury, seeking approval:
...for a 10 mile map in England and Wales showing each river with its tributaries and their watershed lines, and the position of towns whose population is not less than 10,000; to give on such a map as many levels as possible, along the watershed lines, or at the junctions of streams, and as nearly as possible to the 500 feet contour line for the whole, and within the Thames basin, the contour lines of 500, 400, 300 and 200 feet, so far as the information already obtained for the OS will permit.
OS staff were also to note rainfall and gauge streams. James assessed the cost of the map at about £150. Treasury approval was granted on 7.5.1867. An unfinished proof was submitted to the Commissioners on 11.5.1868 and the map was completed that year. The sales edition was in print first. (PRO T1/6746A 19252; T1/7188 6/5; T24/7 8486 433; OS 1/144 70B). Lit: Nicholson (1991)

State 1 (OSR 1868)

Engraved sales edition
   Sheet limits: 6°W-3°E, 50°-55°42!7N. With a graticule at degree intervals within a border graduated at 10' intervals. The additional strip 2°-3°E provides space for the title.
   A map newly drawn and lettered. Berwick upon Tweed is shown on an extrusion. Scotland is blank except main rivers. Isle of Man and Ireland are omitted. In two pieces, divided vertically at 0°, with a table to the right. For characteristics used, see col."F" on p.9.

Ob C17 (304)

State 2 (?1869) (OSC 1873)

Engraved sales edition, with addition to title:
1. 27/- with Case
   NB: Cases were made by HMSO at 5/- (NA OS 5/3007)

SOos

State 3 (1870)

Zincographed edition, with these additions:
1. Between "Walls &c" and "Price 21/-": Rainfall in inches marked thus 27.7 [in circle] from Symons' Rainfall 1867.
2. After "27/- with Case": Zincographed at the Ordnance Survey Office, Southampton; Major General Sir H. James R.E.,F.R.S.,&c Director. 1870
3. Headed: Appendix B.N.
4. Overprinted with the lines of the aqueducts for the Bateman and Hemans and Hassard schemes, and their proposed reservoirs near London
5. Rainfall measurements added

In *Royal Commission on Water Supply: Report of the Commissioners: Appendix, Maps, Plans, and Index* (BPP(HC) 1868-69 [4169-II], XXXIII, Appendix B.N.) (Copies: Lrgs, NTg, Ob)

Ordnance Survey of Scotland: Rivers and their Catchment Basins Scale Ten Statute Miles to One Inch [scalebar 10+50 miles] / Photozincographed and Published at the Ordnance Survey Office, Southampton 1893. / The Altitudes are given in Feet above the Mean Level of the Sea and are indicated thus (2145) / Rainfall in inches marked thus 38.79 [in circle] / Price 12/- Unmounted and Coloured, 21/- Mounted and Coloured, 27/- with Case. / All rights of reproduction reserved

Initiated 2.1885 by an enquiry from Stanford, once the ten-mile map of Scotland was complete. No new survey was required. Treasury approval sought 9.4.1885, and received later that year. The cost of making it was estimated at about £125.
Sources: PRO T1/7501, T1/9699 (lost); OS 1/144 70B

State 1 (OSPR 10/1893)

A newly drawn map. With inset maps of the Shetland Islands and St Kilda. No graticule, but a border graduated at 10' intervals. England and Ireland are in outline, though the rivers in Northumberland that reach the border are drawn. With number references to the table. Hand coloured. For characteristics, see col."H" on p.9.

A large *Table of Reference* was published separately, with print code 150-11/92 (copy Dtc)

CPT; SOos; SOrm

State 2 (1894)

Changes
1. Price deleted, and title altered, after "38.79" to: Reproduced by the kind Permission of the Director General of Ordnance Surveys. / ARRR
2. The map is now colour printed in six colours

In *Twelfth Annual Report of the Fishery Board for Scotland, Being for the Year 1893: Part II: Report on Salmon Fisheries* (BPP(HC) 1894 [C 7428-I], XXIII, 267)

State 3 (1946)

Changes
1. Title as State 1, but lacking price statement line
2. With print code: 146/Ch

Lpro OS 5/109; SOos

Supplement 3. Military issues

The earliest known British map at the ten-mile scale listed by the War Office was published in 1854. From that point on several dozen sheets at the scale appeared showing communication networks and general distribution of various kinds of military organisation. Nicholson (1988b) provides a summary account of the development of British military mapping in the latter half of the nineteenth century, which explains the sequence of interlocking military and government departments responsible for their making. With maps at the ten-mile scale, the part played by the Ordnance Survey was usually peripheral, and most commonly as the source of the base map.

Almost certainly the OS ten-mile Index would have been brought into service as the base map, though with so many military versions printed in outline only, some on sheet lines different to the Index, it is often hard to be certain of this. But some are straight overlays upon OS Index issues. It may also be significant that military issues covering Scotland were first issued in the early 1870s, much the same time as Index issues covering Scotland were printed for the first time.

After 1885, military issues often carry the legend "Transferred from the Ordnance Survey Index Map Jany 1885" (presumably WO copies catalogued IBWO 382, 411 and 412 (north, middle and south sheets)). Although, with qualifications, Sheets 2 and 3 are topographically as States M-3 and S-15 (qv), their railway states differ, which leads to an assumption either that this railway revision was effected on plates used for the New Series Index, but not the Old, and transferred to the military map from there, or that it was added to the military transfers and did not affect the Index at all. In that new railways were added along generalised alignments, the latter is more likely. The work was redone for States NS-M-2 and NS-S-2 much more accurately, and obviously not from the military map, since that includes railways that on them are incomplete or omitted (eg Bradford-Keighley, Ely-Newmarket, and in South Wales).

The starting point for compiling a list of military maps at the ten-mile scale is the *Register of GSGS Maps 1-4795*, and War Office (1889). The latter is certainly incomplete, and can furthermore be misleading as to map title. The descriptive element (usually beginning with "Shewing") only occasionally forms part of a map title: more often it is merely a list of what is in the legend. The catalogue does not distinguish the one from the other. Titles in this list are emboldened only where copies have been located, but with so few copies found, and rarely in circumstances where comparison was possible, the following sequence must be assumed both to be incomplete, and provisional.

---

1. QDWO maps

Distance Map of England and Wales (1854). WO 1889/6. (2 sheets, outline)
Another issue: Compiled in the Quarter Master General's Office, Horse Guards by W.J. Kelly, Draftsman 1870 / Lithographed at the Topographical Dépôt of the War Office under the Superintendence of Capt. C.W.Wilson,R.E. Colonel Sir Henry James,R.E:F.R.S:&c. Director 1870
Ob C17 (301). (WO 1889/32)
Another issue (1895) (IDWO 1095a,b)

I.Supplement 3

2. TDWO maps

England and Wales Recruiting Districts (1865). WO 1889/17. (2 sheets)

England and Wales, showing Engineer and Military Districts (1867). WO 1889/21. (2 sheets)

Military District Map of England and Wales Shewing the Distribution of the Troops (1868). Old Series Index States S-10 and M-P13 (qv), with integral title
Lmd
    Another issue (1869). Old Series Index States S-11 and M-P14 (qv), with title glued over the original
Lmd. (WO 1889/26)

Distance Map of England and Wales, showing Recruiting Districts and Pay Stations (1870). WO 1889/31. (2 sheets)

Military Map of Great Britain (1870) / Lithographed at the Topographical Dépôt of the War Office. Captn C.W.Wilson R.E. Director. 1870 (2 sheets, outline)
Cu Maps 33.87.8. (WO 1889/33)
    Another issue, corrected to 6.75, with a supplementary north sheet
Lbl 1138 (3), 1138 (4) (S sheet)
    Another issue, corrected to 6.76. (3 sheets)
Lbl 1138 (4); Lpro WO 78/626, MR 186, MR 187 (N sheet)
    Another issue (1878), Lith^d at the Intelligence Branch, Quart^r Mr Gen'^{s} Department, under the direction of Captain G.E.Grover, D.A.Q.M.G.
Lbl 1138 (5)

Military Map of Great Britain, showing Military Districts and Dépôt Centres (1872). (3 sheets)
    Another issue, corrected 1873. WO 1889/38

Military Map of Great Britain, showing Military Districts and Head Quarters of Brigade Dépôts, Militia Regiments, Administration, Battalion, Rifle Volunteers, and Volunteer Corps (1872). (3 sheets)
    Another issue, corrected 1874. WO 1889/40

Military Map of Great Britain shewing Head Quarters of Corps of Yeomanry Cavalry (c.1875)
Lpro MR 186 (M,S sheets only)

3. IBWO maps

Military Map of Great Britain, showing Head Quarters of an Army Corps with first division, Head Quarters of a Division, Infantry and Cavalry Brigades, Garrisons and Coast Brigades (1879). WO 1889/47. (3 sheets)

Great Britain showing Military districts, Head Quarters of Military districts, and Head Quarters of Regimental Districts (1882) (IBWO 94,95,96). WO 1889/58

Another issue, corrected to 1.8.83. WO 1889/63

Military Map of Great Britain, showing distribution of troops, 1.5.83 (IBWO 293,294,295). WO 1889/62
    Another issue [10.83] (IBWO 321,322,323)

4. IDWO maps

NB: Maps "Transferred from the Ordnance Survey Index Map. Jany 1885" employ the full Index base, with the index grid deleted. The maps show the topographical revision of the 1884-85 issues, with additional railways (States N-1,M-3,S-15)

Military Map of Great Britain: Shewing the distribution of Head Quarters of Volunteer Corps / Transferred from the Ordnance Survey Index Map. Jany 1885 (IDWO 707, 708,709). Corrected to 1.12.87
Lbl 1190 (66). (WO 1889/71)
    Another issue, corrected to 1.4.88 (GSGS cat.)
    Another issue, corrected to 1.4.91
Lbl 1190 (68). (WO 1889/79)
    Another issue, corrected to 1.4.94. / Published on behalf of the War Office by Edward Stanford, 26 and 27 Cockspur St Charing Cross London S.W. All Rights Reserved
Ob C16 (120)

Military Map of Great Britain / Transferred from the Ordnance Survey Index Map. Jany 1885 (1888) (IDWO 714, 715,716)
Lbl 1190 (67). (WO 1889/72)
    Another issue, corrected 1889. WO 1889/74
    Another issue [undated] / Published on behalf of the War Office by Edward Stanford, 26-27 Cockspur St Charing Cross London S.W. 10/6d the set
Ob C16 (121)

England and Wales (Outline) showing Boundaries of Counties of Military Districts (1890) (IDWO 787)

Scotland (Outline) showing Boundaries of Counties of Military Districts (1890) (IDWO 788)

Military Map of Great Britain: Shewing the distribution of Regular, Militia, Yeomanry and Volunteer units of all arms / Transferred from the Ordnance Survey Index Map. Jany 1885 (1894) (IDWO 979a,b,c) (WO 1889/87)
En Map 14.b.9; Lbl 1190 (69); Lpro FO 925/4194

Map of Great Britain, showing Head-Quarters and Boundaries of Military and Regimental Districts (1897) (IDWO 1320). WO 1889/92. 4 sheets (including Ireland). Prepared at OSO, Southampton, published IDWO

CHAPTER II

TEN-MILE MAPS OF IRELAND IN THE NINETEENTH CENTURY

1. Thomas Larcom's map

The importance of the maps forming the Irish Railway Commissioners' Atlas of 1838[1] has been recognised for nearly forty years now,[2] and there is no reason here to recount the remarkable inventiveness of Henry Drury Harness in introducing to the world the flow map, proportional circles, and his method of treating choropleths which later became known as dasymetric. Richard Griffith's Geological Map in the same atlas is probably also the first of Ireland to be published.

Our subject here is an aspect of these maps that previous writers left as irrelevant to their subject, or treated but superficially for want of evidence: who created the base map which underlies the maps of Harness and Griffith? The atlas comprises six maps, five of Ireland at the ten-mile scale, and one of Great Britain and Ireland at the twenty-mile scale. Of the ten-mile maps, Harness has credit for three, Griffith for one. These four were engraved by J.Gardner, Regent Street, London. The other, the first, was "*Prepared & Engraved under the direction of Lieu^t Larcom Royal Eng^{rs} May 1837*". It has an overlay of railway lines, both built (six miles only) and projected, and of the five it is the most complete as far as the base map detail is concerned. This includes roads and canals, lighthouses, rivers, lakes, offshore banks, names of rivers and coastal physical features, place and county names, inside a border graduated at 10' intervals. Harness's and Griffith's maps are all somewhat smaller than this one (611mm W-E by 750mm S-N neat line measurements), and have a common though different border design, with values given at only degree intervals and no graticule, but nonetheless there is no doubt that the base map of each derives from the Larcom map, with fewer characteristics. The similarities are most obvious in the case of Griffith's Geological Map, which retains much of the Larcom map lettering as well as roads. Lighthouses are missing, however, there are fewer names included, some of them are moved or even respelled (Urris Hd for Erris Hd), and the county boundaries are more emphasised. The Harness maps employed little more than Larcom's outline, and, being probably taken from transfers, are not quite identical. They also have replacement lettering: we will learn why this was so later.

It is natural to assume that the name Thomas Aiskew Larcom, since 1828 in charge of Ordnance Survey Irish affairs at Mountjoy, would mean that this base map was indeed the work of the OSI, but unfortunately there is nothing on the maps themselves to confirm this. OS survey work in Ireland was not so advanced, especially in the south, to make an island map such as this easy to construct. At the time the six-inch Townland survey was very incomplete and one-inch mapping not even begun. It is therefore fortunate indeed that copies of much of the correspondence from the Railway Commissioners (though not to them) and a minute book of their proceedings[3] should have survived, the contents of which prove that this ten-mile map is indeed the work of the Ordnance Survey at Mountjoy. We may therefore herald this as the first ten-mile Ordnance Survey Map of Ireland.

Its history began on 3 December 1836 when the Railway Commissioners had their Secretary Harry D.Jones write to Colonel Colby, requesting him:

..to furnish triangulation of Ireland, and a Copy of the Outline Map of Ireland, shewing the Coast Line,

---

1. Prepared to accompany the *Second Report of the Commissioners appointed to consider and examine a general system of Railways for Ireland* BPP(HC) 1837-38 XXXV, 469-863. Two thousand copies of each map were made.
2. Robinson (1955), Seymour (1980) and Andrews (1983). See also Pilkington White (1888) and Collinson (1903).
3. NA 2D 59 51, 2D 59 52, 2D 59 53 (Board of Works).

Rivers and County Boundaries with the positions and heights of the Principal Stations and of all the Rivers, Villages and Towns, Windmills or remarkable objects which have been fixed and such information as the Survey can afford regarding the Mountain Ranges and natural features of the Country Generally.
Captain Thomas Drummond, the chairman, followed this up five days later with a personal plea to Colby. Colby himself attended a meeting of the Commissioners in Dublin on 13 December 1836, when he was given further explanations "*of the nature of the Maps and information required from the Survey Office*". Colby seems to have put both Larcom and Robert Kearsley Dawson on to this task, because we find on New Year's Eve the Commissioners offering both men until the end of January to complete their joint work on the "10 Mile to an Inch" map (the first time an actual scale was mentioned), if they, the Commissioners, could be provided with a rough copy of the proposed quarter-inch map by 15 January.

By the time of their meeting on 21 March 1837, the Commissioners had become clearer about their intentions, so far as maps were concerned. They wanted three versions of the small map, as they called the ten-mile to save confusion with the quarter-inch, or large map, to accompany the report, "*one shewing all existing and proposed Railway Lines, Canals, &c.*", and the second "*shaded in order to show the distribution of population*". The third would show "*the extent and direction of the Traffic of the country and also the position of the Mines and the courses of the streams navigable from the several Ports*": this later divided into passenger and freight traffic maps. Griffith's Geological Map they had had in their possession since 8 November 1836. The population and traffic maps would go to Gardner in London for final engraving, and the outline map of England and Ireland would be engraved at the Survey Office in Dublin. The following day they wrote again to Colby wanting to know whether the ten-mile map was engraved, and asking him to reduce a map of Ireland to correspond with the twenty-mile map of England sent to him.

By 27 March they had received the engraved ten-mile map from Mountjoy, presumably in proof. Several alterations were required which Richard Griffith undertook to see done. They asked for 24 impressions of the map the following day, and these somewhat curiously arrived on 30 March as four-sheet maps. No mention of this aspect had been made in reference to the 27 March proof, but it is hard to perceive how this could have been a full sheet when two days later quarter sheets were being printed. But whatever the case, the Commissioners wrote again on 31 March asking for the small Map to be engraved in one sheet of copper, and requesting to know how long the engraving of the hill map (the quarter-inch map) would require. Larcom attended a meeting with the Commissioners early in April to receive instructions on these matters. The base map (or groundwork, as it was called then) for the Harness maps was apparently ready first. We learn from Richard Griffith, who delivered it to Harness, why this version of the map was relettered. In order to achieve visual subordination, unnecessary roads and names of divisions smaller than a county were to be deleted; and a lettering for county names was to be chosen "*not to interfere too much with the population shading*".

In spite of difficulties, it seems that the engraving of the maps was completed by 10 May, but for the titles. Larcom asked the Commissioners what was required in this respect, and their response on 13 May puts the case for OS responsibility for the base map beyond any reasonable doubt. The titles were quickly agreed, but in addition they proposed that Larcom inscribe on the maps "*Prepared and engraved at the Ordnance Map Office Phoenix Park Dublin under the direction of Lieu^t Larcom Royal Engineers May 1837*". Colby himself replied on 17 May. This letter is lost, but in it he evidently refused to permit any reference to the Ordnance Survey to appear on the maps, presumably because he doubted their accuracy. The far from perfect methods that must have been adopted in the south to achieve a map at all could not at that time have been the result of properly controlled OS surveying, and probably for this reason he refused to allow his department to be seen as responsible. Jones replied to Colby on 20 May:

I have to acknowledge the receipt of your letter dated the 17th Inst which I have read to the Commissioners who have directed me to inform you that it was their wish that nothing should be written upon the Maps but what met your approbation: they are fully acquainted with the difficulties that have been experienced in the compilation from such very imperfect documents of the South of Ireland as the existing County Maps and in proposing what they did it was with a view that the credit of the labour should fall upon those who deserved it. The proofs which we have just received appear to be very nicely executed and will be a valuable appendage to the 2nd Report and the title now upon it meets with the approbation of the Commissioners viz "Prepared and Engraved under the direction of Lieu^t Larcom Royal Eng^{rs} May 1837".

After this, Colby's name drops from the record altogether, and the later stages of the story, dealing mainly with matters of detail, all passed through Larcom's hands. The ten-mile map was finally engraved in mid-January 1838. The acknowledgement to Larcom that must have received Colby's approbation appeared on only the first of the ten-mile maps, the one including the railway lines, for the engraving of which Larcom was also responsible, and the last, the twenty-mile map. If it ever appeared on those used by Harness and Griffith, it was deleted before Gardner's engraving of them. Even the compilation order of the atlas became Larcom's responsibility, and he was given instructions on this in the middle of March. His suggestion as to its title was approved on 21 April.

Larcom's map is much better known in its second

phase, when it was used in 1845 for the Devon report on Land Tenure. Here again it was to possess an entirely innovative feature, this time provided by the Ordnance Survey, as contours were for the first time added to a ten-mile map. The matter was first raised on 21 May 1844 with a request from the Land Commissioners for a contoured map of Ireland. Apparently recent experiments in Phoenix Park had resulted in a general map being roughly contoured. The contours probably derived from those included on the quarter-inch map Larcom had prepared for the Railway Commissioners, and the "general map" concerned can only have been the Larcom's ten-mile base map. On 28 May 1844 its use was authorised for the Land Tenure Commission.[4] Also added to its base topography were the few railways open or being built at the time. The railway overlay of the 1837 map was deleted, and replaced by another. The map was also put on public sale, through Hodges and Smith in Dublin. Though it was never claimed publicly, OSI provenance becomes undeniable at this stage because sales copies usually carry the initials of James Duncan, the principal engraver, often with the date. This was a feature typical of almost all Phoenix Park produced maps for decades. The railway information was updated at least seven times before the appearance of Sir Henry James's new ten-mile map of Ireland in 1868, and yet again afterwards. It was also used at least twice more for parliamentary papers. Remarkably it remained in OSC until 1920.

2. The projects of Sir Henry James[5]

The ten-mile maps of Ireland that Sir Henry James initiated in 1864 sprung from a highly unlikely and unpromising source. In 1861 the Ordnance Survey Office (OSO), Southampton had zincographed upon the ten-mile Index sheets of England and Wales - still incomplete in the north - a **Plan of the Catchment Basins of the Rivers of England and Wales** to accompany a Royal Commission Report on Salmon Fisheries. The resulting map, which was a very hasty concoction, retained the full topography of the originals, plus the one-inch sheet lines, and to this added the boundaries of river catchment areas, with a computation of the areas involved. It was printed in six colours. On 21 January 1864 Captain Berdoe A.Wilkinson, Superintendent of the Ordnance Survey Office at Mountjoy, was instructed by Colonel John Cameron in Southampton to use this dog's dinner of a map as a model for a River basins map of Ireland on the ten-mile scale. The instructions Cameron sent are interesting. Less detail was required, and though this apparently refers to combining some of the smaller catchment areas in groups rather than to topographical detail, it seems that the topographical element was to be omitted. He enclosed a schema of the new map, with a grid of one-inch sheets already superimposed, since he supposed that it would be essential to draw it by reducing from the one-inch map by pantograph[6]. There were also special instructions about the Liffey basin, enforcing a special drawing at the ten-mile scale entitled **Catchment Basin of the River Liffey.**

A civil assistant named William Harvey was given the task of drawing the new map. It is not entirely clear what features he was to include, but they were probably a graticule, the coastline, rivers, lakes, the catchment area boundaries, and positioning some specified towns. Lettering was probably limited to naming rivers, towns, lakes, islands and coastal and marine physical features. Harvey completed the map by August 1864, though the computations of basin areas and river lengths took until the end of the year. He entitled it **Plan of the Catchment Basins of the Rivers of Ireland**, and coloured it to reflect these. Some 1865 correspondence reveals how it was made: a set of one-inch maps of the whole of Ireland were coloured to show the catchment basins of the entire island, which were then reduced, presumably by pantograph, to the ten-mile scale.

1. The Map of the World

But this was not the only ten-mile map to be in Sir Henry's mind in 1864. He had a much more ambitious project afoot, nothing less than a ten-mile topographical map of the entire world, on a modified form of a projection apparently invented in February 1858 which James called the Rectangular Tangential Projection[7]. North American sheets were already under construction at Southampton. James was in Ireland in 1864, and quite probably saw Harvey's drawing before completion, and he would appear to have given verbal instructions that topography should be added to it to provide the source from which the three sheets of the world map that would be required to cover Ireland could be engraved. James sent as a model one of the American sheets already printed at Southampton. The OSO Dublin final Quarterly Report of 1864 summarises Harvey's additional task:

> *Geographical 10 mile map. On the catchment map the principal roads, the railways, the canals &c.&c. have been reduced & drawn and the additional writing consequent on the alteration completed, so as to agree with the map of portion of America sent as specimen.*

Harvey thus set to work again to overlay a completely new topographical map upon his first. So dominant is this second layer that it makes the unaltered title seem wholly inappropriate. Specification for the new elements was drawn from the American model. The differen-

---

4. NA OSLR 10997 (1844).
5. NA OS 5/3007 is the most important source on James's ten-mile maps of Ireland.
6. For many years the OS incorrectly referred to this instrument as a pentagraph.
7. See Appendices 1 and 2 for information on both projection and project.

ces between America and Ireland in specification terms are slight, given the very different circumstances of the two countries, and they are noted in the lists in Appendix 2. Almost all the lettering and symbols that Harvey drew came to be used on the published maps that followed. The two important exceptions were his use of Egyptian lettering for headlands and Roman for islands. Both of these classes were probably lettered in the first phase of drawing, and it was presumably considered preferable not to alter them in the second, but effect the change to the standard specification by instructions to the engraver.

Wilkinson sent the revised drawing to Southampton for James's inspection on 25 January 1865. His covering letter reveals how pleased he was with it:

> I beg to forward for your inspection the finished Catchment basin Map of Ireland. You have already given me instructions respecting the engraving of this Map as part of the large projection of the World: but as I consider it a good piece of Map drawing and creditable to M*r* William Harvey who executed it I am desirous that you should see it before the Engraving is commenced.

James was equally delighted, and confirmed in his reply on 30 January that it was "*to be engraved on the ten-mile [world] projection and then Electrotyped*". He also made the first mention of it as an island map as he went on to say that it should be "*formed into one plate for the 10 Mile Index Map*" - still not a topographical island map *per se*, but a definite move in that direction.

The Dublin Quarterly Reports through 1865 to 1867 summarise progress towards publication of the "Geographical & Catchment Basin map", as the Ireland sheets of the Map of the World came to be called. First a tracing was taken of Harvey's drawing for engraving. Outline engraving followed, then writing. Wilkinson sent a pull from the Dublin plate to Southampton on 5 December 1866, mindful of the need to assimilate characteristics against those of the American sheets. He need not have worried, since the American sheets were much more delayed at Southampton for want of reliable data and funds. On 14 December Cameron advised him not to add the province boundaries, and to use as a model for waterlining the OS one-inch map. This was a feature still lacking from all the American sheets. Wilkinson had an experimental piece of waterlining done, and sent it to Southampton on 10 January 1867 for comment. Cameron approved it and decided in time to adopt the same style. By the first quarter of 1867, the engraving of sheets 1 and 3 (presumably the unnamed Londonderry and Belfast sheets) was completed, including waterlining. By the second quarter, sheet 2 (Dublin) also was completed.

But by mid-1868, the concept of a topographical map of Ireland as part of a Map of the World had quite clearly been superseded by a map of the island. Even so, the project, though dormant, had not yet died. On 1 November 1869 Cameron in Southampton wrote to Wilkinson in Dublin, saying that he had the unnamed Londonderry and Belfast plates there. He was about to engrave the Hebrides on the one, but had no use for the other "*as the part of England which comes into the Sheet has been engraved on another plate*".[8] Cameron wondered whether to return this to Dublin, or cancel it. Wilkinson was averse to any destruction at that time, and thought it best to keep the plates together, since another electrotype from them might be required in the future. Cameron eventually decided to return the Belfast sheet to Dublin leaving only that of Londonderry at Southampton. Following this all references to them disappear from extant records until 27 May 1891, when Captain T.B.Shaw, for his Officer in Charge of the Irish Division (OID), wrote to the Executive Officer (EO) in Southampton, pointing out that the third plate had still to be returned to Dublin. The correspondence that ensued found the over-zealous Shaw insisting that the three-sheet map of Ireland had still to be formally superseded by an island map then in print for 23 years, and the Keeper of the Southampton Engraved Copper Plate Store recording that nobody had asked for this particular plate in the twenty years he had been there. The upshot was the return of the Londonderry plate to Dublin on 30 June 1891. There is a second copy of the same plate, lacking the Scotland outline, which seems to have escaped documentary record.

2. The island map of Ireland

Cameron's further comments in his letter to Wilkinson on 14 December 1866 in effect begin the story of James's 1868 island topographical map of Ireland:

> Sir Henry wishes the (World) Maps when completed to be joined together by electrotyping so as to form one copper plate for Ireland - and will be glad if you will mount the 3 Sheets together and cut them to the same size as the intended Copper plate -
>
> On the Electrotype plates it may perhaps be desirable to insert the provinces and strengthen the County names.

James asked on 11 January 1867 for the Ireland sheets of the Map of the World to be mounted together and sent to him. He returned them at the end of the month, with a tracing of the sheet lines required for the island topographical map. It took Wilkinson until mid-October to join matrices of the world sheets from which to form an island plate for Ireland. He then requested a tracing from Southampton of the Scotch (*sic*) and Welsh coastline for addition to his new plate, while also pointing out that, since James had apparently encompassed within his sheet lines the whole of Cardigan Bay, the width of map now formed was too great for any paper he knew of, and it might have to be made specially. James was insistent that Holyhead at least be included, with as much of the British coastline as could appear on paper of 28½ins width. This was still too great for Wilkinson, whose

---

8. See Appendix 2 for a discussion of this point.

widest paper was under 28ins. But the wider paper was available at Southampton, and Wilkinson asked for fifty sheets of it. The River Catchments map was also to require paper of the same width. By 12 June 1868, the topographical map of Ireland was published, and the Map of the World sheets effectively superseded. The first OSO Dublin Quarterly Report for 1869 adds a twist to the tail of this account, in recording that "*The sailing routes etc.* (the "etc" being British railways, ports and the electric telegraphs from Howth, Donaghadee and Valentia) *have been engraved on this map*". No such copies have been located, but these features are on a plate adapted as the **Index to the General Map**, which was in print by 30 August 1871. It would seem that the plate so adapted was used entirely for this purpose, though it remains a mystery why this information should be deemed necessary on an index.

New information was engraved from time to time in both topographical and index maps - or removed, as when c.1880 the War Office had labels referring to defence installations deleted. Both plates have survived, and, while they must inevitably have been revised separately, a comparison between the two reveals remarkably little discrepancy. A new electrotype of the index version was introduced c.1890 which for some undisclosed reason was radically revised, with a new title, the reinsertion of the graticule, and some other alterations listed in the cartobibliography. Otherwise, railways and boundaries were the features most revised, with alterations usually made concurrently with the one-inch.[9] A document dated 22 September 1890 provides valuable evidence of the process and demonstrates furthermore that at the time the delay in engraving railway revisions was considerable.[10] It records that the last railway engraved was the Tullow branch (opened 1885) in April 1888, which helps fix the date of State 6 of the topographical map. On 23 March 1890, fifteen more railways were being inserted, all of them present on State 7, and of a group of six additional lines noted as being surveyed, only one was in the event delayed until the next state of the map. The opening dates of these railways were spread through the 1880s. There may well be more states of both maps than those so far identified: by all accounts OSI print runs were short, and done as required. It may be also that a new reprint did not necessarily follow a revision until stocks had reduced. The known overprinted versions also suggest intermediate states, which may or may not have appeared as base map or index printings in their own right. The last, quite extraordinarily, was initiated in 1922, or seventeen years after the 1905 map (III.2) was published. Though this was but an index, the railway state was again updated.

3. Rivers and their catchment basins

Having begun first, the catchment basins map was the last to be published. James wished to see an impression of it on 21 January 1867, and Wilkinson's reply suggests that work had not even begun. He explained on 24 January that his intention was to take a matrix from the world map plates even then mounted together to be shipped at James's instructions to Southampton, once he could get water lines on them completed. From the matrix a few names would be removed and the Catchment basins engraved on a duplicate. In this response, Wilkinson seems surprisingly ignorant of the characteristics required of a river catchments map, and certainly casts doubt on what Harvey's original 1864 instructions were when drawing it. It also suggests that no news on the England and Wales Rivers map, at the time evidently under construction in Southampton, had reached Dublin. James wrote on 18 April 1867 leaving Wilkinson in no doubt of his requirements, and if the detail specifications of the Irish and British maps occasionally differ, the basic thrust is identical:

> *In making the River map of Ireland, I wish you to represent only the Rivers themselves, and the boundaries of their catchment basins, and simply to mark the positions of the towns over 5000 inhabitants on the rivers. No names to be engraved beyond the names of the rivers and these towns, no roads, or divisions of counties or any other detail beyond what I have mentioned, excepting altitudes on the rivers and on the catchment basin boundaries to be inserted.*

These instructions omitted mention of anything outside the coastline, and names of physical features written on the sea areas were left, though occasionally relocated. The lettering used for principal islands and headlands was finer. Aran Islands was relettered in capitals. Engraving began by July 1867, and the map was published by 3 November 1868.[11] Whether this was in advance of the England and Wales map is uncertain. Discussions on colouring followed, and a copy was returned on 7 April 1869 from Southampton to Dublin for use as a model. In view of the more finely engraved boundaries on the Irish map, it was felt that a band of stronger colour should be added around each basin in order to make them more distinct. Then, on 14 May 1869, another copy was sent from Southampton, in a case which had been made for 5/- by HMSO. It was decided to sell the map at 18/6d if in a case. The uncased price was 12/6d, reduced by 1904 to 10/6d. On 5 January 1904, with very low demand for the map, the OID Colonel Sim suggested that future copies should be coloured at Southampton, since he could not have it done economically. This was approved.[12]

---

9. NA OS 6/8540.
10. NA OS 5/8409.
11. NA OS 5/3331.
12. NA OS 6/8633.

Cartobibliography

II.1. Larcom's map, 1838

I. **Map of Ireland: To accompany the Report of the Railway Commissioners 1838. Shewing the different lines laid down under the Direction of the Commissioners and those proposed by Private Parties** [References] Scale of Statute Miles [scale-bar 10+30 miles] Prepared & Engraved under the direction of Lieu^t Larcom Royal Eng^{rs} May 1837

Dimensions:
  Paper: 664mm W-E by 974mm S-N
  Neat line: 633mm W-E by 935mm S-N

Standard features present
B.1   Boundaries
  a   County boundary
B.2   Topographic features
  a   River (with horizontal ruling)
  b   Lake (with horizontal ruling)
  d   Canal
  h   Minor road
  j   Main road
  m   Buildings blocked together in county towns, etc
  t   Town
B.5   Hydrographic features
  b   Sandbank, mudbank
  f   Lighthouse, light vessel
C.1   Roman capitals
  b   City, county town
  c   Large island: Achil
  f   Principal bay, harbour, haven, estuarial river, sea lough, inland lough
C.2   Roman open capitals
  a   Marine name
C.3   Roman
  b   Minor island
  c   Principal headland
  e   Lighthouse, light vessel
  g   Market town
  j   Inland harbour: Shannon Harbour, Richmond Harb.
  k   Canal
  l   Sandbank
C.6   Egyptian capitals
  d   County
C.9   Italic
  b   River
  z   Characteristic of lighthouse, light vessel

Other features: the British coastline, a graticule in the sea areas only, a graduated border at 10' intervals, broken by Clogher Head. Great Blasket is partly inside the border, and is cut off by the map frame.

With an overlay of railways built, for which Acts of Parliament have been obtained, and lines suggested both by the Commissioners and others.

The next four are not OS maps: the outline of Larcom's map was transferred and engraved by Gardner:

II. **Map of Ireland: to accompany the Report of the Railway Commissioners showing by the varieties of shading the comparative Density of the Population.** Constructed under the Direction of the Commissioners by Henry D. Harness Lt Royal Engineers 1837 [Note] [References] Scale of Statute Miles [scale-bar 60 miles] Engraved by J.Gardner, Regent Street London

III. **Map of Ireland, to accompany the Report of the Railway Commissioners shewing the relative Quantities of Traffic in different Directions.** Constructed, under the Direction of the Commissioners, by Henry D.Harness Lt Royal Engineers 1837. [References] Scale of Statute Miles [scale-bar 60 miles] Engraved by J.Gardner, Regent Street, London

IV. **Map of Ireland, to accompany the Report of the Railway Commissioners, shewing the relative Number of Passengers in different Directions by regular Public Conveyances.** Constructed, under the Direction of the Commissioners, by Henry D. Harness Lt Royal Engineers. 1837. [References] Scale of Statute Miles [scale-bar 60 miles] Engraved by J.Gardner, Regent Street, London

Some topographical detail was deleted for:
V. **Geological Map of Ireland to accompany the Report of the Railway Commissioners 1837 Shewing the different lines laid down under the Direction of the Commissioners and those proposed by Joint Stock Companies** [References] Scale of Statute Miles [scale-bar 50 Miles] Engraved by J.Gardner Regent Street London / [signed] Richard Griffith Dublin April 28th 1838

It was reduced to the twenty-mile scale as part of:
VI. **Map of England & Ireland, Explanatory of that part of the Report of the Railway Commissioners, which relates to the communication between London and Dublin, and other parts of Ireland.** Prepared & Engraved under the direction of Lieu^t Larcom Royal Eng^{rs} May 1837

All six maps in *Irish Railway Commission. Maps* [titles] *Presented to both Houses of Parliament by Command of Her Majesty. 1838.* Cover title: *Atlas to accompany 2^d Report of the Railway Commissioners Ireland 1838* (Dtf, Eu, Lbl, Ob). They are also bound into *The Sessional Papers Printed by order of the House of Lords or presented by Royal Command, in the Session 1837-38, (1° & 2° Victoriae)*, (BPP(HL) 1838, XLVII Part II, 32-37)

No I in *Second Report of the Commissioners appointed to inquire into the manner in which Railway Communications can be most advantageously promoted in Ireland* (Minor Edition) London, Printed W.Clowes & Sons, 1838

Map of Ireland: To accompany the Report of the Land Tenure Commissioners 1845. Shewing the Places visited by the Commissioners and the relative proportion of the Surface of each County lying between certain lines of Altitude [References including contour diagram] Scale of Statute Miles [scale-bar 10+30 miles]

Larcom's base map for the Railway Commissioners was electrotyped, trimmed and reused with a new overlay.

Plates (two versions): Dna OS 106
The matrix appears to be that which transformed the Railway Commissioners' map into the Land Tenure map. Its railways are as State 1 below, and the title is lacking. The plate is as the final state below

State 1 (1845)

Dimensions (derived from later states):
  Paper: 690mm W-E by 1015mm S-N
  Matrix: 627mm W-E by 922mm S-N (copper)
  Plate: 625mm W-E by 921mm S-N (copper)
  Neat line: 599mm W-E by 889mm S-N (copper)

Introduction of new standard features
B.2   Topographic features
  mm   Railway
  uu   Contours at 250ft, 500ft, 1000ft and 2000ft, the layers expressly to be coloured yellow, light red, dark red, brown, black
C.3   Roman
  h   Railway company name
C.9   Italic
  bb   Important range of hills: Slievebloom Mountains

Changes (to the 1838 Railway Commissioners' map)
1. The sheet lines are reduced, especially east and north. The Mull of Galloway and Islay just survive
2. The graduated border is replaced by a neat line only which to the west replaces outer line of the frame
3. The graticule is deleted
4. Table of areas added bottom centre, 17mm anl
5. The printed signature of Thos.A.Larcom Capt R^l Eng^r is added bottom right, 9mm anl
6. *Printed from an Electrotype Plate* added bl, 9mm anl
7. Great Blasket, Skillig Rocks are omitted altogether
8. St Georges I^d to Inishmaan, Birr to Parsonstown
9. Dalkey added
10. S^T GEORGES [Channel] moved

The overlay is of "*Places visited by the Commissioners at which evidence was taken*", with placenames in boxes, and "*Places inspected by the Commissioners*", eg King Williams Town, marked by squares.

NB: For a discussion of the principles governing the following lists of railways, see p.15.

Railways, shown as double parallel lines (* named)
BELFAST GREAT VICTORIA STREET-PORTADOWN*   1839-1842
DUBLIN AMIENS STREET-DROGHEDA*   1844
DUBLIN KINGSBRIDGE-THURLES-CASHEL*   1846-1848
    CHERRYVILLE JCN (KILDARE)-CARLOW*   1846
DUBLIN WESTLAND ROW-KINGSTOWN [DUN LAOGHAIRE]*   1834
    KINGSTOWN [DUN LAOGHAIRE]-DALKEY*   1837-1844

In *Index to the Minutes of Evidence taken before Her Majesty's Commissioners of Inquiry into the State of the Law and Practice in respect to the Occupation of Land in Ireland: Part V* (BPP(HC) 1845 [673], XXII, 725). Another copy: Lbl 10820 (18) (id 3 Feb.1847): probably part of the initial stock sold later, with Hodges & Smith details and price (see State 2) lacking.

State 2 (1845) (engraved sales edition)

Change
1. Added bottom centre, 2mm bnl: *Sold for Her Majesty's Government, by Hodges & Smith, 104, Grafton Street, Dublin. Price Two Shillings & Sixpence*, plain

Lrgs Ireland S/G.4 (id 11 Oct.1845) (coloured copy)

State 3 (?1852)

Changes
1. Railway depiction a single pair of parallel lines
2. Washingtongreen to Bagenalstown

Additions to railways (* named)
LONDONDERRY FOYLE ROAD-STRABANE   1847
BELFAST YORK ROAD-CARRICKFERGUS   1848
    GREENISLAND-BALLYMENA*   1848
BELFAST QUEENS QUAY-HOLYWOOD   1848
Belfast Great Victoria Street-Portadown*   ARMAGH*   1848
Dublin Amiens Street-Drogheda*   NEWRY*   1849-1852
    HOWTH JCN-HOWTH   1846
    DROGHEDA-NAVAN [AN UAIMH]   1850
    DUNDALK-CASTLEBLAYNEY*   1849
    NEWRY-WARRENPOINT   1849
DUBLIN BROADSTONE-GALWAY*   1847-1851
Dublin Kingsbridge-Thurles*   CORK*   1848-1849
NB: The unbuilt Thurles-Cashel alignment DELETED
    Cherryville Jcn (Kildare)-Carlow*   KILKENNY*   1848-1850
    LAVISTOWN-THOMASTOWN   1848
    Cork Albert Quay   BALLINHASSIG-BANDON   1849
LIMERICK-TIPPERARY*   1848

Lrgs Ireland G.54 (id 3 Dec.1856) (JWTM 18?54)

State 4 (?1858)

Additions to railways (* named)
Londonderry Foyle Road-Strabane   ENNISKILLEN*   1852-1854
LONDONDERRY WATERSIDE-LIMAVADY*   1852
    LIMAVADY JCN-COLERAINE STN*   1853

II.1

Belfast York Road-Carrickfergus
    Greenisland-Ballymena* PORTRUSH* 1855
        COOKSTOWN JCN-COOKSTOWN* 1848-1856
Belfast Queens Quay-Holywood
    BALLYMACARRETT JCN (BELFAST)-NEWTOWNARDS 1850
Belfast Great Victoria Street-Armagh*
    PORTADOWN-DUNGANNON* 1858
Dublin Amiens Street-Newry* PORTADOWN* 1852
    Drogheda-Navan [An Uaimh] KELLS [CRANANNAS]* 1853
    Dundalk-Castleblayney* NEWTOWN BUTLER*= 1854-1858
    GORAGHWOOD Newry-Warrenpoint= 1854
Dublin Broadstone-Galway*
    MULLINGAR-LONGFORD* 1855
        INNY JCN-CAVAN* 1856
Dublin Kingsbridge-Cork*
    Cherryville Jcn (Kildare)-Kilkenny*
        Lavistown-Thomastown NEWRATH JCN (WATERFORD)* 1853
    PORTARLINGTON-TULLAMORE* 1854
    MALLOW-KILLARNEY* 1853
    CORK ALBERT QUAY Ballinhassig-Bandon* 1851
    CORK ALBERT STREET-PASSAGE* 1850
Dublin Westland Row-Dalkey WICKLOW MURROUGH* 1854-1855
    BRAY-DUBLIN HARCOURT STREET* 1854
Limerick-Tipperary* WATERFORD* 1852-1854
    LIMERICK-FOYNES* 1856-1858
    WATERFORD-TRAMORE* 1853

Dn 16.B.6 (19) (JW) (id 14 Sep.1858)

State 5 (?1860)

Additions to railways (* named)
Londonderry ENNISKILLEN-NEWTOWN BUTLER Dundalk* 1858-59
Belfast Queens Quay-Holywood
    Ballymacarrett Jcn (Belfast)-Newtownards=
        COMBER-BALLYNAHINCH*= 1858
            BALLYNAHINCH JCN-DOWNPATRICK*= 1859
Belfast Great Victoria Street-Armagh* MONAGHAN* 1858
Dublin Amiens Street-Portadown*
    SCARVA-BANBRIDGE 1859
Dublin Kingsbridge-Cork* MIDLETON 1855-1859
    Cherryville Jcn (Kildare)-Kilkenny*
        BAGENALSTOWN [MUINE BHEAG]-BORRIS 1858
    Portarlington-Tullamore* ATHLONE* 1859
    BALLYBROPHY-PARSONSTOWN [BIRR] 1857-1858
    MALLOW-FERMOY 1860
    Mallow-Killarney* TRALEE* 1859
Limerick-Waterford*
    LIMERICK (ENNIS JCN)-ENNIS* 1859
    KILLONAN-CASTLE CONNELL 1858

En C19(104) (TJH 1859) (id 25 Jan.1861)

State 6 (?1863)

Additions to railways (* named)
Londonderry Foyle Road-Enniskillen-Dundalk*
    SHANTONA JCN (BALLYBAY)-COOTEHILL BR* 1860

Belfast York Road-Carrickfergus LARNE HARBOUR* 1862
Belfast Queens Quay-Holywood
    Ballymacarrett Jcn (Belfast)-Downpatrick*
        Comber-Newtownards DONAGHADEE 1861
Belfast Great Victoria Street-Monaghan*
    Portadown-Dungannon* OMAGH* 1861
Dublin Amiens Street-Portadown*
    Drogheda-Kells [Cranannas]* OLDCASTLE* 1863
Dublin Broadstone-Galway*
    CLONSILLA-NAVAN [AN UAIMH]* 1862
    Mullingar-Longford* SLIGO* 1862
        Inny Jcn-Cavan* CLONES* 1862
    ATHLONE-CLAREMORRIS* 1860-1862
    ATHENRY-TUAM* 1860
Dublin Kingsbridge-Cork-Midleton* YOUGHAL* 1860
    Cherryville Jcn (Kildare)-Kilkenny*
        Bagenalstown-Borris BALLYWILLIAM 1870
    QUEENSTOWN JCN [COBH JCN]-QUEENSTOWN [COBH]* 1862
Dublin Westland Row-Wicklow* RATHDRUM 1861-1863
Limerick-Waterford*
    Limerick-Foynes*
        PATRICKSWELL-CHARLEVILLE [RATH LUIRC]* 1862
    Killonan-Castle Connell KILLALOE* 1860-1862

Cu Maps 171.84.1 (TJH 1863), Dtc (both id 30 Apr.1863)

State 7 (?1865)

Additions to railways (* named)
Londonderry Foyle Road-Enniskillen-Dundalk*
    STRABANE-STRANORLAR* 1863
Belfast Queens Quay-Holywood BANGOR* 1865
Belfast Great Victoria Street-Monaghan* CLONES* 1863
    KNOCKMORE JCN (LISBURN)-BANBRIDGE* 1863
    ARMAGH* Goraghwood-Warrenpoint 1864
Dublin Broadstone-Galway*
    Clonsilla-Navan [An Uaimh]*
        KILMESSAN-ATHBOY 1864
    STREAMSTOWN-CLARA* 1863
    Athlone-Claremorris* CASTLEBAR* 1862
Dublin Kingsbridge-Cork-Youghal*
    Ballybrophy-Parsonstown [Birr]
        ROSCREA-NEAR BIRDHILL* 1863-1864
        KINSALE JCN-KINSALE* 1863
Dublin Westland Row-Rathdrum* ENNISCORTHY* 1863

Dna Council Office Map 32 (TJH 1864) (id 18 Apr.1865)

State 8 (?1867)

NB: The railways here are as State 1 of the 1868 map (II.2), except Ballingrane-Newcastle West (absent there)

Additions to railways (* named)
LONDONDERRY GRAVING DOCK-BUNCRANA 1863
    TOOBAN JCN-FARLAND POINT 1863
Londonderry Foyle Road-Enniskillen-Dundalk*
    BUNDORAN JCN-BUNDORAN* 1868

Dublin Broadstone-Galway*
   Athlone-Castlebar* WESTPORT* 1866
Dublin Kingsbridge-Cork-Youghal*
   MARYBOROUGH [PORTLAOIGHISE]* Kilkenny-Newrath Jcn
                         (Waterford)* 1865-1867
    Ballybrophy-Parsonstown [Birr]
      Roscrea-Near Birdhill BIRDHILL* 1864
   CORK CAPWELL-MACROOM* 1866
   Cork Albert Quay-Bandon* DUNMANWAY* 1866
Dublin Westland Row-Enniscorthy*
   WOODENBRIDGE-SHILLELAGH* 1865
Limerick-Waterford*
  Limerick-Foynes*
    BALLINGRANE-NEWCASTLE WEST 1867
  Limerick (Ennis Jcn)-Ennis* GORT ?1869

Lpro MPHH 556 (id 24 Jun.1867), Dn 16.B.4 (16) (id 2 Dec.1867)

State 9 (final state)

Additions to railways (* named)
Londonderry Foyle Road-Dundalk* GREENORE* 1873
Belfast Queens Quay-Bangor*
   Ballymacarrett Jcn-Downpatrick* NEWCASTLE* 1869
Belfast Great Victoria Street-Clones*
   KNOCKMORE JCN (LISBURN)-ANTRIM* 1871
Dublin Broadstone-Galway*
   Clonsilla-Navan [An Uaimh]* KILMAINHAM 1872
   Mullingar-Sligo*
     KILFREE-BALLAGHADEREEN 1874
   Athlone-Westport*
     MANULLA-BALLINA* 1868-1873
Dublin Kingsbridge-Cork-Youghal*
   Cherryville Jcn (Kildare)-Kilkenny*
     Bagenalstown-Ballywilliam MACMINE* 1870-1873
    Ballybrophy-Parsonstown [Birr]
      PARSONSTOWN [BIRR]-PORTUMNA BRIDGE 1868
   Mallow-Fermoy* LISMORE* 1872
Dublin Westland Row-Enniscorthy* WEXFORD* 1872
Limerick-Waterford*
   Limerick (Ennis Jcn)-Gort* ATHENRY* 1869

BFpro OS 18/12A, OS 18/12B

Map of Ireland: To accompany the 18th Annual Report of the Commissioners of Public Works Shewing the position and extent of every Loan made under the Land Improvement Act, 10 Vic.C.32 [symbols] Scale of Statute Miles [scale-bar 10+30 miles] / Ireland_Appendix to 18th Annual Report, Commissioners of Public Works

Not an OS map, but lithographed by Standidge & Co, Old Jewry. The base map is the Larcom map, further cut down, with a new border, no Great Britain, no contours, the overlay of the Land Tenure map removed, and much re-lettering. The inland water is still ruled, and the Blasket Islands are still lacking. The only railways are Dublin-Drogheda and Dublin-Kingstown. The overprint is of green dots in sizes proportionate to their value.

In *Public Works, Ireland: Eighteenth Report from the Board of Public Works, Ireland, with Appendices* (BPP (HC) 1850 [1235], XXV, 569)

Map of Ireland: shewing the sites of the different fixed engines, used for the capture of salmon, during the year 1862, and the boundaries of the several districts, formed by the Commissioners of Fisheries under the Provisions of the Act 11 & 12 Victoria, Cap.92. To accompany the Annual Report of the Commissioners of Irish Fisheries for 1862. Scale of Statute Miles [scale-bar 10+30 miles]

Changes (to the 1845 Land Tenure map)
1. Smaller map area map with no neat line
2. No Great Britain coastline
3. With red overprint for fishery districts and blue for coastguard stations

The railways are as State 6 of the Land Tenure map
The Board of Works requested use of the Land Tenure map, and 850 zincographed copies. In the end 1265 copies of the map were lithographed. The cost to the Board of Works was £20.2.6d. (NA OS 5/2934, 27.4.1863)

In *Report of the Commissioners of Fisheries, Ireland, for 1862* (BPP(HC) 1863 [3225], XXVIII, 125)

General Valuation of Ireland..Scale of Statute Miles [scale-bar 10+30 miles] / Prepared at the Ordnance Survey Office, Phoenix Park, under the direction of Captain Wilkinson R.E. / Colonel Sir Henry James R.E.,F.R.S. Director of the Ordnance Surveys

State 1 (?1867)

The date from railway evidence. A skeleton map, probably transferred from the Larcom map, with the sheet area considerably reduced. Showing county and union boundaries and names, and railways. Its purpose is not known, perhaps for *Report from the Select Committee on General Valuation &c (Ireland)* (BPP(HC) 1868-69 [362], IX, 1-304), and in the event not used.
Not found.

State 2

Undated, certainly post-1922

Dtf

II.2. James's map, 1868

Ordnance Survey of Ireland: Scale Ten Statute Miles to One Inch [scale-bar 10+50 miles] / Constructed and Engraved at the Ordnance Survey Office Phoenix Park Dublin in 1868, under the direction of Captain Wilkinson, R.E. Colonel Sir Henry James,R.E.,F.R.S.,M.R.I.A. &c. Superintendent. The Outline by John West, the Writing by John F.Ainslie, the Water by James Jones

NB: For details of the original drawing, see p.187. Initiated as three sheets of James's ten-mile Map of the World on the Rectangular Tangential Projection. For further information on this, see p.56 and Appendices 1 and 2. Three plates were engraved, from which one for this island map of Ireland was electrotyped. The specification derives from the North American sheets of James's Map of the World, via the three Ireland plates. For details, see col."E" on p.9, and Appendix 2, p.184.

NB: The railway developments that are known only on states of the **Index to the General Map** are included in this sequence for ease of reference. There are states of both maps still apparently lacking, and this carto-bibliography is likely to require revision.

Plate: Dna OS 106
Documentary source: NA OS 5/3007

State 1 (1868)

Dimensions:
    Paper: 723mm (28½ins) W-E by 1047mm S-N
    Plate: 751mm W-E by 1047mm S-N (copper)
    Neat line: 685mm W-E by 965mm S-N (copper)

Changes (to Ireland Map of the World sheets: see p.187)
1. The OSI coat of arms introduced, top left. It is distinctive for its lion with staring eyes, and is a feature also of map covers printed in Dublin
2. Title as above placed 18mm below the coat of arms
3. Scale-bar reduced to 10+50 miles, placed under title
4. Writing in bottom margin condensed
5. Rectangular sheet lines
6. Border broken off the Blasket Islands by waterlining
7. New coastal outline of Great Britain and Isle of Man

Railways (* named) (as on World map plates)
LONDONDERRY GRAVING DOCK-BUNCRANA*   1863
   TOOBAN JCN-FARLAND POINT   1863
LONDONDERRY FOYLE ROAD-DUNDALK*   1847-1859
    STRABANE-STRANORLAR*   1863
    BUNDORAN JCN-BUNDORAN*   1868
      SHANTONA JCN (BALLYBAY)-COOTEHILL BR*   1860
LONDONDERRY WATERSIDE-COLERAINE STN*=   1852-1853
    LIMAVADY JCN-LIMAVADY   1852
BELFAST YORK ROAD-LARNE HARBOUR*   1848-1862
    GREENISLAND-PORTRUSH*   1848-1855
      COOKSTOWN JCN-COOKSTOWN*   1848-1856
BELFAST QUEENS QUAY-BANGOR*   1848-1865
    BALLYMACARRETT JCN (BELFAST)-NEWTOWNARDS=   1850
      COMBER-BALLYNAHINCH*=   1858
        BALLYNAHINCH JCN-DOWNPATRICK*=   1859
BELFAST GREAT VICTORIA STREET-CLONES*   1839-1863
   KNOCKMORE JCN (LISBURN)-BANBRIDGE*   1863
   PORTADOWN-OMAGH*   1858-1861
   ARMAGH-WARRENPOINT   1849-1864
DUBLIN AMIENS STREET [CONNOLLY]-PORTADOWN*   1844-1852
   HOWTH JCN-HOWTH BR   1846
   DROGHEDA-OLDCASTLE*   1850-1863
   SCARVA-BANBRIDGE   1859
DUBLIN BROADSTONE-GALWAY*   1847-1851
   CLONSILLA-NAVAN [AN UAIMH]*   1862
     KILMESSAN-ATHBOY BR   1864
   MULLINGAR-SLIGO*   1855-1862
     INNY JCN-CAVAN-CLONES*   1856-1862
   STREAMSTOWN-CLARA*   1863
   ATHLONE-WESTPORT*   1860-1866
   ATHENRY-TUAM*   1860
DUBLIN KINGSBRIDGE [HEUSTON]-CORK-YOUGHAL*   1846-1860
   CHERRYVILLE JCN (KILDARE)-KILKENNY*   1846-1850
     BAGENALSTOWN [MUINE BHEAG]-BALLYWILLIAM* 1858-1870
   PORTARLINGTON-ATHLONE*   1854-1859
   MARYBOROUGH [PORTLAOIGHAISE]-NEWRATH JCN (WATERFORD)*
                                      1848-1867
    BALLYBROPHY-PARSONSTOWN [BIRR]*   1857-1858
      ROSCREA-BIRDHILL*   1863-1864
   MALLOW-FERMOY*   1860
   MALLOW-TRALEE*   1853-1859
   CORK CAPWELL-MACROOM*   1866
   CORK ALBERT QUAY-DUNMANWAY*   1849-1866
     KINSALE JCN-KINSALE*   1863
   CORK ALBERT STREET-PASSAGE   1850
   QUEENSTOWN JCN [COBH JCN]-QUEENSTOWN [COBH]   1862
DUBLIN WESTLAND ROW [PEARSE]-ENNISCORTHY*   1834-1863
   BRAY-DUBLIN HARCOURT STREET   1854
   WICKLOW MURROUGH STN BR   1855
   WOODENBRIDGE-SHILLELAGH BR*   1865
LIMERICK-WATERFORD*   1848-1864
   LIMERICK-FOYNES*   1856-1858
     PATRICKSWELL-CHARLEVILLE [RATH LUIRC]*   1862
   LIMERICK (ENNIS JCN)-GORT*   1859-1869
   KILLONAN-KILLALOE   1858-1862
   WATERFORD THE MANOR STN-TRAMORE*   1853

Lmd (id 11 Sep.1868) (TJH 1868) (1868 military map, qv); Dki (via J.P.Prendergast from Wilkinson 26.10.1874); Dn 16.K.18 (1) (TJH 1870) (overdrawn with Proclaimed Districts of the Several Counties, at 24.3.1880)

State 2 (1871)

Additions to railways (* named)
Belfast Queens Quay-Bangor*
   Ballymacarrett Jcn-Downpatrick*   NEWCASTLE*   1869

Dublin Broadstone-Galway*
   Athlone-Westport*
      MANULLA-FOXFORD   1868
Limerick-Waterford*
   Limerick-Foynes*
      BALLINGRANE-NEWCASTLE WEST=   1867
   Limerick (Ennis Jcn)-Gort*   ATHENRY*   1869

Not found. But see **Index to the General Map**, State 1

State 3 (?1876)

Introduction of new standard feature
C.8   Egyptian sloping capitals
   c   Extensive district: Connamara, Jar Connaught, Joyce's Country, Cloghan, The Rosses

Changes
1. Province boundaries added (present in North America)
2. Province names added, in Egyptian capitals (C.6e)

Additions to railways (* named)
Londonderry Foyle Road-Dundalk*   GREENORE*   1873
Belfast York Road-Larne Harbour*
   Greenisland-Portrush*
      COLERAINE JCN Stn-Londonderry Waterside*   1860
Belfast Great Victoria Street-Clones*
   KNOCKMORE JCN (LISBURN)-ANTRIM*   1871
   Armagh-Warrenpoint
      KING STREET JCN (NEWRY)-GREENORE*   1876
Dublin Broadstone-Galway*
   Clonsilla-Navan [An Uaimh]*   KILMAINHAM*   1872
   Mullingar-Sligo*
      KILFREE JCN-BALLAGHADERREEN BR   1874
   Athlone-Westport*
      Manulla-Foxford   BALLINA*   1873
Dublin Kingsbridge [Heuston]-Cork-Youghal*
   Cherryville Jcn (Kildare)-Kilkenny*
      Bagenalstown-Ballywilliam*   MACMINE*   1870-1873
   Ballybrophy-Parsonstown [Birr]*
      PARSONSTOWN [BIRR]-PORTUMNA BRIDGE   1868
   Mallow-Fermoy*   LISMORE*   1872
Dublin Westland Row [Pearse]-Enniscorthy*   WEXFORD   1872

Lpro MR 186 (TJHK), showing railways open and projected. (Charles Fort (Kinsale), Camden, Carlisle Forts (Cork), Fort (Dublin) labels are scratched out on this copy)

State 4 (?1878)

Additions to railways (* named)
Dublin Broadstone-Galway*
   NESBITT JCN (ENFIELD)-EDENDERRY BR   1877
Dublin Kingsbridge [Heuston]-Cork-Youghal*
   Mallow-Lismore*   SUIR BRIDGE JCN (WATERFORD)*   1878
      CLONMEL-FETHARD   1879
   Cork Albert Quay-Dunmanway*   SKIBBEREEN*   1877

Dn 16.K.18 (3), used as a military map, Nov.1880 (qv), 16.K.18 (11), used for Resident Magistrates 1881, showing Resident Magistrates' stations and petty sessions to be attended by each

---

State 5 (?1880)

Change
1. Labels to fortifications deleted (Greencastle Fort, Battery (Inch), Fort (Dublin), Camden Fort, Carlisle Fort (Cork), Charles Fort (Kinsale)

Additions to railways (* named)
Dublin Broadstone-Galway*
   Clonsilla-Kilmainham*   KINGSCOURT   1875
   Nesbitt Jcn (Enfield)-Edenderry Br*   NAMED
   Athlone-Westport*   WESTPORT QUAY   1874
Dublin Kingsbridge [Heuston]-Cork-Youghal*
   Mallow-Tralee*
      GORTATLEA-CASTLEISLAND BR   1875
   Mallow-Lismore* (Lismore-Waterford DELETED)
   Clonmel-Fethard DELETED

Dn 16.K.18 (6); Lbl 10805 (78)
See also **Index to the General Map**, State 2

State 6 (1888)

Introduction of new standard feature
A.6   *Price_Two Shillings & Sixpence* (under scale-bar)

Change
1. Fisguard to Fishguard

Additions to railways (* named)
Londonderry Graving Dock-Buncrana*
   Tooban-Farland Point   LETTERKENNY*   1883
Londonderry Foyle Road-Greenore*
   Strabane-Stranorlar*   DRUMINNIN [LOUGH ESKE]*   1882
   ENNISKILLEN-CARRIGNAGAT JCN (SLIGO)*   1879-1881
   INNISKEEN-CARRICKMACROSS BR   1860
Belfast York Road-Larne Harbour*
   Greenisland-Portrush*
      Cookstown Jcn-Cookstown*
         MAGHERAFELT-MACFIN*   1880
         MAGHERAFELT-DRAPERSTOWN*   1883
      BALLYMENA-PARKMORE*   1875-1876
      BALLYMONEY-BALLYCASTLE*   1880
      Coleraine Jcn-Londonderry Waterside*
         Limavady Jcn-Limavady   DUNGIVEN*   1883
   LARNE-BALLYCLARE   1877
      BALLYBOLEY-BALLYMENA*   1878
Belfast Great Victoria Street-Clones*
   Portadown-Omagh*
      DUNGANNON JCN-COOKSTOWN*   1879
   Armagh-Warrenpoint*   NAMED
Dublin Amiens Street [Connolly]-Portadown*
   Scarva-Banbridge   BALLYRONEY*   1880

II.2

Dublin Broadstone-Galway*
   Clonsilla-Kingscourt*
      Kilmessan-Athboy Br*   NAMED
   Mullingar-Sligo*
      Inny Jcn-Cavan-Clones*
         CROSSDONEY-KILLESHANDRA BR*   1886
Dublin Kingsbridge [Heuston]-Cork-Youghal*
   SALLINS-TULLOW*   1885
   Cherryville Jcn (Kildare)-Kilkenny*
      Bagenalstown [Muine Bheag]-Macmine*
         PALACE EAST-NEW ROSS*   1887
   Portarlington-Athlone*
      CLARA-BANAGHER*   1884
   Maryborough [Portlaoighise]-Newrath Jcn (Waterford)*
      CONNIBERRY JN (PORTLAOIGHISE)-MOUNTMELLICK BR 1885
   Ballybrophy-Parsonstown [Birr]*
      Parsonstown [Birr]-Portumna Bridge*   NAMED
   THURLES-CLONMEL*   1879-1880
   Mallow-Lismore*   SUIR BRIDGE JCN (WATERFORD)*   1878
   Mallow TRALEE-NEWCASTLE WEST Ballingrane*   1880 NAMED
      FARRANFORE-KILLORGLIN*   1885
   Cork Albert Quay-Skibbereen*
      CLONAKILTY JCN-CLONAKILTY*   1886
      DRIMOLEAGUE-BANTRY*   1881

Cu Maps 170.86.3
Virtually the state used for IDWO 717 military map (qv)

State 7 (1890)

Introduction of new standard feature
A.14   ARRR bottom left, 11mm bnl

Change
1. *Price Two Shillings & Sixpence* resited bc, 11mm bnl

Additions to railways (* named)
Londonderry Foyle Road-Greenore*
   Strabane-Druminnin [Lough Eske]*   DONEGAL*   1889
   VICTORIA BRIDGE-CASTLEDERG BR*   1884
Belfast York Road-Larne Harbour*
   Greenisland-Portrush*
      BALLYCLARE JCN-BALLYCLARE*   1884
      PORTSTEWART-PORTSTEWART TOWN (TRAMWAY)*   1882
      PORTRUSH-BUSHMILLS-GIANTS CAUSEWAY*   1883-1887
   Larne-Ballyclare DOAGH   1877
BELFAST-WHITEHOUSE (TRAMWAY)*
Belfast Great Victoria Street-Clones*
   Armagh-Warrenpoint*   ROSTREVOR (TRAMWAY)   1878
      NEWRY-BESSBROOK (TRAMWAY)   1885
   TYNAN-MAGUIRESBRIDGE*   1887
Dublin Broadstone-Galway*
   Mullingar-Sligo*
      Inny Jcn-Cavan-Clones*
         BALLYHAISE-DROMOD*   1885-1887
            BALLINAMORE-ARIGNA*   1888
DUBLIN-LUCAN (TRAMWAY)   1881

Dublin Kingsbridge [Heuston]-Cork-Youghal*
   Mallow-Tralee-Ballingrane*
      BANTEER-NEWMARKET BR*   1889
      TRALEE-FENIT*   1887
      LISTOWEL-BALLYBUNION*   1888
   CORK WESTERN ROAD-COACHFORD*   1877-1888
      COACHFORD JCN-BLARNEY TOWER   1887
   Cork Albert Quay-Skibbereen*
      SKIBBEREEN-SKULL [SCHULL]*   1886
DUBLIN-BLESSINGTON (TRAMWAY)*   1888
Limerick-Waterford*   WATERFORD NEW STN   1883
   Limerick (Ennis Jcn)-Athenry*
      ENNIS-MILLTOWN MALBAY*   1887

Not found. But see **Index to the General Map**, State 3

State 8 (?1894)

Additions to railways (* named)
Belfast Queens Quay-Bangor*
   Ballymacarrett Jcn (Belfast)-Downpatrick-Newcastle*
      DOWNPATRICK-ARDGLASS*   1892
Dublin Broadstone-Galway*   CLIFDEN*   1895
   Athlone-Westport Quay*
      CLAREMORRIS-BALLINROBE BR*   1892
      Manulla-Ballina*   KILLALA   1893
      WESTPORT-ACHILL   1894
   ATTYMON-LOUGHREA BR*   1890
   Athenry-Tuam*   CLAREMORRIS*   1894-1895
Dublin-Lucan LEIXLIP (TRAMWAY)   1889
Dublin Kingsbridge [Heuston]-Cork-Youghal*
   Mallow-Suir Bridge Jcn (Waterford)*
      FERMOY-MITCHELSTOWN BR*   1891
   Mallow-Tralee-Ballingrane*
      HEADFORD-KENMARE*   1893
      Farranfore-Killorglin*   VALENTIA HARBOUR*   1893
      TRALEE-DINGLE*   1891
         CASTLEGREGORY JCN-CASTLEGREGORY BR*   1891
   Cork Albert Quay-Skibbereen*   BALTIMORE*   1893
      Drimoleague-Bantry Old Stn*   BANTRY   1892
      Clonakilty Jcn-Clonakilty*
         BALLINASCARTHY-COURTMACSHERRY*   1890-1891
Dublin Westland Row [Pearse]-Wexford*   ROSSLARE HBR   1882
Limerick-Waterford*
   Limerick (Ennis Jcn)-Athenry*
      Ennis-Milltown Malbay*   KILKEE*   1892
         MOYASTA-KILRUSH   1892

Not found. But see **Index to the General Map**, State 4

State 9 (?1896)

Additions to railways (* named)
Londonderry Foyle Road-Greenore*
   Strabane-Donegal*   KILLYBEGS*=   1893
   STRANORLAR-GLENTIES*   1893
Dublin Amiens Street [Connolly]-Portadown*
   DROMIN-ARDEE BR*   1896

Dublin Broadstone-Clifden*
   Athlone-Westport Quay*
      CLAREMORRIS-COLLOONEY*   1895
Dublin Kingsbridge [Heuston]-Cork-Youghal*
   Cork Western Road-Coachford*
     Coachford Jcn-Blarney Tower
       ST ANNES-DONOUGHMORE*   1893
Dublin-Blessington POULAPHOUCA (TRAMWAY)*   1890

Cu Maps b.19.G.48.1

State 10 (?1901) (final state)

Change
1. Fishguard reverts to Fisguard (as is on the plate)

Addition to railways (* named)
Londonderry Graving Dock-Buncrana* CARNDONAGH* 1901

Dn 16.K.18 (2), showing Light Railways constructed under the Acts of 1883, 1889 and 1896, and sites of proposed works under the Marine Works Act, 1902, in red.

   This further railway was added to the topographical, but not the index map (see next). A comparison between the plates demonstrates this, and the slight variations between the two. For instance, the Port Stewart tramway and the line bridging the river at Coleraine are lacking on the topographical plate, the Castleisland branch is not named on the Index, and several names are in slightly different positions. There is an even later railway state on the **Index to Six Inch Sheets** (1922) (qv)

Ordnance Survey of Ireland: Scale Ten Statute Miles to One Inch [scale-bar 10+50 miles]: **Index to the General Map, Published on a Scale of One Inch to a Statute Mile** [this title in sloping letters] / Constructed and Engraved at the Ordnance Survey Office Phoenix Park Dublin in 1868, under the direction of Captain Wilkinson, R.E. Colonel Sir Henry James,R.E.,F.R.S.,M.R.I.A.&c. Superintendent. The Outline by John West, the Writing by John F.Ainslie, the Water by James Jones

   The index was requested on 6.5.1871. The graticule was retained on trial pulls. Wilkinson wanted to remove this, and to lithograph the map with sheet lines in red. James replied on 14.8.1871 approving only the removal of the graticule.
   The index was probably revised in parallel with the topographical map, though slight variations in railway revision grew up over the years. For details of the railway states, refer to the topographical map.

Plate: Dna OS 106
Documentary sources: NA OS 5/3331, 5/3616

State 1 (1871)

Dimensions:
   Paper: 728mm W-E by 1064mm S-N
   Plate: 727mm W-E by 1041mm S-N (copper)
   Neat line: 685mm W-E by 965mm S-N (copper)

Introduction of new standard features
B.5   Hydrographic and marine features
     q Sailing route
     r Electric telegraph, from Holyhead to Howth, from Portpatrick to Donaghadee; Atlantic telegraph from Valentia to Newfoundland
C.6   Egyptian capitals
     g Electric telegraph, Atlantic telegraph
C.9   Italic
     pp Sailing route and distance

Changes (in relation to topographical map)
1. British railways added. These were never revised
2. Port Patrick, Porth [D]inlleyn added
3. Graticule deleted

Railways as topographical map, State 2

Cu Atlas 1.08.16 (TJHK); PC

State 2 (?1880)

Changes
1. Labels to fortifications deleted
2. Extensive districts added
3. Province boundaries and names added

Railways as topographical map, State 5

Dn 16.A.15 (2) (id 16 Mar.1882) (TJHK), 16.A.15 (3)

State 3 (?1890)

Changes
1. The graticule is reinstated
2. Upright lettering replaces sloping for the title
3. *Price Two Shillings & Sixpence* added bc, 11mm bnl
4. ARRR added bottom left, 11mm bnl
5. Porth inlleyn deleted
6. Atlantic Telegraph to Atlantic Cable
7. Fisguard to Fishguard
8. Extensive district names deleted

Railways as topographical map, State 7

Dn 16.K.18 (4) (carrying the stamp of the Congested Districts Board, Ireland, dated 6 Sep.1895), 16.K.18 (5), showing Government Relief works 1895

II.2

State 4 (?1894)

Railways as topographical map, State 8

Dn 16.B.6 (32), shewing Land Inspectors Districts

State 5 (?1896)

Railways as topographical map, State 9. The Dna OS 106 plate is in this state

BFq (3 copies); Do

State 6 (?post-1922)

Changes
1. The coat of arms is deleted
2. Marginalia deleted after "Superintendent"

NTg

NB: The index map, reduced to 27 miles to 1 inch and entitled **Map showing the Hinterlands served by Belfast, Derry, Enniskillen, Sligo, Newry and Dundalk in 1922 as commercial centres for the distribution of dutiable goods** (ARRR, price 1/-, with five overprint colours), is in *Handbook to the Ulster Question* (Dublin, Stationery Office, 1923)

Ordnance Survey of Ireland:...Poor Law Unions (Other title details as the **Index to the General Map**)

For official use only. The Commissioner of Valuation requested 500 copies of the **Index to the General Map** on 9.11.1880, with an overprint in red of Poor Law Union names and boundaries. Treasury authority was received on 4.12.1880. They were lithographed, and the work completed by 3.3.1881. 200 copies were overprinted in black, 300 in red. (NA OS 5/4239)

On 6.7.1883 the Irish Land Commission wanted twelve copies for their own use. They were advised to acquire them from the Valuation Department since the 1881 map had already been removed from stone. (NA OS 5/4525)

NB: All copies so far located have the red overprint

State 1 (1881)

The base is the **Index to the General Map**, State 2

Changes (to this state of the Index)
1. Sheet size reduced, especially to east
2. Bottom marginalia is cleared
3. With an overprint of union names and boundaries
4. With a legend: Reference_The Names and Boundaries of Poor Law Unions are shewn in red

Dn 16.B.4 (2)

State 2 (1890)

Using the valuation figures of the 1881 census. The base used is the **Index to the General Map** prior to its State 4 revisions, but including most of its railways, lacking the Whitehouse, Blessington and Newmarket lines

Changes
1. With a new legend: Coloured by Unions to show Ratio of Valuation to Population
2. The Westport district no longer includes Newport
3. Signed H.Kirkwood Major,R.E., 13 Aug.1890

Dn 16.K.18 (8), used apparently as a model for a fully coloured map: the districts are hand-coloured in varying intensities of red and blue

State 3 (?1890)

Changes
1. The union overprint is new, but apparently in the same state of revision as State 2
2. Bottom right hand corner of the map is cleared as if for a table: Lundy is only half deleted
3. The base is the full size topographical map, State 7

Dn 16.K.18 (9)

State 4 (?1890)

Changes
1. With a table bottom left "showing average valuation per head for each Union"
2. Prepared at the Ordnance Survey Office, Dublin. From Statistics supplied by the Registrar General, H.Kirkwood, Major,R.E.
3. With an additional overprint (as modelled in State 2) showing in eight intensities of red and blue the ratio of valuation to population

Dn 16.K.18 (10)

Map of Ireland: To accompany the Report of the Royal Commission on Public Works: Shewing Rail and Tramways, Fishery Piers and Harbours &c.,&c.1887 / Prepared at the Ordnance Survey Office Phoenix Park under the direction of L^t.Colonel A.B.Coddington R.E. / Scale of 10 English Miles to an inch [Scalebar 10+50 miles] / Colonel Sir C.W.Wilson K.C.B.,K.C.M.G.,F.R.S.,R.E. Director General

On the **Index to the General Map**, with four overprint colours. Many tables, including 187 piers and harbours. No Great Britain.

In *Second Report of the Royal Commission on Irish Public Works* (BPP(HC) 1888 [C 5264], XLVIII, 201). Other copies: Dn 16.K.18 (19), 16.K.18 (20)

Ordnance Survey of Ireland: Scale Ten Statute Miles to One Inch [scale-bar 10+50 miles]: **Resident Magistrates 1888** / Prepared at the Ordnance Survey Office Phoenix Park under the direction of L^t.Colonel A.B.Coddington R.E. / Colonel Sir C.W.Wilson,K.C.B.,K.C.M.G., F.R.S.,R.E. Director General

A newly printed map with the topographical map as base, apparently in the same (c.1887) state throughout. The sheet size is reduced. The overprint shows Resident Magistrate's stations and petty sessions to be attended in red and black, with green county boundaries. There is also a pale hand-coloured wash of counties and blocks of counties. Preceded by an 1881 issue giving the same information, overdrawn on the topographical map, State 3

State 1 (1888)

Dn 16.K.18 (12), 16.K.18 (13)

State 2 (1889)

Changes
1. Title date altered to December 1889
2. Coddington replaced by Major H.Kirkwood R.E.

Dn 16.K.18 (14), 16.K.18 (15), 16.K.18 (16)

State 3 (1899)

Changes
1. Title date altered to June 1899, the overprint title now preceding the scale statement
2. Redesigned with different borders and lettering, and with no marginalia
3. The full sheet size is restored

Dn 16.K.18 (17). *NB*: The next edition on the 1905 map

Ordnance Survey Map of Ireland: Shewing Locality of and Expenditure on Works &c. connected with Fisheries **1891.** Scale Ten Statute Miles to One Inch [scale-bar 10+50 miles]: Prepared from Statistics and information supplied by H.M.Inspector of Irish Fisheries / Prepared at the Ordnance Survey Office Phoenix Park under the direction of Major H.Kirkwood, R.E. / Colonel Sir C.W. Wilson K.C.B.,K.C.M.G.,F.R.S.,R.E., Director General

The base map is grey, with six overprint colours, both hand painted and stamped. With a reference table, and neat line border. The railway revision is similar to that on State 7 of the topographical map. 875 copies were requested. (NA OS 6/8613)

In *Report of the Inspectors of Irish Fisheries on the Sea and Inland Fisheries of Ireland for 1890* (BPP (HC) 1890-91 [C 6403], XX, 329). Another copy: Lbl Maps 57.a.11

Map of Ireland for Relief of Distress Report Scale Ten Statute Miles to One Inch [scale-bar 10+50 miles]. Undated, but c.1894

An unattributed and reduced area lithograph, lacking Great Britain. With graticule and a plain border, and new title lettering. Open railways are overprinted red, those under construction are blue, ie mostly the rail-railways additional to State 8 of the topographical map. Its purpose is unknown: the title suggests a parliamentary report though in the event none such was completed

Dn 16.B.8 (7)

Map of Ireland to accompany Report on Loan Fund Societies / Reference: The circles around the towns in which the offices of the Loan Societies are situate, indicate the area within which each Society is authorized by the rules to operate. A double circle implies that two Societies are working within the same radius.

An unattributed and reduced area lithograph, lacking Great Britain. With graticule and a plain border, and title lettering as the last. With one colour overprint. Railways are similar to State 7 of the topographical map

In *The Loan Fund Board for Ireland: Report of the Committee appointed to inquire into the Proceedings of Charitable Loan Societies in Ireland, established under the Act 6 & 7 Vic.,Cap.91* (BPP(HC) 1897 [C 8381], XXIII, 417). Other copies: Dn 16.B.5 (15), 16.B.36 (6)

Ordnance Survey of Ireland: Scale Ten Statute Miles to One Inch [scale-bar 10+50 miles]: **Index to Six Inch Sheets** / Constructed and Engraved at the Ordnance Survey Office Phoenix Park Dublin in 1868, under the direction of Captain Wilkinson R.E. Colonel Sir Henry James R.E., F.R.S.,M.R.I.A.&c. Superintendent. The Outline by John West, the Writing by John F.Ainslie, the Water by James Jones

State 1 (1922)

Specially prepared for the Commander in Chief. 300 copies were printed and supplied to the Stationery Office, GHQ, Portobello, on 30.11.1922. The Ministry of Industry and Commerce wanted a copy: they were advised to apply to the military authorities. (NA OS 6/20612) Not found.

State 2 (?1922)

Graticule, price and ARRR as State 10 of the topographical map, and "Fisguard" reappears. Six-inch sheet lines and county boundaries are added in black. It is possible that this state is part of the State 1 stock.

Railways (additional to those on State 10 of the 1868 topographical map above) (* named)
Londonderry Graving Dock-Carndonagh*
   Tooban-Letterkenny BURTONPOINT* 1903
LONDONDERRY VICTORIA ROAD Strabane-Killybegs* 1900
   STRABANE-LETTERKENNY* 1909
   DONEGAL-BALLYSHANNON* 1905
Belfast Great Victoria Street-Clones*
   ARMAGH-CASTLEBLAYNEY* 1909-1910
Dublin Amiens Street [Connolly]-Portadown*
   Scarva-Ballyroney* NEWCASTLE 1906
Dublin Kingsbridge [Heuston]-Cork-Youghal*
   DUBLIN LOOP LINE 1877-1901
   Cherryville Jcn (Kildare)-Kilkenny*
     ATHY-WOLFHILL BR* 1918
   Bagenalstown [Muine Gheag]-Macmine*
     Palace East-New Ross* WATERFORD* 1904
   Maryborough [Portlaoighise]-Newrath Jcn (Waterford)*
     KILKENNY-DEERPARK BR* 1919
   Ballybrophy-Parsonstown [Birr]*
     Parsonstown [Birr]-Portumna Bridge CLOSED
   Thurles-Clonmel*
   GOOLDS CROSS-CASHEL BR 1904
   Cork Albert Street-Passage CROSSHAVEN* 1902-1904
Limerick-Waterford* ROSSLARE STRAND JCN 1906
   KILLINICK-FELTHOUSE JCN 1906

Dtf; Mg

State 3 (1924)

Changes
1. ARRR replaced by CCR(!), bottom left
2. Reprinted in 1924 added, bottom left
3. Price replaced by 4/-, bottom centre, 22mm bnl
4. Black overprint replaced by red

Dn 16.A.15 (8)

State 4 (1937: still in OSI 1949 catalogue)

Changes
1. CCR replaced by CR, bottom left, 10mm bnl
2. Reprint date altered to 1937, bottom left, 14mm bnl

BFq; Dtf; NTg

NB: For OSI skeleton map on this base, see p.91.

Supplement 1. Rivers and their Catchment Basins

Plan of the Catchment Basins of the Rivers of Ireland Scale Ten Statute Miles to One Inch [scale-bar 10+ 50 miles] Reduced and Drawn by William Harvey Augt 1864 / at foot: Ordnance Survey Office Phoenix Park Dublin August 1864 [signed] Berdoe A.Wilkinson Captain RE

The original coloured drawing on cloth-backed paper, drawn in two phases, first as a map of rivers and their catchment basins, achieved by August 1864, secondly with topographical information, completed in January 1865. With a graticule, but no border. See pp.56,187.

Dimensions: 673mm W-E by 963mm S-N

Dos. Its transfer to Dna is planned
Documentary source: NA OS 5/3007

Ordnance Survey of Ireland: Rivers and their Catchment Basins Scale Ten Statute Miles to One Inch [scale-bar 10+50 miles] / Constructed and Engraved at the Ordnance Survey Office Phoenix Park Dublin, in 1867-8, under the direction of Captain Wilkinson, R.E. Colonel Sir Henry James, R.E.,F.R.S.,M.R.I.A.&c Superintendent. The Outline by John West, the Writing by John F.Ainslie

A map of Ireland derived from the **Plan of the Catchment Basins of the Rivers of England and Wales** of 1861 was prepared in 1864 following a request from the Irish Fishery Commissioners. The map cost £208.17.3d to prepare and engrave. Treasury authority is not recorded.

Plate and matrix: Dna OS 106
Documentary sources: NA OS 5/3007, 5/4713; PRO OS 1/144

State 1 (OSR 1868)

Dimensions:
   Paper: 725mm (28½ins) W-E by 1091mm S-N
   Plate: 711mm W-E by 1025mm S-N (copper)
   Matrix: 708mm W-E by 1023mm S-N (copper)
   Neat line: 654mm W-E by 954mm S-N (copper)

Published 1868. No Great Britain. Showing river systems, spot heights near their sources, water names, towns on rivers with more than 5000 inhabitants, coastal names. No graticule, but with a border graduated at 10' intervals. With references to a large table at the foot. For characteristics used, see col."G" on p.9.

BFpro OS 15/4/1; Cu Maps 171.86.7; Dki (via J.P.Prendergast from Wilkinson 26.10.1874); Dn 16.E.16; Lbl Maps 7.c.33 (c.1872), 10820 (6) (id 24 Nov.1879); Lrgs Ireland G.35; Ob C19 (225); SOos. The BFpro copy is in an original slip case. Impressions were probably run off and hand coloured as needed

State 2 (1923)

Changes
1. Printed at the Ordnance Survey Office, Dublin in 1923
2. Colour printed

This is Map 4 in *Coimisiún na gCanálach agus na mBóthar Uisce Intíre (Canals and Inland Waterways Commission): Report: July 1923* (IPP 1923, III, 611 (250 11.23)). Report title and map number in top margin

State 3 (1927)

Changes
1. Printed at the Ordnance Survey Office, Dublin, 1927
2. The canal commission overprint of State 2 removed
3. Crown Copyright Reserved added, bottom centre
4. Price 10/6d added, bottom centre below CCR

BFq

State 4 (1945)

Changes
1. The title below "Superintendent" deleted
2. Crown Copyright Reserved to Copyright Reserved
3. After "Dublin, 1927": Reprinted in 1945

Lrgs (uncoloured copy)

State 5 (1958)

Changes
1. Marginalia rewritten: bottom centre: *Altitudes are referred to Ordnance Survey Datum, which is a Low Water of Spring Tides observed in Dublin Bay on 8th. April 1837, this Datum being 8.218 feet below Mean Sea Level. Published by the Director at the Ordnance Survey Office, Phoenix Park, Dublin. Copyright Reserved*; bottom right: *Prepared and Printed at the O.S.O. Phoenix Park, Dublin 1958*
2. Price 10/6d moved bottom left
3. Colour printing rationalised
4. Queenstown to Cobh, Kingstown to Dún Laoghaire, Birr or Parsonstown to Birr

BFq; Cu; Dtf; Lse

State 6

Changes
1. The map is uncoloured
2. Price 10/6d deleted

On sale (1992): in OSI 1968 Catalogue

Map of Ireland: Showing Drainage Districts...Rivers and Main Streams. Scale Ten Statute Miles to One Inch [scalebar 10+50 miles]

An unattributed, but presumably authorised, use of the **Rivers** map, dated 11.1886, in five base and three overprint colours for boundaries, with black railways

In *Appendix to First Report of the Royal Commission on Irish Public Works: Minutes of Proceedings, Evidence, and Index* (BPP(HC) 1887 [C 5038-I], XXV, 815). Another copy: Dn 16.B.5 (4)

Ordnance Survey of Ireland: Rivers and their Catchment Basins Showing Rainfall Distribution and Rainfall Stations / Constructed and Engraved at the Ordnance Survey Office Phoenix Park Dublin, in 1867-8, under the direction of Captain Wilkinson, R.E. Colonel Sir Henry James, R.E.,F.R.S.,M.R.I.A.&c Superintendent. Scale Ten Statute Miles to One Inch [scale-bar 10+50 miles]

State 1 (1921)

Scale statement and scale-bar are moved to the foot of the map to make way for the extended title. With catchment area boundaries in thick and thin black lines, with rainfall stations named and their locations marked with blue dots. An extra reference table is on the right hand side. An isohyetal map prepared by Dr Crowley from data by Dr H.R.Mill and (for the Dublin area) W.J.Lyons.
Requested 22.4.1918. A separate plate of the rivers map was made ready by 12.11.1918. 750 copies were printed by 11.7.1921 (NA OS 6/18000). See also p.79.

In *Board of Trade: Report of the Water Power Resources of Ireland Sub-Committee* (Dublin, HMSO, 1921)

State 2

Lacking overprint: a plain **Rivers and their Catchment Basins** map in all but title

NTg

Water-Power Sites in Ireland: Index Map...[scale-bar 10+50 miles] / Printed at the Ordnance Survey Office, Dublin in 1922

On the **Rivers** map with the table at the foot removed and the south section of the border raised 213mm. Water power sites are in red, with text reference numbers

In *Coimisiún Fhiafruighthe maoin is tionnscal éireann. Report on Water Power. January, 1922.* (Dublin, Commission of Inquiry into the Resources and Industries of Ireland, 87 Grafton Street)

II.2.Supplement 1

Ordnance Survey of Ireland: Rivers and their Catchment Basins Scale Ten Statute Miles to One Inch [scalebar 10+50 miles] / Printed at the Ordnance Survey Office, Dublin, 1924

State 1 (1924)

Showing fishery districts in red. Railways added in black, the sea is coloured blue. This is the **Rivers** map with reduced sheet size and a simple border.

In *The Angler's Guide to the Irish Free State* (Dublin, Stationery Office, 1924 (3000 7/24))

State 2 (1930)

Changes
1. Title altered to **Ordnance Survey of Ireland: Rivers, Lakes and Fishery Districts**
2. Prepared at the Ordnance Survey Office Dublin, 1924. Revised in 1929
3. The fishery district boundaries revised at the border with Northern Ireland
4. Railway company labels revised
5. Irish-speaking districts are coloured yellow: this is a simplified version of the overprint that first appeared in the 1926 Gaeltacht Report (see III.2)

In *The Angler's Guide to the Irish Free State.... Second Edition* (Dublin, Stationery Office, 1930)

State 3 (1937)

Reprinted in 1937 in *The Angler's Guide to the Irish Free State....Third Edition* (Dublin, Stationery Office, 1937 (3005 3/38))

State 4 (1948)

Changes
1. The east neat line brought in, forcing the County Down coastal area on to an extrusion
2. Further revision of railway company labels

Reprinted in 1948 in *The Angler's Guide....Fourth Edition* (Dublin, Stationery Office, 1948 (5000 10/48))

Supplement 2. Military issues

The base map for most if not all the maps in the list below was the island topographical map published by James in 1868, though most of them were stripped of their topographical detail and were published in strongly amended forms, usually by other agencies. Sometimes the topographical map itself was used, overdrawn in some way. As will be deduced from the list below, the War Office received copies of the new map within days of publication. A few copies were requested by the Quarter Master General's Office, Dublin Castle on 2 November 1868, and ten more (perhaps of the new index version) on 30 August 1871 (NA OS 5/3331). In 1888 a lithograph was "Transferred from the Ord^e Survey Index Map. Feby 1888": in fact it was transferred from State 6 of the topographical map immediately prior to its completion.

NB: For other general comments, see I.Supplement 3

1. TDWO maps

Ireland, showing distribution of regular troops and Military Districts (1867). WO 1889/12

Ordnance Survey of Ireland: Shewing the distribution of the Troops in 1868
The 1868 topographical map, State 1 (qv)
Lmd (acquisition date 15.9.1868). (WO 1889/17)
Documentary source: NA OS 5/3331

Map of Ireland (1868). Printed by Kell Bros
In *Ireland: Handbook of Railway Distances* (WO, 1869)
Ob GA Irel. 4° 38
Another issue (1874). Printed by Harrison and Sons
In *Ireland: Handbook of Railway Distances* (London, Harrison and Sons). Map print code: (600 8/74 1980)
Ob GA Irel. 4° 39

Ireland. Telegraphic Map (1870)
Another issue, corrected to 1872. WO 1889/18

Distance Map of Ireland (1873). Compiled in the Quarter Master General's Office, War Office and by W.J. Kelly, Draftsman. 1873. Lithographed at the Topographical Dépôt of the War Office. Major C.W.Wilson R.E. Director
Dn 16.B.6 (13) (TDWO stamp 8.11.1873) (WO 1889/20)
Another issue, ([IBWO] 88)
Dn 16.B.8 (32); Ob C19 (153)

Ireland showing Head Quarters and Numbers of Brigade Dépôts (1874). WO 1889/21

Ireland, showing Head Quarters of Brigade Dépôts and Infantry Militia Regiments (1874). WO 1889/22

Ireland: Lithographed at the Topographical Dépôt of the War Office. Major C.W.Wilson R.E. Director. (1874)
Another issue, corrected to 6.76
Lpro WO 78/626
Another issue, corrected to 1877. WO 1889/26

Ireland, showing Head Quarters of Brigade Depots and Pay Stations (1879). WO 1889/29

Ordnance Survey of Ireland: [Shewing cavalry, infantry and marine headquarters, battery of artillery, forts and towers, detachments, unoccupied barracks, brigade depots, boundaries]. Additional railways are in manuscript, and the three regular telegraphs that appear on the index map, plus a fourth, from Greenore to Abermawr.
The 1868 topographical map, State 4 (qv)
Dn 16.K.18 (3) (in use November 1880)

2. IBWO maps

Ireland, showing Military Districts, Head Quarters of Military Districts and Boundaries and Numbers of Regimental Districts] (1881) (IBWO 93)
    Another issue, corrected to 1.7.82. WO 1889/35
    Another issue, corrected to 8.83. WO 1889/36

Ireland, showing the Distribution of Troops, 1.3.82 (IBWO 134)
    Another issue, corrected to 1.5.83 (IBWO 291). WO 1889/37
    Another issue, corrected to 1.10.83 (IBWO 320)

3. IDWO maps

Military Map of Ireland for Basis Map (1888) (IDWO 689)

Ireland: Military Districts / All Rights Reserved / Published on behalf of the War Office by Edward Stanford, 26 and 27, Cockspur St Charing Cross London S.W. Price 3/6 / Transferred from the Ord^e Survey Index Map. Feby 1888 (IDWO 717)
    In fact the topographical, not the index map, with State 6 railways (except that Palace East-New Ross is lacking), with marginalia altered to the above
Ob C19 (150). (WO 1889/40)

Ireland (Outline) showing Boundaries of Counties of Military Districts (1890) (IDWO 789)

Map of Ireland, showing Head-Quarters of Military and Regimental Districts (1897) (IDWO 1320 Sheet 4). WO 1889/47. Prepared at OSO, Southampton, published IDWO

Londonderry Coast Defence Scheme / Secret / Map 1 (August 1901) (IDWO 1551(a)). An extract of the 1868 map (II.2), south to 53°N. Showing naval and port war signal stations, cables, coasts practicable for landing. Not found.
    Another issue: 1904 Revision (November 1904)
    In *Belfast Defences* (full title probably as below) (WO A.929). Copy at PRO WO 33/323
    Another issue: 1906 Revision (May 1906)
    In *Ireland: Londonderry Coast Defence Scheme: Part I: Lough Swilly Defences: Revised to 1st January, 1906* (WO A.1099 (9/1906)). Another copy in *Part II: Belfast Defences: Revised to 1st January, 1906* (WO A.1100 (9/1906)) (both classified "Secret": copies at PRO WO 33/401)
    Another issue, retitled **North Irish Coast Defence Scheme: 1909 Revision** / Secret / Map 1 / [Printed at the] War Office Octr 1909
    In *Ireland Defence Scheme: North Irish Defended Ports: Part II: Belfast Defences. Revised to May, 1909* (WO A.1362 (29 10/09). Another copy in *Remarks on the North Irish Coast Defence Scheme, Revised to May, 1909* (WO A.1381 (57 5/10) (both classified "Secret": copies at PRO WO 33/484)

4. Ordnance Survey map

Map of Ireland: Showing Districts of Army Corps / Printed at the Ordnance Survey Office, Southampton, 1900 (with manuscript additions 1903)
Lpro MPGG 22 (ex-WO 78 2150): miscatalogued, not found.

CHAPTER III

THE JOHNSTON TEN-MILE MAPS

1. Great Britain

1. A map of the three kingdoms?

It was implied in Director General (DG) Sir Charles Wilson's written evidence to the Dorington Committee of 1892 that the OS was authorised to produce a topographical map at the ten-mile scale, though the date of that authority seems never to have been established.[1] Henry Tipping Crook argued before the same committee that the ten-mile Index was the ten-mile map, and that it had been out for years, the inference being that it was the statutory map.[2] Wilson, in his written evidence, denied this. Attempts at making one had in fact begun by 1887, following the completion in 1884 of the north sheet of the Index and the refurbishment of the middle and south sheets to accommodate New Series sheet lines. The proposal was to create a composite ten-mile map of Great Britain and Ireland. On 2 March 1887 the EO was told:

At present we possess a set of plates containing Index map of England and Scotland. From matrices of these plates I (?O.Trig) can scrape off sheet lines and numbers and so produce duplicates on which these lines and numbers will not appear. This had already been done for the Southern sheet No 3 when the Index to the New Series map was made.

For Ireland there is already such a plate in existence. The difficulty is to construct a set of plates to form a combined map of Great Britain. I have considered two methods...[3]

The next question was whether:

....the 10 mile Index map of the three kingdoms can be combined with sufficient accuracy to make one grand geographical map on that scale of Great Britain and Ireland, the lines of latitude and longitude to be supplied by your (O.Trig) Dept.

O.Trig thought it could:

The degree lines of latitude and longitude can be constructed with reference to the one-inch sheet lines on dry proofs and transferred to the plates before these sheet lines were erased. The production of the degree lines to the margins of the new plates and the marginal graduations to be a subsequent operation.

As the existing maps were on different projections, this opinion would seem to be highly over-optimistic.

The plan was recommended to Wilson on 15 March 1887. He was advised that, with its mixture of projections, it was a less than perfect scheme, but the alternative of creating a new map on a single central meridian, and the inclusion on it of the results of the still incomplete cadastral survey would take years to implement, be more expensive and require Treasury approval. Nonetheless, Wilson chose the latter path. He replied on 17 March, concerned about present expense and the current commitments of the engravers, and he questioned the wisdom of publishing outdated information, especially as regards roads. He recommended awaiting the results of the cadastral survey, completion of which was expected in two years. So this plan lapsed, and it was this situation

---

1. *Report of the Departmental Committee...* Q.1444.
2. *Report of the Departmental Committee...* q.1443.
3. PRO OS 1/144 70B. This file was the work of Harold Jolly, OS Research Officer, in 1934. Winterbotham asked him to investigate the history of the ten-mile map with special emphasis on the supposed projection of the south sheet of the nineteenth century map on the Airy "Balance of Error" projection. Jolly's research took him to Index map file 1879-1898 (CC 880/87), now lost, but his transcript of its contents provides the most important evidence of late nineteenth century developments. The remainder of this section in part depends upon it. See also Appendix 1.

reported by Wilson to the Dorington Committee in 1892:

It was decided about four years ago that the 10 mile map was not to be brought out until the material from the Cadastral Survey was available.[4]

This in effect meant awaiting the completion of the one-inch map, and then the quarter-inch map, both necessary steps in the process of reducing large-scale surveys to a very small-scale map. As OSR 1892 put it:

The map on a scale of 10 miles to an inch has been engraved as an index to the 1-inch map. It will eventually be revised and published as a general map of the Kingdom.

It was not until the end of the century that the OS could at last foresee the arrival of that circumstance. At that point they requested Treasury approval.

2. Johnston's ten-mile map

It fell to the Secretary of the Board of Agriculture, J.H.Elliot, to supply the evidence supporting the submission on 29 December 1898 to the Treasury for a new OS ten-mile map. After referring to the ineffectiveness of photozincography in picking up the detail of a small-scale map, which was better achieved through engraving on copper plates, he went on to point out:

The 1-inch Revision of the whole of England is now on the point of completion; the 1-inch Revision of Scotland has been completed within the last two or three years; and the present is therefore by far the most favourable time for producing the 10-mile map of Great Britain which has always formed part of the Scheme of the Ordnance Survey, but which has hitherto had to be postponed because there have been no 1-inch maps for either country, of which the details were approximately up to date.

The Board of Agriculture's submission was:

It is proposed to complete the 1/10 inch map of Great Britain in 3 years from [31st March 1899] at a total cost of about £2,000 of which £300 would be provided in the year 1899-1900.

At the same time they requested £3000 for the making of the Scotland quarter-inch map, £1000 of which was required in the first year. In January 1899 the Treasury gave its approval in the 1899-1900 Estimates, unfortunately muddling the figures in the written declaration by awarding £300 to the Scotland quarter-inch and £1000 to the ten-mile map, with more in subsequent years. The amounts were no doubt reversed in practice as intended.[5]

The original idea was still to erase, amend and add to the old plates of the Index, but gradually the major advantages of drawing anew at the quarter-inch scale became more marked:

The impressions would have become so confused as to throw an impossible burden on the engravers, and, worse still, the inaccuracy disclosed by a comparison between one of the Index sheets and a New Series One Inch reduced by pentagraph showed that a large measure of re-drawing would have to take place in any case.

The decision, therefore, to redraw at the quarter-inch scale was recorded in a minute from Director General D.A.Johnston on 9 March 1900. But the map would still have to be an amalgam, of the England and Wales quarter-inch on Cassini's Projection, and the Scotland quarter-inch on Bonne's. A specification for the map was quickly drawn up, and in July, the **Characteristic Sheet for the Revised Ten-mile Map of Great Britain** was engraved. Perusal of this (illustrated on p.75) and the specification table on p.9 will explain more to the interested reader than two pages of text, but, briefly, a few additional points need to be made, to set the specification of this map between those of its predecessor and successor. It is much the simplest of the three, with little information offshore other than lightships. The coastal outline is black, only giving way to the blue inland water plate at the points where rivers become single lines. Black stipple was used for sand and mudbanks. Some specification elements seem strongly influenced by the 1868 Ireland map, with an ordered sequence of circular symbols for the various categories of town classification, and a narrower gauge for railways and roads, though a double line was still used for main roads. The number of roads shown was reduced, but railway tunnels and stations were marked, also battle sites, lightships, lighthouses and some spot heights. A handful of ancient sites were shown, including, remarkably, a Roman signal station at Kaims Castle, Rufus's Stone in the New Forest, and Standing Stones on Mainland, Orkney. Ancient and other notable buildings were also marked, with a separate symbol for ruined examples.

OS officers continued a fruitless search through the summer to find a way of applying a graticule to a map of composite Cassini and Bonne projections, and finally recommended redrawing Scotland on Cassini's. But on 14 September 1900, Johnston decided to bypass the problem by omitting the graticule altogether. He announced that the revised ten-mile map would be a simple reduction from the four-mile maps and therefore virtually from the one-inch scale. The projections would not be altered, and the sheet lines would not be quite parallel to the east and west lines of the Scotland one-inch map. Proposals would soon be made as to imprint and marginal writing, which should be as simple as possible.

The map was drawn on impressions of the quarter-inch sheets printed in blue on Whatman paper, on which the draughtsman inked in only the detail required at the ten-mile scale. On photographing the result, the blue detail disappeared, leaving only the black. Heliozincography[6], a process the OS had been using since 1893, was

---

4. *Report of the Departmental Committee... Q.1444.*
5. PRO T1/9335c.
6. See James (1875, revised 1902 179ff) for a description of this process.

ORDNANCE SURVEY
CHARACTERISTIC SHEET
for the
REVISED TEN-MILE MAP
of
GREAT BRITAIN.

Outline

Roads
1st Class
2nd Class

Railways
Passenger Lines, only
    do   with stations & tunnel
County Boundaries
Canals
River and streams

Ornament
Large Parks
Trigonometrical station & altitude
Sands
Large Towns
Medium
Smaller
Villages

Character of Writing

County names	C x
marginal do	HAMPSHIRE
Extensive Districts	G O W E R x
County Towns	COLCHESTER x
Large do	BRADFORD x
Medium do	TETBURY
Small do	BRIDLINGTON
Villages, Mansions, etc.	Gavrickhead
Large Forests & Moors	D A R T M O O R x
Principal Lakes, Bays & Navigable Rivers	WINDERMERE, SWANSEA BAY, R I V E R T H A M E S
Minor Lakes, Bays, Rivers, & Channels	Hawes Water, Bogbury Bay
Large Islands	ISLE OF MAN x
Medium & Minor do	Fair Isle x Coquet I.d
Principal Headlands	LANDS END
Minor do	Rame Head x
Important Ranges of Hills	GRAMPIAN MOUNTAINS x
Minor do	Pentland Hills x
Important Valleys	Glen Tilt x
Railways	CALEDONIAN RAILWAY
Important Sands	Goodwin Sands
Antiquities	Fyvie Castle

x *Gauge according to size and importance.*

All rights of reproduction reserved.
Ordnance Survey Office, Southampton, July 1900.
Price Sixpence.

Plate 6. Ordnance Survey Characteristic Sheet for the Revised Ten-Mile Map of Great Britain, 1900. Original dimensions are 31cm by 47cm.

used for reducing the map from four- to ten-mile scale. Two versions were to be published: one in outline with a vignetted blue sea, the other with the addition of light sienna colouring in the first class roads (burnt sienna from about August 1904), blue inland water, and hills shaded. As latterly on the one-inch, hill plates were made separately.

The outline map was designed to cover Great Britain in twelve quarter-sheets, in two columns of six, without overlaps. It is not known why a quarter-sheet size was chosen. It is possible that it stemmed from the limitations of colour printing at the time of inception, and though, by the time of publication of the hillshaded version, some combined sheets were used, a much larger format still would have been possible, as is apparent from the decision in June 1904 to publish Ireland on one, not two sheets. Remarkably, scheming the sheet limits of the northern sheets took place directly on the copper plates of the nineteenth-century index. Appearing on final states of these maps, usually as ticks where they cross one-inch sheet lines, are the joining lines of Sheets 1 and 2 with 3 and 4, and the vertical joining line one mile (0.1") east of the Delamere Meridian.[7] The sheet lines used were conventional, and not common with those of any other current series, save the joining line between Sheets 7 and 9, and 8 and 10, which the New Series one-inch map largely shared. The standard sheet size was 20ins E-W by 13ins S-N (511mm by 329mm), so giving a sheet coverage of 26,000 square miles. For no obvious reason, Sheets 9-12 are five miles deeper S-N. The map itself claims reduction from the four-mile map, which itself was of course a reduction of the one-inch map. No revision dates are given.

The outline map appeared in 1903 with an illuminated title **Ordnance Survey of Great Britain** on Sheet 6, obviously with an eye to the wall map market. Ireland was given similar treatment on publication in 1905. The hillshaded map was introduced in 1904. Combination of the top eight sheets into four by the elimination of sea areas, and incidentally the title, reduced the number required to eight, which for some years retained combined sheet numbers. According to Winterbotham[8], this rationalisation of sheet lines was known as the "C" Series. Late in 1905, thin paper copies of the hillshaded map, unmounted or mounted on cloth, were sold in covers. A dissected format followed in about 1906. Mounted and dissected outline copies were advertised, but have not been recorded. The hillshaded edition was advertised from 1907 flat on thick paper. Also available from 1905 was a set of sheets (whether outline or coloured is unclear) especially printed for mounting as one map, together with a title sheet and marginal lines. The only examples seen in this style have been hillshaded, with sheet lines the same as the eight-sheet version, and blank paper additions to fill out the gaps.

The title from the outline edition, with scale-bar and symbols panel (evidently from the outline edition as well since the main road symbol is uncoloured), were assembled in the North Sea, with an additional line of text between scale-bar and symbols panel: "Published by the Director General at the Ordnance Survey Office, Southampton, 1904". Stanford produced similar examples in Ansell-fold. The set of sheets could also be obtained mounted on holland to form one map bound with ribbon, and, if desired, fitted with moulding and roller. Prices for maps in these forms, both British and Irish, were raised in OSPR 1916/3, and again on 1 January 1920.

Eventually logic prevailed, and the hillshaded map was renumbered 1-8. It has proved frustratingly difficult to determine precisely when this occurred. It is tempting to assume it formed part of the rationalisation process approved in March 1914 to reconstitute Scotland and Ireland quarter-inch maps as large sheet series (OSR 1913-14). But at least one sheet (7,8) retained its combined number following the price increase which took place in April 1915, which therefore either suggests this was overlooked, or that the renumbering occurred later than this. In fact no sheet can be said with certainty to have been renumbered before 1921, and the situation is hardly clarified by contemporary indexes, with DOSSSM [1919] showing an eight-sheet index, and OSC 1920 the combined twelve- and eight-sheet variety.

It also remains unclear whether combined sheets, let alone renumbering them, ever affected the ten-mile outline map. A move towards it was announced in OSPR as early as 9/07 with a combined reprint of Outline Series Sheets 5 and 6, and the comment that all future reprints would be done on combined sheet lines. This state has not yet been located, and its absence from the copyright collections may suggest that it was not published. There is also no evidence for the continuation of the policy, at least among known later reprints. But with none located so far dated later than 1910, further comment would be speculative.

Sheets were further revised in ways consistent with all contemporary small-scale families, though, as with the others, decisions to make changes were not always reflected in changes to the sheets themselves. Reprint codes were added, initially in the number printed/year format, then from 1905 in a month/year format. Finally in 1923 a different number printed/year format was used. The change from "All rights of reproduction reserved" to "Crown Copyright Reserved" seems to have been implemented in mid-1912. Price changes affected the ten-mile map in 1905 with the introduction of mounted and folded copies, probably in 1906 when dissected copies were added to the range, and then after April 1915 when the price of dissected copies was increased. The general increase in prices introduced on 1 January 1920 caused the removal of the price from the map. Naturally these changes

---

7. I am most grateful to Brian Adams for pointing out the relevance of these lines.
8. Winterbotham (1936) 134.

Column numbers: as far as possible I have retained a common function for each column in lists throughout the book. No edition of the ten-mile map has entries in all columns. Sometimes they are irrelevant, and some would list unchanging features which are better noted as "Standard attributes". A few types of information change columns for convenience of organisation. The most important of these is the sheet price. Overprinted matter is marked ■. This includes most entries in columns 13 and 14: details in other columns occasionally supplied on overprint plates are here marked ■, and preceded by ■ in the lists. For further details, see p.82 (III), p.117 (IV), p.155 (V).

1. **Sheet number**, or **N**-orth, **S**-outh, **NS** back to back, **n**-orth of England strip map, **E**-ngland & **W**-ales sheet (some ■)
2. **State** (I): of **S**-outh, **M**-iddle, or **N**-orth sheet, **P**-artially completed state, **N**-ew **S**-eries
   Print code (III,IV,V). Reprint code (III), or OSI re-publication date (III.2). Print code (IV,V), in the form: base map:■overprint elements, ■overprinted code, or **p**-roof (perhaps with proof number or letter, or print code)
3. **Publication date**. Dates in I,II are estimated. III,IV,V: printed publication date of base map, or **n**-o **d**-ate. The place by default is **S**-outhampton to 1946 and from 1968, **C**-hessington 1948-1967, Dublin on Ireland maps. Exceptions are preceded **S** or **C**. For ■IV.2 dates included here, see col.13. V: **P**-rinted and published was used until 1958, **M**-ade and published from 1958. 1958 dates and exceptions are preceded **p** or **m**
   Legend codes. A double letter system is used. See p.82 for **a** (III) codes, p.117 for **b**, **c** (IV.1), **d** (IV.2), **q** (IV.2R), **e**, **f**, **X** (IV.3), **s**, **t** (IV.4) codes, p.155 for **g** (V.1), **p** (V.2), **h**, **j**, **k**, **r** (V.3), **u** (V.4) codes
   Sheet price, or **n**-o **p**-rice (IV,V). *NB*: I,III price is in col.8
4. **Railway revision**: see also col.5
5. **Reservoir revision**. *NB*: IV.3 information appears in col.4. Reservoir labels were deleted following a security directive in 1957: those in V.1 which precede or survive it are shown as + in col.8
6. **Minor correction date** (III,IV.1,V)
   Map revision (IV.3,4): see p.109 and p.140 for details
   Graticule intersections: + when present (IV.3), - when lacking (IV.3,V)
7. **E**-mbossed **P**-rinting **D**-ate (I: the earliest one known is quoted)
   Printed in Southampton date (III): derived from **f**-our **m**-ile, or **q**-uarter **i**-nch (given as ¼-inch) map
8. **Copyright statement** (I,III,IV.4,V). IV,V: **C**-rown **C**-opyright **R**-eserved, or **C**-rown **C**-opyright 19xx. This is ignored for IV.1,2,3. *NB*: ■IV.3 appear in col.13. V: ■ dates are quoted if later than the base map CC date
   Sheet price, or **n**-o **p**-rice (I,III). III: + or - the relevant altitude statement. V: **u**-nivers lettering, or **d**-igitally produced sheets from data stored on computer; additional + or - are explained in cols.5,6. *NB*: IV,V price is in col.3
9. **Grid information**: **A**-lpha **N**-umerical squares, **g**-ra-**t**-icule, **I**-rish **G**-rid, (War Office) **C**-assini **G**-rid, **CGg**: CG+ graticule, **N**-ational **G**-rid, **NGg**: NG+graticule, **NGm**-ilitary system, **NGmg**: NGm+graticule, National **Y**-ard **G**-rid, **YGan**: YG with marginal AN system, (some ■). Grid information is implicit in col.3 codes, so is omitted from IV.3,4,V. On IV.3 maps printed in 1942 the National Grid was referred to as the **O**-rdnance **S**-urvey **G**-rid
10. **Magnetic variation date** (some ■). V.1 Sheet 1: + corrected figures in "Difference from Grid North" panel
11. **Colour of base map**. Second letters (if used): on **f**-ilm, **h**-eavy (usually 152gsm) paper, **t**-racing paper, or the colour of grid if different from the base map: an upper case letter implies a grid in 100km squares. III,IV: additional symbols (+,-,x) concern the water plate if uncharacteristic: see p.119, and overleaf on colour codes
12. **Depiction of height**: **H**-ill **S**-haded, **O**-utline with **H**-ills, **O**-utline, **R**-elief map, or number of coloured layers, the number underlined if there are uncoloured layers below
    Colour of sea. This may be uncoloured paper. See note on colour codes on the reverse of this sheet
    ■**Number of overprint colours**. IV: + or - contours, if uncharacteristic: see p.119 for details
■13. **Overprint publication date. Price.** See col.3 for details and abbreviations, which may also be relevant here. IV.3 with **c** are ■copyright dates. Maps not for publication may be marked **f**-or **i**-nternal (or **o**-fficial) **u**-se **o**-nly. *NB*: ■III.1 and ■IV.2 details are given in col.3
■14. **Date of overprinted information**
15. **Print run. Date.** The information is taken as far as possible from the job files, and by default gives the number of good copies printed, and the date of printing. Less positive evidence comes from **E**-stimate, **R**-equest, History **C**-ard (issue date), **D**-ispatched information in job files, and PRO files, usually OS 1/432, 1/433, 1/999. An approximation is of course often implicit in the print code
16. **Cover type** by reference to my lists in Browne (1991) and Sheetlines Supplement 31, **x** if in unlisted cover
17. **O**-rdnance (or **G**-eological) **S**-urvey **P**-ublication **R**-eport (yyyy/q [quarterly], m/yy [monthly] reports). Since 1988 OSPR publication has postdated that of the maps: map publication dates are given. GSPR reports suffixed **g**
18. **Location of copies** (see list of abbreviations: "on sale" as in 1992)
19. **Notes**

Abbreviations

This reference system for ten-mile map families is maintained throughout the book. It appears at the top of each page:-
I.   Index to the one-inch maps: 1. South sheet, c.1817. 2. Middle sheet, c.1824. 3. North sheet, c.1881
II. 1. Ireland, 1838. 2. Ireland, 1868
III. 1. Great Britain, 1903. 2. Ireland, 1905
IV. 1. Three-sheet map, 1926. 2. Two-sheet map, 1932,1937. 3. 1:625,000, 1942. 4. 1:1,250,000, 1947, to "D"
V.   1. "Ten Mile" Map, 1955. 2. Physical Map, 1957. 3. RPM, from "B", 1965. 4. 1:1,250,000 "E", 1975
Appendix 1 to 6. (Appendix 2 concerns James's Map of the World. Appendix 3 concerns Indexes)

Titles as given here may be taken from different areas of the map: these elements are separated by /. Subdivisions of a title, if expedient, are shown by :. Overprinted title elements are marked ■: overprinting always ceases at /, and, if necessary elsewhere, with >. + titles are members of the Planning Series. Upper case and italic writing have been ignored except in specification tables and lists of changes where they may be relevant to specification change

Abbreviations
    Map libraries. Capital letter abbreviations for places: AB: Aberystwyth, B: Birmingham, BF: Belfast, BS: Bristol, C: Cambridge, D: Dublin, DR: Durham, E: Edinburgh, EX: Exeter, G: Glasgow, L: London, LD: Leeds, LE: Leicester, LV: Liverpool, M: Manchester, N: Nottingham, NT: Newcastle upon Tyne, O: Oxford, R: Reading, S: Sheffield, SO: Southampton, Y: York; also CPT: copyright collection, PC: private collection, RH: the author's collection (reprints only listed)
    Combined with lower case letters for collections, thus: BFpro: Public Record Office, BFq: The Queen's University; Cjc: St John's College, Cambridge; Da: Royal Irish Academy, Dki: King's Inns Library, Dna: National Archives, Do: Oireachtas, Dos: Ordnance Survey of Ireland, Dtc: Trinity College, Dtf: Freeman Library, Trinity College, Ebgs: British Geological Survey, Lbl: British Library, Lgh: Guildhall Library, Lkg: King's College, Lmd: Ministry of Defence Map Library, M.C.E., Tolworth, Lnmm: National Maritime Museum, Lpro: Public Record Office, Lraf: Royal Air Force Museum, Hendon, Lrgs: Royal Geographical Society, Lsa: Society of Antiquaries, Lse: London School of Economics, Ob: Bodleian Library, SOos: Ordnance Survey, SOrc: Royal Commission on Historical Monuments (England), SOrm: Record Map Library, Ordnance Survey. With g: university or college geography department, n: national (ie ABn of Wales, Dn of Ireland, En of Scotland, C-On: Ottawa, Canada), p: public, u: university
    Official organisations: BGS: British Geological Survey, DAS: Department of Agriculture for Scotland, DEP: Department of Employment and Productivity, DHS: Department of Health for Scotland, DOE: Department of the Environment, GS: Geological Survey (Great Britain), GSGS: Geographical Section, General Staff, GSGS (AM): Geographical Section, General Staff (Air Ministry), GSI: Geological Survey (Ireland), IBWO, IDWO: Intelligence Branch (later Department, then Division), War Office, IGS: Institute of Geological Sciences, LUS: Land Utilisation Survey of Britain, MAF[F]: Ministry of Agriculture [and] Fisheries [and Food], MFP: Ministry of Fuel and Power, MHLG: Ministry of Housing and Local Government, MOS: Ministry of Supply, MOT: Ministry of Transport, MOW: Ministry of Works, MTCP: Ministry of Town and Country Planning, MWT: Ministry of War Transport, OS: Ordnance Survey (Great Britain), OSI: Ordnance Survey (Ireland), QDWO: Quartermaster General's Department, War Office, SDD: Scottish Development Department, TDWO: Topographical and Statistical Department, War Office, TSGS: Topographical Section, General Staff, WO: War Office
    Railway Companies: B&D Jcn: Birmingham & Derby Jcn, CR: Caledonian, EC: Eastern Counties, EU: Eastern Union, GE: Great Eastern, GW,GWR: Great Western, I&B: Ipswich & Bury, L&B: London & Birmingham, L&C: London & Croydon, L&M: Liverpool & Manchester, L&NW: London & North Western, L&SW: London & South Western, L&Y: Lancashire & Yorkshire, LM&SR: London, Midland & Scottish, LT&S: London, Tilbury & Southend, M&C: Maryport & Carlisle, MD: Metropolitan District, MR: Midland, MS&L: Manchester, Sheffield & Lincolnshire, N&E: Northern & Eastern, NB: North British, SE: South Eastern, TV: Taff Vale, VN: Vale of Neath, WH&FR: Welsh Highland & Festiniog, WHR: Welsh Highland, WL: West London, WM&C: Wilsontown, Morningside & Coltness, WM&CQ: Wrexham, Mold & Connah's Quay, Y&N: Yarmouth & Norwich; also Ry: Railway, Ty: Tramway, Br: Branch, Ln: Line, Jcn,Jn,J: Junction
    Watermarks: APSL: A.P.& S.London, APSLB: A.P.& S.London Bodleian, HC: Hodgkinson & Co, JW: J.Whatman, JWTM: J.Whatman Turkey Mill, MK: Monckton Kent, RT: Ruse & Turners, TEW: T.Edmonds Wycombe, TEWNB: T.Edmonds Wycombe Not Bleached, THSL: Thos H.Saunders London, TJH: T.& J.Hollingsworth, TJHK: T.& J.H.Kent
    Others: anl,bnl,lnl,rnl: above, below, left of, right of neat line; tl,tc,tr,bl,bc,br: top (or bottom) left, centre, right; m-onth, q-uarter, y-ear; ARRR: All rights of reproduction reserved, CCR: Crown Copyright Reserved; gsm: grams per square metre; i-nitialled d-ate; NA: Notice to Airmen. Bibliographical and documentary abbreviations are on p.197

Colour codes appear in columns 11 and 12. The following are used throughout the cartobibliography: b-lue, g-rey (g* see IV.3), j-et (= black), k-rystal (or crystal) black (earlier broken black), o-range, r-ed, s-ienna (= brown), w-hite, y-ellow. These last two in effect mean the colour of the paper, and are noted when no sea spray colour is used.

With developments in printing colour combinations and screens, the number of overprint colours actually used is ever more difficult to assess until the general introduction of the trichromatic process in 1978, after which all colours have probably been derived from magenta, yellow and cyan, with black. It is now known as four-colour process printing.

would not physically have appeared on a map until it was next revised, and clearly they were often overlooked.

The map revisions themselves must form a future study. Here I was concerned solely with attempting to locate all the states of the maps I could find and set them in sequence. Without this and side by side comparison, map revisions are notoriously difficult to itemise, and the sources are far flung enough to make the job doubly difficult, especially when before this study began there was no clue as to how many states of these maps there were. The list is still not complete, but at least now the gaps are more obvious, and the work of comparative analysis, assisted by a list of consecutive map states, is more approachable. What has always been clear is that the OS organised its revision of the ten-mile as with the other small-scale families, and divided minor corrections, presumably of a topographical nature, from railway revision, and dated each alteration by month and year as they were effected. But because the OS dated these alterations, it was not necessary for me to identify them in order to list the states of the map. It may be, of course, that revisions were made without notice, but the *prima facie* situation here is quite the reverse to that of the nineteenth century map. This is true also for the map of Ireland.

2. Ireland

Since the initial moves towards the new ten-mile map of Ireland took place in Southampton, documentation is incomplete. We learned above of the abortive attempt in 1887 to make a composite ten-mile map of Great Britain and Ireland. The earliest known reference to a new map of the island lies in an EO's memorandum dated 1 October 1902,[9] wherein instructions were given that a graticule should be put on it, but there may well have been developments earlier than this. Application for Treasury approval was received on 1 December 1902.[10] Outline and hillshaded maps were both requested, provision for them being required in the OS estimates for 1903-4. Unfortunately the subsequent Treasury file has also disappeared, though permission was evidently granted. Almost the same specification was used as for Great Britain: the differences were slight, and caused by matters already traditional on the Irish map. Beside the graticule it had a graduated border, and province boundaries were necessary. As with the British map, it was reduced from the quarter-inch, itself a reduction of the one-inch. The original plan was to engrave it on two sheets, dividing just north of 53°20'N.[11] As in the case of Great Britain, this may have been to do with the exigencies of colour printing at the time, and limitations on the size of paper it was possible to employ.

It was more than a year before the news about the new Ireland map actually filtered through to Dublin. Colonel Sim (OID) decided for himself late in 1903 that the 1868 ten-mile map was desperately in need of refurbishment, and he submitted his proposals in a long memorandum to Southampton on 13 November for freshening up both topographical and index versions.[12] It was a somewhat embarrassed Hellard (EO) who had to break the news to Sim that he had been wasting his time. Drawing the new map was complete, and engraving of both its intended sheets well advanced. Water plates had been engraved in October, and the hills drawing, at least for Sheet 2, was completed by 29 September 1903. But Sim still required direction on the index version. Hellard told him that when engraving had been completed there would be no objection to a second duplicate being made, and to engrave on it the one-inch sheet lines and numbers if he considered there would be a satisfactory sale for it.

Sheet 2 was proved on 15 April 1904. Up to this point the Welsh coast had not been included: it was added to the proof in red, with directions to engrave it on to the plate. After this, we hear no more about the map until 30 June 1904, when Lieut. D.Champion Jones, for OE, reported to Hellard that:

The Director General told me this morning that he would like the Ten Mile Map of Ireland published in one sheet instead of two sheets.

It is at present in two sheets - but when matrices have been made from the originals - these matrices could be tentatively joined together by O. Stores, and the position of the plates relative to each other could be tested by OTT, and if found correct, a duplicate could be electrotyped.[13]

The index version was also to be on one sheet. Sim was certain there was enough demand for it, but suggested saving money by transferring it to zinc from the copper plate, before engraving sheet lines and numbers. Johnston's view was that few people would buy a 2/6d index in order to locate one or two 1/- maps, and that decisions should be delayed on it. Sim provided him with sales figures, and some fifty copies a year seems to have been adequate to persuade Johnston to proceed with the map - and on copper, not zinc.

The union of the two plates seems to have taken much longer than expected. It was achieved by taking matrices of the two original plates, joining them, and from this making a duplicate plate. Captain de Vitre (OE) was able to record success on 15 November 1904.[14] It

---

9. PRO OS 1/144 70B.
10. PRO T3 series, volume for 1892.
11. See the illustration on p.78.
12. NA OS 6/8540.
13. NA OS 6/9070. But OSR 1904 still showed the two-sheet format.
14. NA OS 6/9254. The plate itself is dated 12.11.1904.

No. 14.
INDEX
TO THE
ORDNANCE SURVEY OF IRELAND,
On the Scale of 10 Miles to an Inch.
Showing State of Preparation, 31st March, 1904.

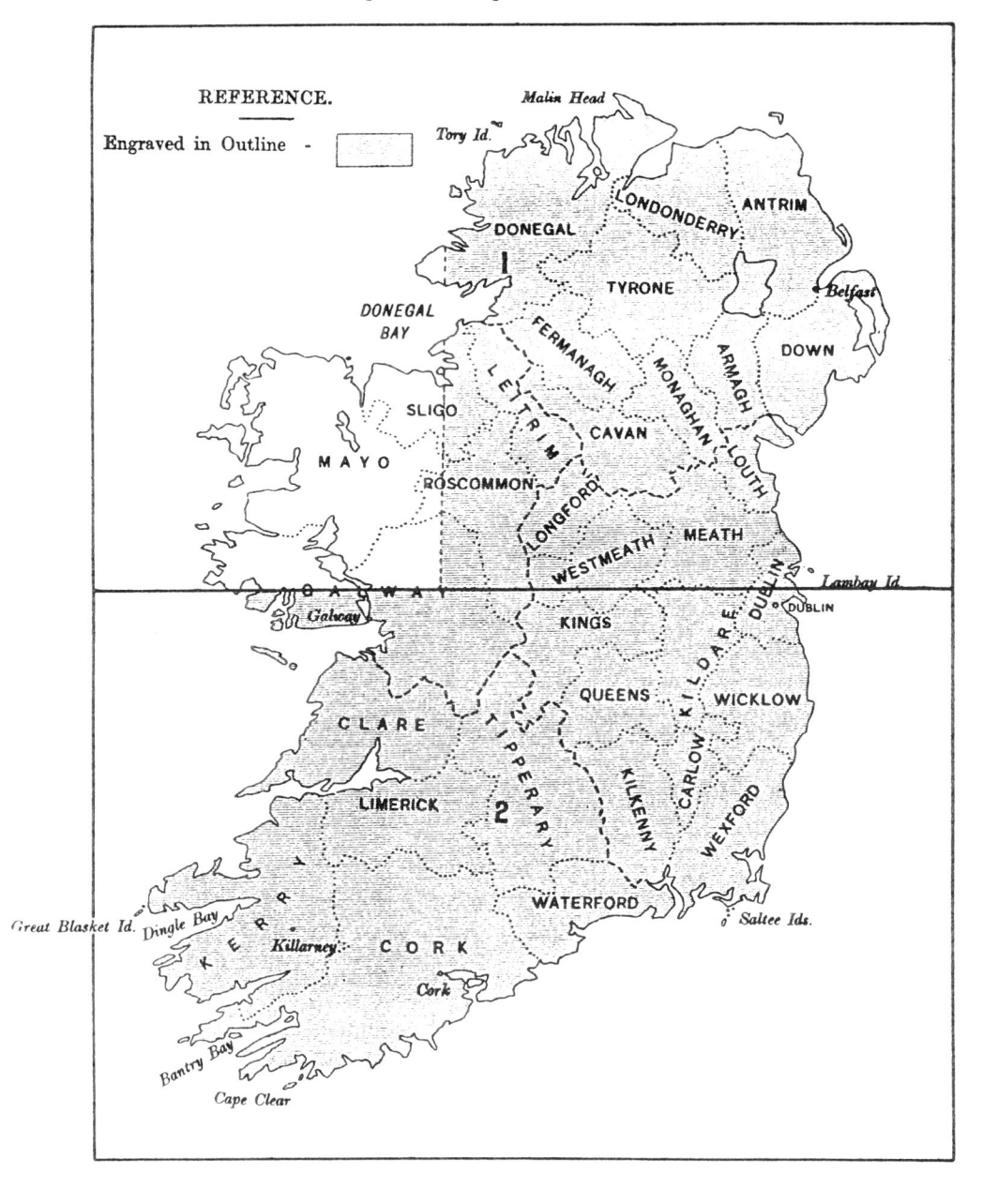

Plate 7. Intended sheet lines and numbers for the 1905 ten-mile map of Ireland. From OSR 1904.

was decided to use this plate for printing. A further duplicate was required from the matrices for the index version. This was completed on 16 December 1904, and a printing duplicate from it was made later. Meanwhile, transfers were taken of the topographical map plate, and it was sent to Dublin for printing on 24 November, arriving on 2 December. But a problem arose when OID noticed some altitudes on the new plate within a War Department prohibited area. He had them removed from the printing plate, but it caused a lengthy correspondence before similar corrections could be made on materials remaining in Southampton. Probably because Dublin lacked the facilities for large sheet colour printing, the hillshaded map was printed at Southampton.

As with the British map, this is not the place for a detailed study of map revision, since, at least in theory, alterations to the plate are noted in minor correction and railway revision dates. Publication of the map was almost coincident with the completion of the Irish railway network, and the additions made in November 1906 (Strabane-Letterkenny-Burtonport, Donegal-Ballyshannon, Armagh-Castleblayney, Ballyroney-Newcastle, Goolds Cross -Cashel, Waterford-Rosslare, Monkstown-Crosshaven) left only the Athy-Wolfhill and Kilkenny-Deerpark colliery lines outstanding. These revisions were usually present though in fact unrecorded. After the minor corrections of 1916, the only topographical revision seems to have been the occasional alteration to Irish placenames. Then in 1949 Shannon and Dublin Airports were drawn on **Biological Subdivisions** complete with runway layouts!

Ten-mile maps for the Geological Survey (GSI)[15]

Richard Griffiths's remarkable ten-mile geological map made in 1836 for the Railway Commissioners was noted on p.54. Then, in February 1869, a map showing all the large bogs in Ireland was mooted. Wilkinson suggested using either river catchments or topographical map base. Sir Henry James approved, but nothing seems to have been achieved[16]. First World War demands led Grenville Cole of the Geological Survey (GSI) to approach the OSI on 10 July 1917, as intermediary for the Department of Agriculture and Technical Instruction (Ireland) (DATII). He asked for about 600 copies of an outline ten-mile map, showing coalfields in black stipple, bogs in brown, railways in bold black, canals in blue and water partings in blue broken lines, heights to be added where necessary to show the fall of rivers. The map was for the DATII journal. The OSI accepted the commission, though the map could not be contoured as required. The resulting map would become known as the fuel map.

On 22 April 1918, Hill of DATII confirmed with Cole his need for six maps: the fuel map, one of river catchment areas and rainfall, probably in five tints of blue, based on the 1868 rivers map, and four of town areas. Hill wanted 3000 copies of each map, with more of the first, presumably for the DATII journal. Cole passed on the order to the OSI the following day, with his drawing of bogs on a ten-mile blue. He recommended using "Tint No 9 of the OS standard series" for overprinting bogs. Cole extended his order for the fuel map by a further 600 copies, with minerals information overprinted in a strong carmine red, for a memoir on the distribution of fuel supplies and economic minerals. The version with this additional overprint is known as the minerals map.

Proofs of the fuel map on black and grey bases were sent to Cole on 29 July 1918. He preferred the black, and suggested selling the map flat as a GSI publication. On 30 October, Cole supplied to the OSI a ten-mile index map with indications of mineral resources for them to overprint. In turn the OSI supplied Cole with a further proof of the fuel map on 12 November. He recommended printing 3000 copies for folding, 600 for sale flat by GSI, and 600 copies more for further overprinting as the minerals map. A price of 5/- was approved by Director General Sir Charles Close. Cole delayed the printing of the fuel map by retaining his proof until one with the minerals overprint was ready. Green and red versions were sent him on 22 March 1919: he preferred the green. Proceedings were then delayed for a year while the OSI awaited the return of proofs. Early in 1920 they took advantage of the delay by adding both the Athy-Wolfhill (opened 1918) and Kilkenny-Castlecomer lines. This was opened in 1919 and extended in 1920 to Deerpark Colliery in the south of the same coalfield.

It was decided on 6 March 1920 to treat both maps as GSI publications, and on 24 March 1920 the OSI announced printing costs. 3000 copies of the fuel map on 96lb paper with another 600 copies on 140lb paper were printed by 14 May at a cost of £30.8.8d. The estimate for an additional 600 on 96lb paper and 20 on 140lb paper, further overprinted in green with minerals information, totalled £14.11.8d. In the event it seems that all 620 copies were printed at the heavier weight, 120 of which were sent to GSI on 25 June 1920. Cole found the contrast between black base and green overprint unsatisfactory, and the paper too heavy. He wanted another 620 printed on a lighter paper for the memoir. OID replied on 26 July, suggesting a grey base, and selling the initial printing to the public straightaway. To him this became a first edition as soon as Cole asked for a further eight names to be added to the plate for the memoir version. OID supplied three more proofs on 70lb paper on 6 August, with minerals overprinted in black, carmine and green. Cole opted for the carmine and requested 620 copies. These were sent to GSI on 17 September 1920.

The situation was a distinctly confused one, and OID wrote to DATII on 11 July 1921 reviewing the stock situation, in an attempt to unravel it. Apparently neither map had yet reached the public, and though officially on sale since June 1920, did not do so until November 1921.

---

15. This section is based on NA OS 6/18000.
16. NA OS 5/3514.

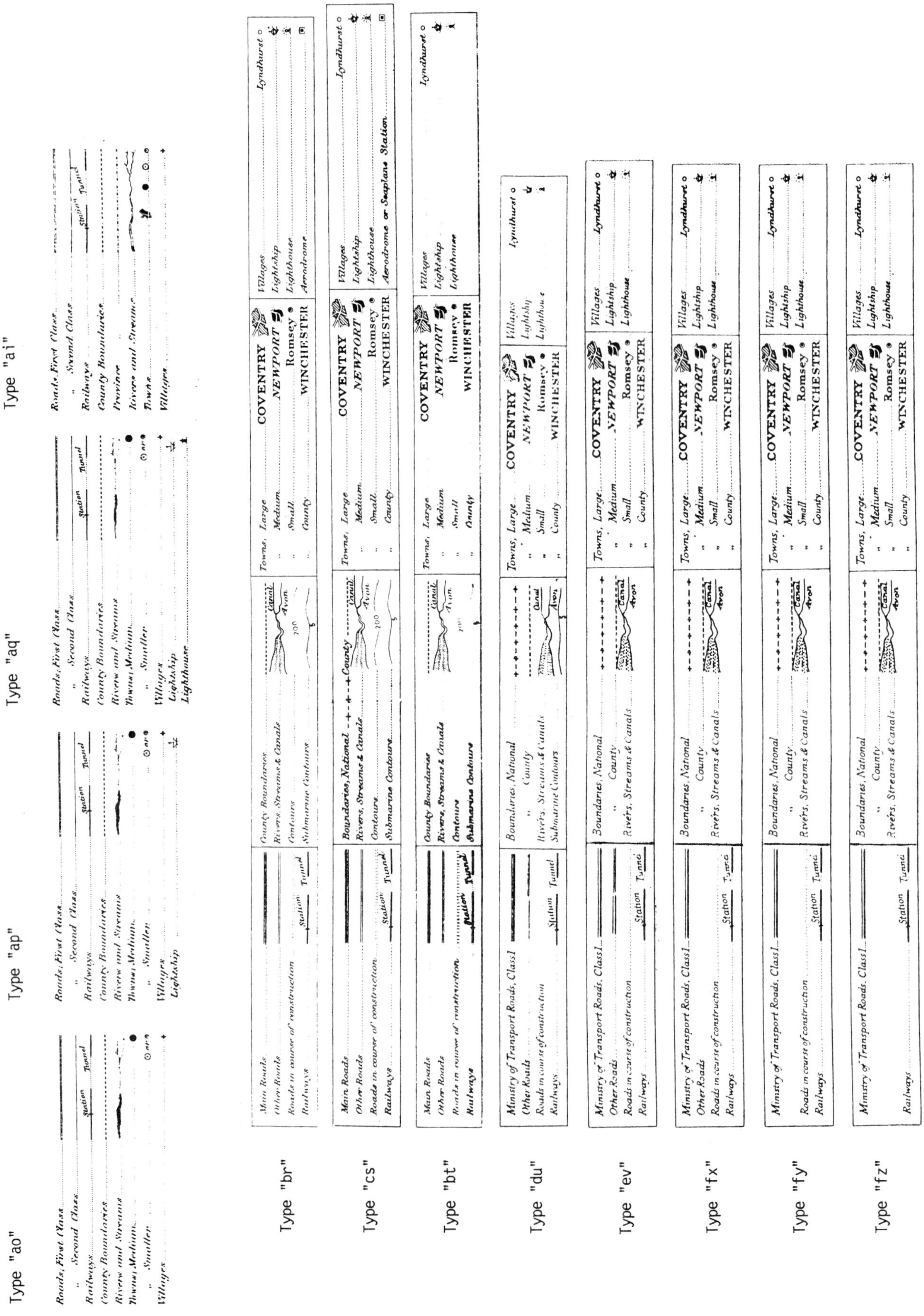

Plate 8. Legends on Ordnance Survey ten-mile maps, 1903-1946. Reduced to 90% of original size.

III   81

Standard attributes: Great Britain (III.1), Ireland (III.2) maps. *NB*: These may vary on OSI issues

Map dimensions: III.1 Sheets 1-8: 20ins (508mm) W-E by 13ins (330mm) S-N. Mapped area per sheet: 26,000 square miles
                III.1 Sheets 9-12: 20ins (508mm) W-E by 13½ins (343mm) S-N. Mapped area per sheet: 27,000 square miles
                III.2: 26½ins (673mm) W-E by 36ins (915mm) S-N. Area mapped: 95,400 square miles
Irish Grid co-ordinates: III.3: 10km E - 370km E; 10km N - 470km N. Area mapped: 165,600 square km

1. III.1: Neat line border, suitable for mounting together as a wall map. III.2: Map border showing latitude and longitude ticks and values at 10' intervals
2. Illuminated title: ORDNANCE SURVEY OF GREAT BRITAIN (III.1: Sheet 6, Outline Series), ORDNANCE SURVEY OF IRELAND (III.2, top left). Scale Ten Statute Miles to one Inch, and scale-bar (10+50 miles) are below titles
3. In the top margin:
    a. Top left corner: ORDNANCE SURVEY OF GREAT BRITAIN (III.1 only)
    b. Top right corner: SHEET 1 (or 2-12) (III.1 outline, rare on III.2). Combined sheet number (eg SHEETS 7. 8. (shown here as 7,8)) from 1904, then sheets renumbered 1-8, apparently after 1915 (probably III.1 hillshaded only)
4. In the bottom margin:
    a. Bottom left:
    1. Unboxed legend: "ao","ap","aq" (III.1), "ai" (III.2) types
    2. Collective reprint codes *m.yy m.yy* etc
    b. Next right: *Diagram shewing the Numbers of the adjoining Sheets* (III.1 only)
    c. Bottom centre:
    1. Price statement: III.1: usually, but not invariably, nearest neat line. Original sheet price: *unmounted 1/-*; first reprints (1905) added *mounted 1/6d*, and from c.1906 *or folded in Sections 2/-, net*. The folded in sections price rose to 2/6d, probably after April 1915. The price rises on 1 January 1920 to 1/6d, 2/6d and 3/6d caused the deletion of the price from the map. III.2 price (when present): 2/6d, furthest from neat line
    2. Scale of Ten Statute Miles to One Inch 1/633600 (III.1 only: see illuminated title above for III.2, where the scale statement has no representative fraction)
    3. Scale-bar in ten and unit miles (10+50 miles) (III.1 only: III.2 scale-bar is below title and scale statement)
    4. *Altitudes* statements measured at Liverpool (III.1), Dublin Bay (III.2) (sometimes lacking on overprinted maps). Some islands (eg Bardsey, Scilly) have additional altitudes statements of their own
    5. Copyright statement: altered from All rights of reproduction reserved, to Crown Copyright Reserved, not necessarily consistently, from mid-1912. After 1922, III.2 versions employ various OSI forms, initially Copyright Reserved, achieved by deleting "Crown" and leaving the remaining words off-centre
    d. Bottom right: revision and publication details and dates:
    1. *Published by the Director General, at the Ordnance Survey Office, Southampton* (III.1) (or *Dublin* (III.2)), *19xx*. OSI issues may read: *Published by the Director at the Ordnance Survey Office, Phoenix Park, Dublin. 19xx*
    2. *Reduced from the Four Mile Map* ("fm"); altered c.1914 to ¼-*inch Map* ("qi") (III.1)
    3. *Printed at the Ordnance Survey Office, Southampton in 1904* (III.1 rare, III.2 sometimes lacking)
    4. *Railways revised* (later *inserted* (III.1)) *to Month, 19xx*. The actual changes have not been listed
    5. *Minor corrections to Month, 19xx*. The actual changes have not been listed
    6. Standard *right of way* warning (often deleted from overprinted maps)
5. III.1: no graticule or reference systems; III.2: graticule across the map at one degree intervals
6. Symbols: see col."J" on p.9. Not all symbols used appear in the legend or **Characteristic Sheet**, 1900
7. Geographical, national and county names spanning two sheets are completed in the margin (III.1)
8. Coastal outline of Ireland and France (III.1) and Great Britain (III.2 - more generalised than on the 1868 map). The detail includes coastal placenames, with appropriate symbols, and coastal physical features

Colour plates (III.1,2):
1. Black for everything except:
2. Blue: rivers, canals, lake boundaries (not open water or coastline)
3. Blue or blue-green vignetted wash: coastal waters and inland open water
4. Light sienna (after August 1904 burnt sienna): main roads
5. Brown: hillshading
Outline editions: plates 1,2, in black only (III.2), 1,2,3 (III.1). Maps coded "OH" in col.12: plates 1,3,5

Lettering: As the **Characteristic Sheet**, 1900, except that the lettering for Roman sites and principal hydrographic names were not listed there. See col."J" on p.11 for a more detailed analysis

Covers types recorded: H.3.2a1, H.3.2b1, ?H.3.2b4, H.5, ?H.7.1b, H.7.2b. Some OSI covers also recorded. The outline map was advertised from 1907 as available dissected in covers, but none have been located. Probably no first states of the British hillshaded map exist in covers, with the earliest references to covered maps in the OSPR's coinciding mainly with the dates of the first reprints. The bottom four (and presumably the top four) maps were also mounted together and sold in covers entitled Great Britain North (or South) (?H.7.1b, H.7.2b).

Variable attributes (III.1,2,3)

1. Sheet number, **n**-orth (or **s**-outh) section
2. Collective reprint codes (m.yy: latest only quoted). Pre-1905 and post-1921: print run/year. III.2 is inconsistent
3. Publication date. Legend type "a" (III.1: "ao","ap","aq"; III.2: "ai": see illustration on p.80):
   - **ao** 8-line legend: "Roads; First Class", "Roads; Second Class", "Railways" (showing Station, Tunnel), "County Boundaries", "Rivers and Streams", "Towns; Medium", "Towns; Smaller" (two symbols), "Villages"
   - **ap** 9-line legend with "Lightship"
   - **aq** 10-line legend with "Lighthouse". The order of "Lightship" and "Lighthouse" symbols is reversible
   - **ai** 8-line legend, Irish system, with "Province" boundary. Four types of town symbol are depicted on one line, the three British types, and the general infill type not shown in the British legend
4. Railway revision date.   6. Minor correction date.   7. Southampton imprint date. Reduced from **fm** (or **qi**) map
8. Copyright statement: **A**: All rights of reproduction reserved, **B**: Crown Copyright Reserved, **C**: Copyright Reserved, and later OSI variants (III.2,3). + or − the relevant altitude statements (III.1: usually listed after price)
   Price: **v**: 1/−, **w**: 1/−,1/6d, **x**: 1/−,1/6d,2/−, **y**: 1/−,1/6d,2/6d, **z** (or **np**): no price (III.1); in figures (III.2)
9. Grid information (III.2,3).   11. Colour of base map. + or − blue water plate, if abnormal
12. **H**-ill **S**-haded, **O**-utline, **O**-utline with **H**-ills. Colour of sea. ■Number of colours in overprint
13. Issues printed **f**-or **o**-fficial **u**-se **o**-nly (III.2). ■13. Overprint publication date (III.3)
14■ Date of overprinted information.   16. Cover type.   17. OSPR.   18. Location of copies.   19. Notes

---

III.1. Great Britain, 1903

1	2	3	4	6	7	8	11	12	■14	16	17	18	19

Ordnance Survey of Great Britain [Outline Series]

1	-	1903ao	-	-	-:fm	Av+	j	Ob0	-	notes	3/03	CPT,Lrgs,Rg	
2	-	1903ao	-	-	-:fm	Av+	j	Ob0	-	notes	3/03	CPT,Lrgs,Rg	
3	-	1903ao	-	-	-:fm	Av+	j	Ob0	-	notes	3/03	CPT,Lrgs,Rg	

The Islands of St Kilda, Boreray, Soa and Dune are taken from Admiralty Surveys. They are outside the neat line

4	-	1903ao	-	-	-:fm	Av+	j	Ob0	-	notes	3/03	CPT,Lrgs,Rg	
5	-	1903ao	-	-	-:fm	Av+	j	Ob0	-	-	3/03	CPT,Lrgs,Rg	
5	2.06	1903ao	10.05	-	-:fm	Aw+	j	Ob0	-	-	-	Lse	
5,6							j	Ob0	-	note	9/07	not found	

"*5.6 now available "combined," printed on thin paper for mounting and folding. All the sheets of the Outline Series will be published, when reprinting, in a similar manner*" (OSPR 9/07). There are later reprints of Sheet 6, and others, on original sheet lines, so it remains unproven whether this, or the later renumbering of sheets, ever occurred

6	-	1903ao	-	-	-:fm	Av+	j	Ob0	-	notes	3/03	CPT,Lrgs,Rg	
6	1.10	1902ap	1.10	-	-:fm	Ax+	j	Ob0	-	notes	-	Lse	
7	-	1903ao	-	-	-:fm	Av+	j	Ob0	-	-	3/03	CPT,Lrgs,Rg	
7	1.06	1902ao	12.05	-	-:fm	Aw+	j	Ob0	-	notes	-	Lse	

III.1   83

1	2	3	4	6	7	8	11	12	■14	16	17	18	19
8	-	1903ao	-	-	-:fm	Av+	j	Ob0	-	-	3/03	CPT,Lrgs,Rg	
8	8.04	1903ao	-	-	-:fm	Av+	j	Ob0	-	notes	-	Cu	
8	12.08								-	notes	-	not found	
8	8.11	1903ao	8.11	8.11	-:fm	Ax+	j	Ob0	-	notes	-	Lse	
9	-	1903ao	-	-	-:fm	Av+	j	Ob0	-	notes	5/03	CPT,Lrgs,Rg	
9	1.11	1903ao	1.11	1.11	-:fm	Ax+	j	Ob0	-	notes	-	Lse	
10	-	1903ao	-	-	-:fm	Av+	j	Ob0	-	-	5/03	CPT,Lrgs	
10	9.03								-	-	-	not found	
10	5.06	1903ao	3.06	-	-:fm	Ax+	j	Ob0	-	notes	-	Lrgs	
11	-	1903ao	-	-	-:fm	Av+	j	Ob0	-	notes	5/03	CPT,Lrgs	
11	1.14	1903aq	12.13	-	-:fm	Bx+	jh	Ob0	-	notes	-	Lpro OS 1/10/4	
12	-	1903ao	-	-	-:fm	Av+	j	Ob0	-	-	5/03	CPT,Lrgs,Rg	
12	2.05	1903ao	7.04	-	-:fm	Av+	j	Ob0	-	notes	-	Lpro MR 934	

Ordnance Survey of Great Britain [hillshaded, or "C" Series]

1,2	-	1904ao	-	-	-:fm	Av+	j	HSb0	-	-	2/04	CPT,Lrgs	
1,2	5.07	1904ao	-	-	-:fm	Ax+	j	HSb0	-	notes	7/07	Sg	
1	5.07	1904ao	-	-	-:fm	Ay+	j	HSb0	-	notes	-	EXg	
3,4	-	1904ao	-	-	-:fm	Av+	j	HSb0	-	-	2/04	CPT,Lrgs	
3,4	5.07	1904ao	4.07	-	-:fm	Ax+	j	HSb0	-	notes	7/07	Sg	
3,4 2	5.11	1904ao	4.11	4.11	-:fm	Ax+	j	HSb0	-	notes	-	EXg,Lrgs	

The Islands of St Kilda, Boreray, Soa and Dune are taken from Admiralty Surveys. Sula Sgeir, Rona and the St Kilda group are outside the neat line. At least one later printing is probable

5,6	-	1904ao	9.03	-	-:fm	Av+	j	HSb0	-	-	2/04	CPT,Lrgs	
5,6	9.06								-	-	-	not found	
5,6	6.07	1904ao	4.07	-	-:fm	Ax+	j	HSb0	-	notes	7/07	Lrgs	
5,6	1.14	1904aq	12.13	12.13	-:qi	Bx+	j	HSb0	-	notes	-	not found	
3	1.14	1904aq	12.13	12.13	-:qi	By+	j	HSb0	-	notes	-	EXg,PC	

Sanda Island breaks the neat line. States from 6.07 have no eastern neat line

7,8	-	1904ao	10.03	-	-:fm	Av+	j	HSb0	-	-	2/04	CPT,Lrgs	
7,8	10.05	1904ao	3.05	-	-:fm	Aw+	j	HSb0	-	notes	12/05	Og	
7,8	11.07	1904ao	10.07	-	-:fm	Ax+	j	HSb0	-	notes	-	Lrgs,Og	
7,8	5.09	1904ao	4.09	-	-:fm	Ax+	j	HSb0	-	notes	-	Lse,PC	
7,8	5.09	1904ao	4.09	-	-:fm	Ay+	j	HSb0	-	notes	-	PC	
4	200/23	1904ao	4.09	-	-:fm	Ay+	j	HSb0	-	notes	-	EXg	
9	-	1904ao	8.03	-	1904fm	Av+	j	HSb0	-	-	2/04	CPT,Lrgs	
9	10.05	1904ao	-	-	1904fm	Aw+	j	HSb0	-	notes	11/05	Og	
9	8.08	1904ap	7.08	-	1904fm	Ax+	j	HSb0	-	notes	-	Sg,RH	
9	1.11								-	notes	-	not found	
9	3.13	1904aq	3.13	1.11	1904fm	Bx+	j	HSb0	-	notes	-	RH	
5	3.13	1904aq	3.13	1.11	1904fm	By+	j	HSb0	-	notes	-	EXg	
10	-	1904ao	9.03	-	-:fm	Av+	j	HSb0	-	-	2/04	CPT	
10	400/1904	1904ao	9.03	-	-:fm	Av+	j	HSb0	-	-	-	Og	
10	1.06	1904ao	12.05	-	-:fm	Aw+	j	HSb0	-	notes	6/06	PC	
10	3.08	1904ao	2.08	-	-:fm	Ax+	j	HSb0	-	notes	-	Sg	
10	8.11	1904ao	8.11	-	-:fm	Ax+	j	HSb0	-	notes	-	PC	

1	2	3	4	6	7	8	11	12	■14	16	17	18	19
10	3.14	1904aq	3.14	-	-:qi	Bx+	j	HSb0	-	notes	-	PC	
6	3.14	1904aq	3.14	-	-:qi	By+	j	HSb0	-	notes	-	PC	
6	12.21	1904aq	3.14	-	-:qi	Bz+	j	HSb0	-	notes	-	EXg,RH	

Smith's Knoll and Outer Gabbard lightships break eastern neat line

11	-	1904ao	9.03	-	1904fm	Av+	j	HSb0	-	-	2/04	CPT,Lrgs	
11	10.05	1904ao	4.05	-	1904fm	Aw+	j	HSb0	-	notes	11/05	Sg	
11	7.08	1904ao	8.08	-	1904fm	Ax+	j	HSb0	-	notes	-	RH	
11	1.14	1904aq	12.13	-	1904fm	Bx+	j	HSb0	-	notes	-	RH	
7	200/23	1904aq	12.13	-	1904fm	Bz+	j	HSb0	-	notes	-	PC	

A Sheet 11 or 7 "By" version is possible, but not found

12	-	1904ao	9.03	-	-:fm	Av+	j	HSb0	-	-	2/04	CPT,Lrgs	
12	5.05	1904ao	8.04	-	-:fm	Av+	j	HSb0	-	notes	11/05	Og,Sg	
12	9.07	1904ao	9.07	-	-:fm	Ax+	j	HSb0	-	notes	-	RH	
12	6.10	1904ao	6.10	-	-:fm	Ax+	j	HSb0	-	notes	-	RH	
8	6.10	1904ao	6.10	-	-:fm	Az+	j	HSb0	-	notes	-	EXg,PC	

A Sheet 12 or 8 "Ay" version is possible, but not found

Extracts of the hillshaded map published, eg of Sheet 3,4 in OS Indexes c.1905-6, and Close (1913), or of Sheet 10 (renumbered 6) in DOSSSM [1919],1920,1921,1923 and Cox (1924)

Indexes to these maps

Index to the Ordnance Survey of Great Britain on the Scale of Ten Miles to an Inch: 31st December, 1903 / Published in 12 sheets, 1/- each. (Scale 85 miles to 1 inch, in OSC 1903, showing the twelve-sheet layout only)

Index to the Ordnance Survey of Great Britain on the Scale of Ten Miles to an Inch: 31st December, 1904. Other issues dated 31st December, 1905, 31st December, 1906, and undated. (Scale 85 miles to 1 inch, in OSC 1904, etc, showing the twelve-sheet layout, with pecked lines of the eight-sheet layout superimposed)

Index to the Ordnance Survey of Great Britain on the Scale of 1/10-inch to 1 Mile (Scale 85 miles to 1 inch, in OSC 1914, showing the twelve-sheet layout, with pecked lines of the eight-sheet layout superimposed)

Index to the Ordnance Survey of Great Britain on the Scale of 10 Miles to 1 Inch (Scale 85 miles to 1 inch, in DOSSSM [1919],1920,1921, and with title reset in DOSSSM 1923 and OSC 1924, showing the eight-sheet layout)

A larger and more detailed index was advertised in DOSSSM [1919]-1923, and OSC 1924, at 2d (not found). Preparation diagrams are in OSR 1902-04.

Supplement 1. Special sheets

■Royal Commission on Canals and Waterways: Plate No 1 / ■**Map of the Canal Systems and Navigable Rivers of England and Wales**

[1]	-	1906ao	-	-	-:-	Az+	gh	Ob5	nd	-	-	PC	1
[1]	-	1906ao	-	-	-:-	Az+	g	Ob5	31.7.06	-	-	BPP	2
2	-	1906ao	-	-	-:-	Az+	gh	Ob5	nd	-	-	PC	1
2	-	1906ao	-	-	-:-	Az+	g	Ob5	31.7.06	-	-	BPP	2

In two sections, dividing at Crewe. Parts of Sheets 5-12. With an inset map at 1:126,720 of Birmingham

1. Draft maps with red title instead of black and darker coloured base map. Its county boundaries were deleted for the final version, and the national border altered to a blue hatched line
2. In *Royal Commission on Canals and Waterways: Volume I, Part II* (BPP(HC) 1906 [Cd 3184], XXXII, 613,615)

Royal Commission on Canals and Waterways: Plate No 2 - see Ireland (III.2)

■Royal Commission on Canals and Waterways: Plate No 3 / ■**Map of the Canal Systems and Navigable Rivers of Scotland**

| - | - | 1907ao | - | - | -:- | Az- | s | OHb3 | nd | - | - | BPP | |

III.1.Supplement 1   85

| 1 | 2 | 3 | 4 | 6 | 7 | 8 | 11 | 12 | ■14 | 16 | 17 | 18 | 19 |

Parts of Sheets 3-8. In *Royal Commission on Canals and Waterways: Volume III: England and Wales, and Scotland* (BPP (HC) 1907 [Cd 3718], XXXIII Part 1, 1303)

■Royal Commission on Canals and Waterways: Plate No 4: **Map of the Canals & Navigable Rivers in the Catchment Basins of England and Wales**
[1]   -   1908ao   -   -   -:-   Az+   g   OHb6   nd   -   -   BPP
 2   -   1908ao   -   -   -:-   Az+   g   OHb6   nd   -   -   BPP
 In two sections, dividing at Crewe. Parts of Sheets 5-12. With an inset map at 1:126,720 of Birmingham. Full title on north sheet only. In *Royal Commission on Canals and Waterways: Volume IV* (BPP(HC) 1907 [Cd 3719], XXXIII Part 2, after 510. *NB*: This volume has no parliamentary pagination)

■Royal Commission on Canals and Waterways: Plate No 5: **Map of the Canal Systems and Navigable Rivers of England and Wales: showing Railway Owned, Railway Controlled and Independent Canal Systems, and Tidal Portions of Navigable Rivers**
[1]   -   1908-   -   -   -:-   Az+   g   Ob3   nd   -   -   BPP
 2   -   1908-   -   -   -:-   Az+   g   Ob3   nd   -   -   BPP
 In two sections, dividing at Crewe. Parts of Sheets 5-12. With an inset map at 1:126,720 of Birmingham. No title on Sheet 2. In *Royal Commission on Canals and Waterways: Volume IV* (BPP(HC) 1907 [Cd 3719], XXXIII Part 2, after 510)

■Royal Commission on Canals and Waterways: Plate No 6: **Plan of Canal Routes**
■Royal Commission on Canals and Waterways: **Plan of Canal Routes A, B, C, and D** [BPP version]
 -    -   1909-   -   -   -:-   Az+   g   Ob6   nd   -   -   PC   1
 -    -   1909-   -   -   -:-   Az+   g   Ob6   nd   -   -   BPP   2
 Parts of Sheets 7-10
 1. Draft map version. Unlike the final version, the canal routes are numbered
 2. In *Royal Commission on Canals and Waterways: Volume VII* (BPP(HC) 1910 [Cd 4979], XII, 163)

Map No 1: Appendix No XLIV
 -    -   1909-   -   -   -:-   Az+   j   Ob4   nd   -   -   BPP
 Title, with tables, to the left of the map. Parts of Sheets 7-12. In *Royal Commission on Coast Erosion and Afforestation: Volume III Part II: Appendices* (BPP(HC) 1911 [Cd 5709], XIV, 709)

Map of Scotland showing Places in which Properties of Congregations have been allocated, by The Churches (Scotland) Act Commission
 -    -   1910-   -   -   -:-   Az-   j   Ob2   nd   -   -   BPP
 Parts of Sheets 3-8 (1 and 2 as insets). In *Churches (Scotland) Act Commission: Report of the Royal Commissioners appointed under the Churches (Scotland) Act 1905: Volume II* (BPP(HC) 1910 [Cd 5061], XIII, 551)

Map showing Controlled Canals in England & Wales: Canal Control Committee: Board of Trade
 -    -   1918-   -   -   -:-   -+:np   g   Ob3   11.18   -   -   book
 In *Canal Control Committee (Board of Trade) Handbook on Canals* (Second Edition, London, HMSO, November 1918 (4000 12/18)). Parts of Sheets 7-12. Back to back with the corresponding map of Ireland (qv)

Map of Eastern District of Scotland: to accompany Report of Rural Transport (Scotland) Committee
 -    -   1919-   -   -   -:-   ---:np   j-   HSb4   nd   -   -   BPP
Map of Central and Southern Districts of Scotland: to accompany Report of Rural Transport (Scotland) Committee
 -    -   1919-   -   -   -:-   ---:np   j-   HSb4   nd   -   -   BPP
Map of Northern District of Scotland: to accompany Report of Rural Transport (Scotland) Committee
 -    -   1919-   -   -   -:-   ---:np   j-   HSb5   nd   -   -   BPP
 The three maps (parts of Sheets 3-6, 5-8, 3-6) are in *Committee on Rural Transport (Scotland): Report of the Rural Transport (Scotland) Committee* (BPP(HC) 1919 [Cmd 227], XXX, 95,109,127)

Map of Orkney and Shetland: to accompany Supplementary Report of Rural Transport (Scotland) Committee
 -    -   1901ao   -   -   -:fm   Az+   j-   HSb0   -   -   -   BPP
 Parts of Sheets 1-4. This map presents a classic example of the "beware OS dates" maxim, 1901 being three years prior to the publication date of the parent sheets. It was actually published in 1920 in *Committee on Rural Transport (Scotland): Supplementary Report of the Rural Transport (Scotland) Committee* (BPP(HC) 1920 [Cmd 987], XXIV, 925)

1	2	3	4	6	7	8	11	12	■14	16	17	18	19

■Water & Power Resources: Map of Area 3: Plate IV

| - | - | ■1920- | - | - | -:- | --:np | g | Ow2 | - | - | - | book | |

Part of Sheet 9. In *Board of Trade: Final Report of the Water Power Resources Committee* (London, HMSO, 1921)

Fire Brigades: Map prepared by the Royal Commission on Fire Brigades and Fire Prevention showing Fire Brigades maintained in Counties (London, Lanark and Renfrew) and Boroughs and Urban Districts

| [n] | - | - | - | - | -:- | - | g | Ow2 | nd | - | - | BPP | |
| [s] | - | 1922ao | - | - | -:- | Bz- | g | Ow2 | nd | - | - | BPP | |

In two sections, dividing at Manchester. Parts of Sheets 5-12. The title is on the top half, publication details on the bottom. Prepared from Returns furnished by the Local Authorities in 1921-22. The scale-bar is on the title area. In *Report of the Royal Commission on Fire Brigades and Fire Prevention* (BPP(HC) 1923 [Cmd 1945], XI, 441,443)

Aeroplane Raids

| - | 3084/30 | 1930- | - | - | -:- | --:np | j | Ow2 | 1914-1916 | - | - | book | |

Map No 1 in H.A.Jones *The War in the Air* Volume 3, Maps (OUP, 1931). Parts of Sheets 10,12

■Airship Raid maps

| - | 3084/30 | 1930- | - | - | -:- | --:np | j | Ow1 | 1915-1916 | - | - | book | 1 |
| - | 3084/30 | 1930- | - | - | -:- | --:np | j | Ow2 | 1915-1916 | - | - | book | 2 |

In H.A.Jones *The War in the Air* Volume 3, Maps (OUP, 1931). Various areas, dealing with 1915-1916 raids
1. Map Nos 2,4,5,6,7,8,9,10,11,12,14,15,16,17,20,22,23,24,25,26,27,28,29,32,33,35,36,37,40,42
2. Map Nos 13,18,19,21,30,34,38,39,41

■Airship Raid maps

| - | 3100/35 | 1935- | - | - | -:- | --:np | g | Ob2 | 1917-1918 | - | - | book | |

Map Nos 1,3,4,7,8 in H.A.Jones *The War in the Air* Volume 5, Maps (OUP, 1935). Various parts of Sheets 8,10,12

■Aeroplane Raid maps

| - | 3100/35 | 1935- | - | - | -:- | --:np | g | Ob2 | 1917 | - | - | book | 1 |
| - | 3100/35 | 1935- | - | - | -:- | --:np | g | Ob4 | 29.9.1917 | - | - | book | 2 |

In H.A.Jones *The War in the Air* Volume 5, Maps (OUP, 1935). Various areas, dealing with 1917-1918 raids
1. Map Nos 10,13,14,15,16,19. Parts of Sheets 10-12.   2. Map No 17. Parts of Sheets 10,12

Supplement 2. Military issues

Ordnance Survey of Great Britain: Shewing Military Districts

1	-	1904-	-	-	-	A-:3/-	j	Ow1	nd	-	2/04	CPT	
2	-	1904-	-	-	-	A-:3/-	j	Ow2	nd	-	2/04	CPT	
3	-	1904-	-	-	-	A-:3/-	j	Ow2	nd	-	2/04	CPT	

Great Britain covered in three sheets, as with the military maps based on Old Series mapping (see p.52). This is an outline map, showing towns, rivers and railways, but lacking roads and altitude representation. (WO 1889/102)

Ordnance Survey of Great Britain / [Telegraph and Telephone Map]

-	-	nd:ao	-	-	1906-	Az+	j	Ow3	nd	-	-	Lbl,Lpro	1
-	-	nd:ao	-	-	1906-	Az+	j	Ow3	nd	-	-	Lbl,Lpro	1
-	-	nd:ao	-	-	1906-	Az+	j	Ow3	nd	-	-	Lbl,Lpro	2
-	-	nd:ao	-	-	1906-	Az+	j	Ow3	nd	-	-	Lbl,Lpro	2

With lists of Principal Head-Offices, Telegraph and Telephone Transmitting Offices, and Principal Testing Stations With an inset map at 1:253,440 of the Glasgow and Edinburgh area. The PRO copies are at CO 700 Misc.26. (WO 1889/108)
1. TSGS 2027. Composites of Sheets 1-4, and Sheets 5,6 with parts of Sheets 7,8. Scotland only is overprinted
2. TSGS 2026. Composites of parts of Sheets 5,6 with Sheets 7,8, and Sheets 9-12. England only is overprinted

NB: Two sheets of the OS map reportedly published in 1916 overprinted with reference squares has not been found

III.2   87

III.2. Ireland, 1905

| 1 | 2 | 3 | 4 | 6 | 7 | 8 | 9 | 11 | 12 | 13 | ■14 | 16 | 17 | 18 | 19 |

NB: Some fourteen plates and other source materials survive. The plates are at Dna OS 106. Most were made for the map as originally envisaged, in two sheets, and comprise water plates dated 10.1903, topographical and hills plates dated 8.1905, ie after the publication of the map in one sheet! There are also four full plates following the fusion of the two halves, for outline, hills and index versions. Their dimensions are 818mm W-E by 1065mm S-N.

Still within OSI (in 1991) are some of the original Sheet 2 drawings and proof sheets, including a model from the 1868 map, a blue reduced from the quarter-inch map, a hills drawing by Malings dated 29.9.1903, and a graticule and border sheet. One of the proofs, dated 15.4.1904, has the Welsh coast added in red, with instructions to engrave.

NB: Modern island maps, mostly at 1:500,000 or 1:575,000, are derived from the ten-mile map, but are not listed here

Ordnance Survey of Ireland [hillshaded map]
```
-    -       1905ai  -      -      1905fm  A+:2/6  gt  j   HSb0  -   -   notes  5/05  CPT,Lrgs,Sg
-    2.07    1905ai  11.06  -      1905fm  A+:2/6  gt  j   HSb0  -   -   notes  -     Lrgs
-    11.16   1904ai  11.06  -      1904fm  A+:2/6  gt  j   HSb0  -   -   notes  -     NTg,RH
-    250/40  1904ai  11.06  -      1904fm  A+:np   gt  j   HSb0  -   -   x      -     EXg
1    -       1944ai  1944   -      -:fm    A+:2/6  gt  j   HSb0  -   -   x      -     Lrgs,Sg
-    -       1948ai  1947   -      1904fm  A+:np   gt  j   HSb0  -   -   x      -     PC
```
Published by OS, and OSI since 1922. Price increased to 4/- on 1.1.1920. Also issued in OS and OSI covers. The edition with "Sheet 1" derives from the outline plate. Inland open water on OSI issues is on the blue plate

Ordnance Survey of Ireland [outline map]
```
1    -       1904ai  -      -      -:fm    A+:2/6  gt  jh  Ow0  -  -  notes  6/05  CPT,Lrgs
1    1919    1904ai  -      -      -:fm    B+:2/6  gt  jh  Ow0  -  -  -      -     Lpro Air 5/769  1
1    1920    1904ai  -      -      -:fm    B+:3/-  gt  jh  Ow0  -  -  -      -     Lpro Air 5/769  2,3
1    1920    1904ai  -      -      -:fm    B+:np   gt  jh  Ow0  -  -  -      -     note            4
1    1921    1904ai  -      -      -:fm    B+:np   gt  jh  Ow0  -  -  -      -     note            4
1    1922    1904ai  -      -      -:fm    B+:3/-  gt  jh  Ow0  -  -  -      -     BFq
-    11.16   1904ai  11.06  -      1904fm  A+:np   gt  jh  Ow0  -  -  -      -     BFpro OS 8/13  5
-    250/40  1904ai  11.06  -      1904fm  A+:np   gt  j+  Ow0  -  -  -      -     Lrgs
1    -       1942ai  -      -      -:fm    A+:2/6  gt  j   Ow0  -  -  -      -     NTg
```
Published by OS, and OSI since 1922. Conceived as a two-sheet map, hence the appearance of a sheet number (those without derive from the hillshaded map plate), but it was decided to publish on one sheet, and to make an index version. Also issued dissected in covers. Price increased to 3/- on 1.1.1920. The Northern Ireland portion of this map also appears divided on to three pages in *The National Trust Maps* (London, National Trust, 1957), with green overprint
  1. Showing (ms) brigade, divisional boundaries.   2. Showing (ms) various headquarters, dropping stations
  3. This base map also used for the murders maps (qv).   4. Known with military additions (see Supplement 1)
  5. ?Printed 1939 (this copy has an office record copy stamp of the Ministry of Finance, Ordnance Survey, 13.2.1939)

Ordnance Survey of Ireland: Index to the General Map Published on a Scale of One Inch to a Statute Mile
```
1    -       1905ai  -      -      -:fm    A+:2/6  gt  j   Ow0  -  -  -      6/05  CPT,Dn,SOrm
```
Uncoloured outline, with one-inch sheet lines. Lbl copy in Outlines 1899-1902

Chart showing Positions of Lighthouses, Fog Signals and other Sea Marks, under the Jurisdiction of the Commissioners of Irish Lights / Ordnance Survey of Ireland [top left 3mm anl]
```
-    -       1905ai  -          -  -:fm  A+:np   -  j  Ob2  fouo  nd    x  -  Dn ILB 6279
-    ?1911   1905ai  ?11.06     -  -:fm  ?B+:np  -  j  Ob2  fouo  1911  -  -  not found      1
-    1925    1905ai  -          -  -:fm  B+:np   -  j  Ob2  fouo  1926  x  -  Lnmm           2
```
With a neat line border, extended 78mm W, 284mm E, 41mm N, 7mm S, to enclose an extensive title and tables of sea marks, etc. Graticule, British coastline and original references to lights deleted, Atlantic Ocean moved, scale statement and scale-bar moved to bottom centre, and other bottom marginalia moved outwards. The Irish Light Commissioners requested the map of OSI in 11.1904, before completion of the base map (NA OS 6/9254). 1911 would be early for CCR on a map, but an improbable alteration in 1926. Later editions are by Geographia, c.1967, OSI (at 1:575,000), 1985
  1. ?"1911" added at end of title. ?With revised tables. ?With the 11.06 railway revision
  2. "1926" added at end of title. With revised tables and headings

1	2	3	4	6	7	8	9	11	12	13	▪14	16	17	18	19

▪Royal Commission of Congestion in Ireland: A.B.C. Map referred to in the evidence of Mr Henry Doran
- - nd:ai - - 1906- A-:np gt j Ow2 - nd - - BPP

In *Royal Commission on Congestion in Ireland: Appendices to the First Report* (BPP(HC) 1906 [Cd 3267], XXXII, 959)

▪Viceregal Commission on Arterial Drainage, 1907 / Ordnance Survey of Ireland: Catchment Basins
- - nd:ai - - 1907- A+:np gt g OHb7 - nd - - BPP

With many tables. Features of the title are from the 1868 map. In *Vice-Regal Commission on Arterial Drainage (Ireland): Appendix to the Report of the Arterial Drainage Commission (Ireland), 1905* (BPP(HC) 1907 [Cd 3467], XXXII, 489)

Ordnance Survey of Ireland: Shewing Locality of and Expenditure on Works, etc., connected with Fisheries, 1891
- - 1904ai - - 1907fm A+:2/6 gt j Ow6 - nd - - BPP

With a table, and a note: Reference added for the Royal Commission on Congestion 1906-7. In *Royal Commission on Congestion in Ireland: Appendix to the Fourth Report* (BPP(HC) 1907 [Cd 3509], XXXVI, 171)

▪Royal Commission on Canals and Waterways: Plate No 2 / ▪Map of the Canals and navigable rivers of Ireland
- - nd:ai - - 1907- A+:np gt s OHb4 - 7.07 - - BPP

In *Royal Commission on Canals and Waterways, Volume II Part II: Ireland* (BPP(HC) 1907 [Cd 3717], XXXIII Part 1, 403)

Ireland:.....Electoral Divisions scheduled as Congested
- - nd:ai - - 1907- A-:np gt j Ow1 fouo nd - - Dn 16.B.5 (13)

The map and margin dimensions are reduced, and the legend is brought within the plain border. ARRR is in the bottom left hand corner. All the electoral divisions scheduled as congested are coloured blue. Another copy at Dn 16.B.5 (14)

▪Royal Commission on Congestion in Ireland / Map of Ireland: Showing the Travelling, the principal Tours of Inspection, and the Country Sittings of the Commission
- - nd:ai - - 1908- A-:np gt j Ow3 - nd - - BPP

With a table. In *Royal Commission on Congestion in Ireland: Final Report* (BPP(HC) 1908 [Cd 4097], XLII, 857)

Ordnance Survey of Ireland: Showing County Boundaries, Rural Districts
- - 1908ai - - -:fm A+:1/6 gt jh Ow0 note - - - Dn 16.B.16 (5)

One hundred copies supplied in error with topography to the Local Government Board 10.1908. They were put on sale, appearing in OSC 1912 at 1/6d, increased on 1.1.1920 to 2/6d ("showing Administrative Boundaries"). See p.90. There is another copy at Dn 16.K.18 (7)

Ordnance Survey of Ireland: ▪Resident Magistrates. September 1909
- - 1909ai - - -:fm A+:np gt j Ow3 fouo 9.09 - - Dn 16.K.18 (18)

Earlier editions of this map were on the 1868 base. See II.2 on p.68.

Map showing Controlled Canals in Ireland: Canal Control Committee: Board of Trade
- - nd:- - - 1918- --:np gt g Ob3 - 11.18 - - book

In *Canal Control Committee (Board of Trade) Handbook on Canals* (Second Edition, London, HMSO, November 1918 (4000 12/18)). Back to back with the corresponding map of England and Wales (qv)

Ordnance Survey of Ireland: ▪[Map showing the numbers murdered in each county up to 1st December 1920]
1 1920 1904ai - - -:fm B+:3/- gt j Ow3 fouo 1.12.20 - - not found

Requested by the Lord Lieutenant on 2.12.1920 and delivered on 24.12.1920 (NA OS 6/19806). Three copies were made, shaded white through blue to red by counties to show the numbers of murders to have occurred there

Ordnance Survey of Ireland: ▪Map showing places where police, military, civilians have been murdered up to 1st December 1920
1 1920 1904ai - - -:fm B+:3/- gt j Ow4 fouo 1.12.20 - - Do

General note as above. Three copies were made with black outline flags at the places where police (coloured red), military (blue) and civilians (green) were murdered. Each flag carries the number of murders that had taken place. This was probably the first copy made, since pencil circles ring each location, linked by a line to the edge where the information to be added was detailed. Additional placenames not present on the base map were applied with a stamp. Otherwise, the flags, title and legend are entirely hand drawn and coloured on this copy, and probably on the others

III.2   89

1   2   3   4   6   7   8   9   11 12   13   ■14   16   17   18   19

   Geological Survey of Ireland: Map showing the distribution of Peat-bogs and Coalfields in Ireland
-   1920   1905ai -   -   -:-   B+:5/- gt j   Ow3   -   1907   -   -   on sale
     Published by GSI. Known as the Fuel Map. On the **Index to the General Map**, with scale statement and scale-bar moved
to the bottom margin, and the western border brought in 25mm. Bog outlines etc, inserted by GSI, 1918. Canals and
railways enhanced (including legend), Athy-Wolfhill and Kilkenny-Deerpark added, though this latter is sometimes to be
found incompletely deleted. 3000 copies were printed on 96lb, and 600 on 140lb paper (copies Dn,Eg,Og)

   Geological Survey of Ireland: Map showing the distribution of Peat-bogs and Coalfields in Ireland: Also the principal localities of Minerals of economic importance
-   1920   1905ai -   -   -:-   B+:5/- gt jh   Ow4   -   1907   -   -   Dn,Lrgs,Lse,Ng   1
-   1920   1905ai -   -   -:-   B+:5/- gt g   Ow4   -   1907   -   -   book   2
-   1920   1905ai -   -   -:-   B+:5/- gt j   Ow4   -   1907   -   -   IPP   3
-   1920   1905ai -   4.56 -:-   B+:5/- gt j   Ow4   -   1907   -   -   Dtf,Lrgs   4
     Known as the Minerals Map. General details as above
     1. With silvermines overprinted in green. 620 copies printed on 140lb paper, unwanted by GSI, put on public sale
     2. With silvermines overprinted in carmine. Eight more locations named. In *Memoir and Map of Localities and Minerals of Economic Importance and Metalliferous Mines in Ireland* (Dublin, Stationery Office, 1922). 620 copies printed
     3. With silvermines overprinted in green. This is Map 3 in *Coimisiún na gCanálach agus na mBóthar Uisce Intíre (Canals and Inland Waterways Commission): Report: July 1923* (IPP 1923, III, 609 (250 11.23)). Report title and map number are in the top margin. This version contains the eight extra locations recorded on the carmine edition
     4. With silvermines overprinted in magenta. Also accompanying the 1956 Memoir reprint. Title elements reordered

   Ordnance Survey of Ireland
-   -   1923-   -   -   -:-   --:np   gt j   Ow5   -   nd   -   -   book
     Printed at OSO, Dublin. Sheet size reduced: scale and scale-bar survive in the bottom margin. The counties are
coloured, with emphasised boundaries. In *Handbook of the Ulster Question* (Dublin, Stationery Office, 1923)

   Ordnance Survey of Ireland: Map showing inland navigations and railways
-   -   1923-   -   -   -:-   --:np   gt j   Ow3   -   nd   -   -   IPP
     With no inner measured border, and reduced sheet area. Showing canals and inland navigation systems, with names,
and enhanced railways. This is Map 1 in *Coimisiún na gCanálach agus na mBóthar Uisce Intíre (Canals and Inland Waterways Commission): Report: July 1923* (IPP 1923, III, 605 (250 11.23)). Report title and map number are in the top margin

   Bogs in Ireland: Map of Districts Selected by the Commissioners Appointed to Inquire into their Nature & Extent and the Practicability of Draining and Cultivating Them 1810
-   -   1923-   -   -   -:-   --:np   gt j   Ow8   -   1810   -   -   IPP
     General base map details as above. The map shows bog areas surveyed on behalf of the 1809 Bog Commission, and has a
table of Bog Districts surveyed and reported on. Canals are shown. The map was prepared by Senator Sir John Purser
Griffith (Chairman) for the Irish Peat Committee of 1917: publication then is unproven. Here it is Map 2 in *Coimisiún na gCanálach agus na mBóthar Uisce Intíre (Canals and Inland Waterways Commission): Report: July 1923* (IPP 1923, III, 607 (250 11.23)). Report title and map number are in the top margin

   Map of Ireland: Showing the Irish speaking districts and the partly Irish speaking districts as defined by the Commission / Map No.3 / Coimisiún na Gaeltachta
-   -   1926-   -   -   -:-   -:np   gt j   Ow3   -   nd   -   -   book
     General base map details as above. Based initially on the 1911 census figures, and then a special survey carried
out by the Gárdai in July and August 1925. The Angler's Guides show the same overprint without the subdivision. This
is Map No 3 in *Coimisiún na Gaeltachta: Report* (Dublin, Stationery Office, [1926] (1000 8/26))

   Ordnance Survey Map for Official Handbook of Saorstát Eireann
-   1931   1904ai -   -   -:fm   C+:np   gt j   HSb1 -   nd   -   -   book
     The original sheet size is reduced. With a table of Principal Steamship Routes and Distances to and from Saorstát
Ports. The national border and one-inch alpha-numeric grid are overprinted in red. Queenstown is altered to Cobh, and
Kingstown to Dún Laoghaire. Two versions of the map are known, the first is on thin waxed paper (copy Dn 16.B.7 (13)),
the second, on the usual mat paper, appears in *Saorstát Eireann: Irish Free State: Official Handbook* (Dublin, Talbot
Press, 1932, another edition: London, Ernest Benn, 1932). This state has considerable deterioration in the hillshading

1	2	3	4	6	7	8	9	11	12	13	■14	16	17	18	19

Ordnance Survey of Ireland: Biological Subdivisions

-	-	1949ai	1949	-	-:-	A+:4/-	gt	j	Ow2	-	3.49	x		Da,RH	1
-	-	1949ai	1949	-	-:-	A+:4/-	gt	j	Ow2	-	3.49			Da,Dn	2
-	-	1949ai	1949	-	-:-	A+:4/-	gt	j	Ow2	-	3.50			Da,En,NTg,Ob	3

Published by OSI. With base map revisions: Gaelic county names and placenames incorporated (eg replacing King's County, Queen's County, Maryborough, Kells, Navan, Kingstown, Parsonstown, Charleville, Queenstown, Edgeworthstown), and showing Shannon and Dublin airports, complete with runway layouts. Lit: Harrison (1949), Praeger (1950)

1. With a legend not found on the third version: "*Boundary between West Cork and Mid Cork shown as altered and definition of boundaries where they occur at broad bodies of water, both as approved by the Fauna & Flora committee of the Royal Irish Academy in March 1949.*" The Da copy was a donation from the OSI Director General on 7.10.1949, the author's copy is in an OSI cover

2. The boundary round the west of Lough Mask is relocated through the centre. The Da copy was acquired 11.11.1949, the Dn copy is at 16.B.10 (14)

3. The above legend lacking. The boundary to the north of the Aran Islands is relocated to the south, and that around the west side of Londonderry is adjusted closer to the city

---

Supplement 1. Skeleton maps

Viceregal Commission on Irish Railways including Light Railways 1906

| - | - | nd:- | - | - | 1907- | -:np | - | j | Ow0 | - | - | - | - | BPP | |

Showing coastal outline, canals and railways including stations. In *Vice-Regal Commission on Irish Railways, including Light Railways: Appendix to the First Report* (BPP(HC) 1907 [Cd 3633], XXXVII, 575). *NB*: The three maps in the Second Report (BPP(HC) 1908 [Cd 3895], XLVII, 971) and the Fifth and Final Report (BPP(HC) 1910 [Cd 5247], XXXVII, 133, 641) are based on Railway Clearing House maps at nine miles to an inch. They were printed by the OS in Southampton

Ordnance Survey of Ireland: Showing County Boundaries: Rural Districts
Ordnance Survey of Ireland: Showing County Boundaries: Rural Districts (former) [since 1957]

-	-	1908-							Ow0	fouo		-	-	not found	
-	1921	1908-							Ow0	-		-	-	not found	
-	1922	1908-	-	-	-:-	B:2/-	-	j	Ow0	-	-	-	-	Do	
-	1944	1908-	-	-	-:-	C:2/-	-	j	Ow0	-	-	-	-	Mg	
-	1952	1908-	-	-	-:-	C:2/6	-	j	Ow0	-	-	-	-	NTg	
-	10-57	1957-	-	-	-:-	C:3/6	-	j	Ow0	-	-	-	-	Lse,Ob	
-	4-62	1957-	-	-	-:-	C:3/6	-	j	Ow0	-	-	-	-	BFq,Dtf	
-	-	1979-	-	-	-:-	C:np	-	j	Ow0	-	-	-	-	Dtf	

Published by OS, and OSI since 1922. An island and four province maps were requested by the Local Government Board on 8.7.1908. One hundred of each were made in October with a topographical base, but the Board required skeleton maps. Three hundred of each were supplied on 14.11.1908. The 1921 issue was a reprint with alterations. Showing coastline, county and rural district boundaries, and names. Great Britain deleted from 1957. A version of the map was reduced in 1957 to 1:1,013,760 (copy Dtf). (NA OS 6/11727)

Ireland [1905 illuminated title]

-	-	1928	-	-	-:-	C:1/6	-	j	Ow0	-	-	-	Dtf	
-	1941	1928-	-	-	-:-	C:1/6	-	j	Ow0	-	-	-	Lrgs	
-	10-56	1928-	-	-	-:-	C:2/6	-	j	Ow0	-	-	-	Ob	
-	10-58	1928-	-	-	-:-	C:2/6	-	j	Ow0	-	-	-	Dn,Lse	
-	5-59	1928-	-	-	-:-	C:2/6	-	j	Ow0	-	-	-	Lse	
-	2-64	1928-	-	-	-:-	C:2/6	-	j	Ow0	-	-	-	Cu	
-	-	1928-	-	-	-:-	C:np	-	j	Ow0	-	-	-	BFq	

Printed at OSI. Showing coastline, county and national boundaries, county and town names

Ordnance Survey of Ireland: Showing County Boundaries: Barony Boundaries

-	-	-	-	-	-:-	:	-	j	Ow1	fouo	nd	-	-	not found	1
-	-	1938-	-	-	-:-	C:2/-	-	j	Ow1	-	nd	-		on sale	
-	5-59	1938-	-	-	-:-	-:-	-	o	Oy1	-	nd	-	-	book	2

III.2.Supplement 1

| 1 | 2 | 3 | 4 | 6 | 7 | 8 | 9 | 11 | 12 | 13 | ■14 | 16 | 17 | 18 | 19 |

Published by OSI. Showing coastline, county (black), barony (red) names and boundaries. Now usually monochrome
1. Prepared for the Taoiseach (OSI Annual Report 1937-38 (unpublished typescript))
2. Prepared by Robert Johnston. Rivers are added in blue. In Robert C. Simington *The Civil Survey AD 1654-1656*: Volume X: *Miscellanea* (Dublin, Stationery Office, 1961)

Railway Map of Ireland [derived from 1868 base map (II.2)]
-      1944   -   -   -:-   ?fouo   -   not found   1
- -   1944-   -   3.49   -:-   -:np   -   j   Ow0   -   -   -   on sale
- 11-56   1944-   -   3.49   -:-   C:5/-   -   j   Ow0   -   -   -   Cu

Published by OSI. Showing coastline and railways, with stations. With plans of Belfast, Dublin and Cork districts
1. Perhaps originally for official use. Not in OSI 1946 catalogue: the 3.1949 issue may have been the first

Ordnance Survey of Ireland: Showing County Boundaries
- 7-56   nd:-   -   -   -:-   C:2/6   -   j   Ow0   -   -   -   Dn,Ob

Published by OSI. Showing coastline, and county boundaries only, with names. No Great Britain

Ireland [Index to six-inch sheets]
- -   nd:-   -   -   -:-   1962:np   -   j   Ow1   -   nd   -   BFq,Dtf

Prepared and printed by the OSI. With coastline only in black, and six-inch sheet lines in blue. Another version at 1:1,000,000 has county boundaries in addition (copy Dtf)

Supplement 2. Province maps and other special sheets

Connaught / Ordnance Survey of Ireland: showing County Boundaries: Rural Districts
- -   1908ai   -   -   -:fm   A+:9d   gt   jh   Ow0   note   -   -   -   Dn 16.B.16 (4)
- -   1908-   -   -   -:-   -   j   Ow0   fouo   -   -   -   not found   1
- 1918   1908ai   -   -   -:fm   B+:9d   gt   jh   Ow0   -   -   -   Dn   2
- 1925   1908ai   -   -   -:fm   B+:1/-   gt   j   Ow0   -   -   -   Lse,NTg

Leinster / Ordnance Survey of Ireland: showing County Boundaries: Rural Districts
- -   1908ai   -   -   -:fm   A+:9d   gt   jh   Ow0   note   -   -   -   Dn 16.B.16 (2)
- -   1908-   -   -   -:-   -   j   Ow0   fouo   -   -   -   not found   1
- 1918   1908ai   -   -   -:fm   B+:9d   gt   jh   Ow0   -   -   -   Dn   2
- 1925   1908-   -   -   -:-   B:1/-   -   j   Ow0   -   -   -   NTg   1
- 1.56   1908ai   -   -   -:fm   C+:1/-   gt   j   Ow0   -   -   -   Lse

Munster / Ordnance Survey of Ireland: showing County Boundaries: Rural Districts
- -   1908ai   -   -   -:fm   A+:9d   gt   jh   Ow0   note   -   -   -   Dn 16.B.16 (3)
- -   1908-   -   -   -:-   -   j   Ow0   fouo   -   -   -   not found   1
- 1917   1908ai   -   -   -:fm   B+:9d   gt   j   Ow0   -   -   -   Dn   2
- 1922   1908ai   -   -   -:fm   B+:1/-   gt   j   Ow0   -   -   -   NTg
- 1953   1908ai   -   -   -:fm   C+:1/-   gt   j   Ow0   -   -   -   Lse,Ng

Ulster / Ordnance Survey of Ireland: showing County Boundaries: Rural Districts
- -   1908ai   -   -   -:fm   A+:9d   gt   jh   Ow0   note   -   -   -   Dn 16.B.16 (1)
- -   1908-   -   -   -:-   -   j   Ow0   fouo   -   -   -   not found   1
- 1918   1908ai   -   -   -:fm   B+:9d   gt   jh   Ow0   -   -   -   Dn,NTg   2

Three hundred skeleton maps were printed for each province in 1908, and a further hundred were printed in error with topography. The latter were put on sale and appeared in OSC 1912 at 9d, increased to 1/- on 1.1.1920. All known reprints except one (Leinster, 1925) have topography. See notes on the similarly entitled island map on p.90.
1. Skeleton maps. 2. The Dn copies of these reprints are in 16.B.17 and 16.B.18

Map showing the market areas served by the following towns in the north of Ireland:- Londonderry, Strabane, Lifford, Omagh, Ballyshannon, Enniskillen, Clones, Monaghan, Castleblayney, Dundalk, Newry, Castlewellan, Downpatrick, Armagh, Dungannon, Augher, Clogher & Fivemiletown, Cookstown, Magherafelt, Draperstown, Coleraine, Limavady, Dungiven
- -   1904ai   -   -   -:fm   B-:np   gt   j   Ow4   -   nd   -   book

The map extends south to Louth. In *Handbook of the Ulster Question* (Dublin, Stationery Office, 1923)

1	2	3	4	6	7	8	9	11	12	13	■14	16	17	18	19

Map of Northern Ireland

-	-	nd:-	-	-	1926-	B-:np	gt	j	2w2	-	nd	-	-	book	1
-	2500/32	nd:-	-	-	1926-	B-:np	gt	j	2w2	-	nd	-	-	book	2
-	1750/38	nd:-	-	-	1926-	B-:np	gt	j	2w2	-	nd	-	-	book	3

Lower levels are green, those over 500 feet are yellow. Railways are coloured red, canals black
1. In *Ulster Year Book* (Belfast, HMSO, 1926). 2. In *The Ulster Year Book* (third issue) (Belfast, HMSO, 1932)
3. The print code (and 2500/32) is outside the border. In *The Ulster Year Book* (fifth issue) (Belfast, HMSO, 1938)

Supplement 3. Military issues

Ordnance Survey of Ireland: Shewing Military Districts

| - | - | 1904- | - | - | - | A-:3/- | - | j | 0w2 | - | nd | - | 2/04 | CPT | |

An outline map, showing rivers, railways and towns. No roads, no altitude. (WO 1889/50)

Ordnance Survey of Ireland / [Telegraph and Telephone Map]

| 1 | - | 1904ai | - | - | 1906fm | A+:np | gt | j | 0w4 | fouo | nd | - | - | Lbl | |

TSGS 2028. With lists of Head-Offices and Testing Stations. (WO 1889/56)

Ordnance Survey of Ireland / Irish Command Scheme

| - | - | nd:ai | - | - | 1908- | A+:np | gt | j | 0b2 | fouo | nd | - | - | book | |

TSGS 2360. Showing regular depots, naval and port war signal stations, coastline sections with their concentration centres, and cable landing places. In *Irish Command Scheme: May 1908* (London, War Office, A.1287 (33 9/08)). There is a copy at PRO WO 33/459. The map is overprinted Secret, and numbered 38

Ordnance Survey of Ireland / Divisional Boundaries, Brigade Boundaries

| 1 | 1920 | 1904ai | - | - | -:fm | B+:np | gt | jh | 0w2 | - | nd | - | - | Lpro Air 5/769 | |
| 1 | 1921 | 1904ai | - | - | -:fm | B+:np | gt | jh | 0w2 | - | nd | - | - | Lpro Air 5/769 | |

3. Ireland maps at 1:625,000, 1960

1	2	3	4	6	7	8	9	11	12	■13	■14	16	17	18	19

■Monastic Ireland
■Monastic Ireland (2nd Edition) [from 1965]

-	-	nd-	-	-	-:-	C:np	IG	k	3b2	1960	nd	x		CPT	
-	■5-65 7000	nd-	-	-	-:-	1965:np	IG	k	4b2	1965	nd	x		CPT	1
-	■7-77	nd-	-	-	-:-	1977:np	IG	k	4b2	1977	nd		-	BFq	
-	■8-79	nd-	-	-	-:-	1979:np	IG	k	4b2	1979	nd	x	-	on sale	

Published by OSI. Compiled by R.N.Hadcock. With letterpress. The second edition has more overprinted in red
1. Copies were also issued with A.Gwynn and R.N.Hadcock *Medieval Religious Houses: Ireland* (London, Longman, 1970)

[Blank base map]

| - | - | - | - | - | -:- | -:np | IG | k | 0w0 | - | - | - | - | BFq,Dtf | |

No publication information. The base map of **Monastic Ireland**, varying in colour from black to grey

Ireland: Irlande: Ierland: Irland: Irlanda

| - | - | nd- | - | - | -:- | C:np | AN | - | Rb0 | - | - | - | - | on sale | |

Printed by OSI, and carrying their copyright. A repayment service multicoloured map, published by Estate Publications [1990]. With photographs and mileage chart

■Average Annual Rainfall (in Millimetres) International standard period 1941-70 / ■North of Ireland

| - | Met.0.886(NI) | - | - | - | -:- | 1976:np | IG | k | 4b1 | - | 1941-70 | - | - | Bg,BFq,Og | |

Published by HMSO, London, for the Meteorological Office. The Irish Grid co-ordinates are 110km E - 380km E; 240km N - 470km N. There are companion sheets for Great Britain

CHAPTER IV

THE TEN-MILE MAP BETWEEN THE WARS AND ITS METRIC CONVERSION

1. The three-sheet map

1. Topographical editions

There is no documentation known that gives details of the origin of the 1926 three-sheet ten-mile map. It has no obvious connection with the 1903 map, and, as the last of the replacement post-war small-scale families to appear, was almost certainly made by reduction from the new quarter-inch map. It can be no coincidence, as with the 1903 map before it and the 1964 **Route Planning Map** (RPM) after it, that the completion of the contemporary quarter-inch map was awaited before the ten-mile came to be made. Winterbotham[1] wrote that the decision to redraw was made in 1924, which hardly tallies with the one documentary reference to its making so far unearthed, that drawing was under way on 28 March 1923.[2] Elsewhere, he stated that the new edition was begun in 1922, which is altogether more probable.[3] It seems unlikely that Treasury approval for it was obtained, because when the matter was investigated in 1941, the acknowledged predecessor of the 1942 map was that published in 1903![4]

The new ten-mile map has several features in common with the other post-1919 small-scale families. Layers were preferred to hillshading for the representation of altitude. The map carries a grid of two-inch squares supported by an alpha-numerical reference system. There are obvious similarities between borders. Much of the lettering on the map and in the margins is similar in style if not size. The blue water plate includes the coastline, coastal water features and fathom lines, though unique to the ten-mile is the inclusion also of marine names. Some of the characteristic symbols, if not the descriptive wording, are identical with those of the quarter-inch as it was originally drawn, though on the quarter-inch these details were later enlarged.

Such assimilation into a corporate specification as this led inevitably to changes from the 1903 map. The question of lettering is dealt with in the table on p.11. The coastline's move to the blue plate has been noted already. Also added to it were submarine contours at five and ten fathoms, and sand, in blue, not black, stipple. Inland many more roads were marked, with minor roads reverting to a double line, and a wider gauge for major roads, with one side enhanced. A wider gauge was also used for railways. Altitude was shown by contours and layers, and additional spot heights. Battle sites were dropped, and there was a very different selection of antiquities. Town sizes were reclassified, though the classification controls were not defined beyond the identification of county towns. Large towns were distinguished from medium by the use of hatched or solid black infill. Location dots were rationalised, and the cross for parish finally disposed of. A symbol for the new feature of aerodrome, later aerodrome or seaplane station, was introduced. Most waterfalls were named on the blue plate, though, perhaps in error, The High Force appears in black. The fact that the French landed at Carreg Gwastad Point in 1797 was duly noted.

On later printings of Sheet 2, the overlap on Sheet 3 was marked, though oddly not that on Sheet 1, and the latitude measurement 53° 30' was corrected. Topograph-

---

1. Winterbotham (1936) 80. H.St J.L.Winterbotham was Director General of the OS from 1930 to 1934.
2. PRO OS 1/15/1.
3. Winterbotham (1936) 134.
4. PRO OS 1/789 27A: "*It* (ie the 1942 map)...*is a suitable base map for the overprinting of National statistical information. As a provisional measure it takes the place of the map next mentioned*" (ie the 1903 map). From *Extracts from Surveys of Great Britain Civil Estimates April 1948* in Treasury file E43631, now apparently lost.

Plate 9. Unused Ordnance Survey map covers: Jerrard's hand-lettered **Monastic Britain** and Martin's **Ten-mile Map**. From the collections of Mr R.A.Jerrard and Dr T.R.Nicholson.

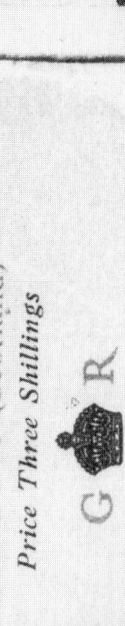

Plate 10. Martin's covers with the additional word **(Scotland)**. The author's collection.

ical information, especially with regard to roads, railways and reservoirs, was updated regularly. Many roads were reclassified "main" in 1935 and so coloured sienna. Many of these were changed again or deleted by 1937 with Ministry of Transport (MOT) classification. This explosion of activity affected also the railway coverage on Sheet 3, when for a short while many lines open only to freight traffic were added to the map. Port Merion and Mersey Road Tunnel were added in 1935. Aerodromes were inconsistently removed. The Isles of Sheppey and Thanet names were curiously deleted in 1933 minor corrections. From 1936, the name "Hebrides or Western Isles" was relocated between the Outer Hebrides and Skye. The letter spacing of this name was increased on the later two-sheet map, when the Sheppey and Thanet names reappeared.

It is not certain that all reprint states of the topographical map have been listed, but sales figures dated 18 November 1934,[5] presented in the knowledge that the new two-sheet edition was forthcoming, suggest that this is probably the case. Sheet 3 was selling at the rate of 900 annually, and in stock were 991 copies: hence the reprint of 1000 copies (1035) in 1935. The Sheet 2 situation was more desperate, with annual sales of 880, but a stock of only 775: hence the reprint of 1900 copies (1935) in 1935. With Sheet 1, the situation was quite the reverse, exaggerated as it was by the duplicate special **Map of Scotland** sheet. With a stock of 1409 and annual sales of but 500, no reprint was needed.

Ellis Martin produced a special cover for the three-sheet map (H.17). He also experimented with a border of alternate red roses and black thistles (known on a proof for Sheet 2, so perhaps intended only for this Anglo-Scottish sheet), but in the event a simple black border was used. The same cover was adapted for the 1937 two-sheet version. Complaints from Scottish customers that there was no longer a ten-mile map of Scotland led to "Scotland" being added to later Sheet 1 covers of both topographical and Road Map in order to promote sales.

2. Air editions

Documentation on the early history of the air editions of the ten-mile map has also disappeared, and one is left to glean information from surviving copies of the maps and chance remarks made in later files. Thus we learn from C.R.Brigstoke of the Air Ministry that:[6]

In 1927...the opportunity was taken of issuing copies of the ¼" to 1 mile and 10 mile to the inch maps, Royal Air Force (RAF) edition, in a form modified for civil use.

In fact, no pre-1934 ten-mile maps specifically for the RAF have as yet been located, though, if the above be true, they must in fact have been the predecessors of the "Special Air Edition (Provisional)" of 1928-29.

It would seem to be the progress towards publication of this apparently civilian map and its successor editions, derived presumably from the 1927 RAF Edition, that was the subject of statements in OSR 1927-28, 1928-29 and 1929-30. These confusingly all refer to the Royal Air Force Map. But, as Nicholson (1988a) suggests, perhaps the new map was not aimed specifically at either service or civilian user, but rather to serve the needs of both. On these editions, the layer below 200ft is uncoloured rather than green as on the topographical map, and the yellow tints to 600ft are browner. This left (uncased) green available for woodland, and yellow rather than burnt sienna for main roads. The air information is dated, and shows aerodromes, seaplane stations, both separate and joint, (emergency) landing grounds, prohibited flying areas, danger areas and air navigation lights. A compass rose was displayed. Provisionally in red, the overprint colour used for air information was changed in 1930 to blue, and the title to "Air Edition". The graticule remained purple. From about this time, some copies were sold inside covers specially designed by Ellis Martin (H.19).

The first suggestion of a new demarcation between civilian and RAF printings appears in OSR 1931-32 with the remark that:

....sheets 1, 2 and 3 [of the RAF 10 mile map] *now brought up to date with the latest information available, and with the British Modified Grid.*

Examples of this date with this grid have not been located, so again further comment would be speculative, except to say that it would have been inconceivable on a civilian map. The demarcation is confirmed by a 1932 map, now called "Civil Air Edition", and certain evidence that by this time the two branches were treated separately. This map, like the Air Editions before it, is overprinted in blue.

The Air Ministry was never happy with the ten-mile air map, which had been largely used by RAF bomber squadrons. Brigstoke (*op. cit.*) wrote that it:

... must be regarded as a very poor adaptation of an existing ordnance survey series and it compares most unfavourably with air maps specially produced by other countries.

But some improvements were in hand. Following Winterbotham's recommendations on 26 July 1933, a wide range of new air information was added both to Civil Air and RAF Editions, as shown in the comparative illustrations on p.97. The numerals on the compass rose were enlarged and redrawn radially. Since it had proved difficult to show a combination of layer colouring and woods distinctly, an additional layer between 200ft and 400ft was cleared of colour in the hope that woodland would become more apparent. Contour lines and figures were also deleted from Civil Air issues, though quickly restored.

From 1936, air information revision was recorded, as always to a specific date, but in addition according to the latest "Notice to Airmen". There was also some rev-

---

5. PRO OS 1/144.
6. PRO OS 1/456 39A, 28 July 1934. Other information in this section derives mainly from PRO OS 1/55.

AERODROME	⬤
LANDING GROUND	⊙
SEAPLANE STATION	⊛
AERODROME & SEAPLANE STA.	⦿
PROHIBITED FLYING AREA	▨
DANGER AREA	⸨ ⸩
AIR NAVIGATION LIGHT	☼

Height in feet of Aerodrome or
Landing Ground above M.S.L. 100

AERODROME	⬤
HEIGHT IN FEET ABOVE M.S.L.	
LANDING GROUND	○
HEIGHT IN FEET ABOVE M.S.L.	○
SEAPLANE STATION	⚓
SEAPLANE MOORING AREA	⚓
AIRSHIP STATION	⬭
MOORING MAST FOR AIRSHIPS	L
AIRWAY BEACON	✹
MARINE LIGHT	★
AERONAUTICAL W/T COMMUNICATION STN.	✥
AERONAUTICAL W/T D/F OR BEACON STN.	✥
CUSTOMS FACILITIES	CROYDON
AERODROME WITH BEACON	☀
AERODROME & SEAPLANE STN.	⬤⚓
AERONAUTICAL W/T COMMUNICATION AND D/F OR BEACON STATION	✥
PROHIBITED FLYING AREA	▭▨
EXPLOSIVES AREA	X
DANGER AREA	⬭

AIRWAY OBSTRUCTION OR OBSTRUCTION OVER
200 FEET (60 METRES) ABOVE GROUND LEVEL.

UNLIGHTED............... ⚑   LIGHTED............ ⚐

NOTE.—IF THE HEIGHT OF THE OBSTRUCTION ABOVE GROUND LEVEL EXCEEDS 300 FEET THE MAXIMUM HEIGHT IN FEET ABOVE MEAN SEA LEVEL IS INDICATED BY FIGURES.

H.T. CABLES OF PARTICULAR
DANGER TO NAVIGATION .... ∿∿∿∿

CONTROLLED AREA ....

Plate 11. Aviation symbols on air editions of Ordnance Survey ten-mile maps, 1930 and 1934. The author's collection.

ision in the classification of air information, including a change from "Air Beacon" to "Air Light", and the division of "Marine Light" between lightship and lighthouse. Another later change was the relocation of the layer box to the face of the map to avoid removal when mounting into the new long covers, with its illustration derived from Ellis Martin's c.1930 design. The earlier adherently covered map had proved hopelessly impracticable while in the air, since the user had had to cope with three folds in one direction after fully extending its ten panels in the other, which was virtually impossible in an open cockpit. Also, though cloth backed, the map was not waterproof. Inside the new long covers, the map was folded only once along its longer axis. This meant that it was unnecessary to open more than two of its eight panels above or below this.[7] The northern location diagram on the front cover was printed upside down to assist in rotating the map correctly. But the new format necessitated not only the removal of the map margins, but also the cropping of some of the generous overlap between the sheets, so reducing their dimensions to 98cm W-E by 59cm S-N. New indexes showing no overlap at all were drawn for these covers, though an explanatory note makes reference to ten-mile overlaps between the sheets. Mounted copies were waterproof sprayed, and for the first time flat copies printed on Place's waterproof paper. We are informed that on 17 April 1934:[8]

10 mile sheets 1 and 2 are at the Air Ministry for proofing. Sheet 3 is ready paper flat, and mounted and folded and waterproofed. Sheets on Place's paper are now being waterproofed.

But all this proved but a temporary palliative to the RAF, who remained very dissatisfied with the ten-mile map, the scale of which they found too small, with irrelevant detail and insufficient emphasis on important features such as water and railways. Nonetheless, the new map went through several reprints incorporating new air information between 1934 and 1937. But in reality the 1934 map was doomed before it appeared, because even in 1933 serious consideration was being given to following the French standard scale of 1:500,000 for aviation maps. Winterbotham's recommendations of 26 July 1933 mentioned above were countered by Major Fryer of the Map Section, Air Ministry, who formally proposed a 1:500,000 map on 11 December 1933. The decision to pursue it was made at a meeting at the OS the following day. The scale was authorised in 1935, and in 1938 the ten-mile map was finally superseded by the fifteen sheets of the Civil Air and RAF Editions of the 1:500,000 map.

3. The military edition

As with the 1934 quarter-inch map (GSGS 3950), air information was also applied to the little known and short-lived ten-mile Military Edition GSGS 3955: indeed, but for the main road colour (the burnt sienna of the topographical map rather than yellow as on the air map), it is to all intents and purposes identical to the 1934 RAF version, with the same blue air information, including compass rose, and the same purple Cassini grid. It too was superseded in 1938, by the two-sheet GSGS 3993.

2. The two-sheet map[9]

1. The Road Map

Winterbotham's vision of a ten-mile road map was governed by the principle that it would normally be used in the car, where it would be impossible to open it fully. He favoured the Ansell method of back to back folding which he had experienced for himself. This involved dissecting the map into sixteen sections (4 by 4) on heavy card, with the bottom half of Sheet 1 and the top half of Sheet 2 on the one side, backed in each case by the other half of the other sheet set upside down. This permitted continuous use, with a single panel uppermost, in any direction with minimum refolding. As a preliminary, therefore, to taking final decisions on the format of his planned road map, he wrote on 3 September 1931 to Stanford, who, he knew, had possessed rights in this method of map mounting:

The 10 mile to the inch scale goes uncommonly well in the style of folding which I send to you with this letter. I want to know if anybody has any commercial rights in this type of folding. I used it first in the war and then again during the great strike for certain confidential maps of London. It has great possibilities for a car map.....The map would be so section mounted than any particular section you wished to use by itself could be folded onto the top of the map. I believe that a type of map like this would sell well. Two questions then. Is the method of folding protected, and what do you think of its sale possibilities.

Edward Stanford replied to Winterbotham on 9 September:

..the method of mounting was patented by a Major G.K.Ansell of the Inniskilling's and was purchased by us in January 1906. Major Ansell is now dead.

We have mounted maps in this way ever since that date, although in recent years we have not kept a stock of them, only doing them to order...

One of the chief things that rendered this form of mounting more or less obsolete was the introduction of overlaps in map sheets as published which necessitated cutting to waste and extra expense in joining. Also, when the new Stanford-Bridges mounting was invented it cut out virtually all other forms of patent mounting, because one could mount individual sheets just as they were published, and

---

7. This method of folding is called Michelin fold.
8. PRO OS 1/456 10.
9. The primary documentary source for this section is PRO OS 1/144.

also in a car or for that matter walking when there was wind, one could go all over the map without opening it further than an ordinary book.

During the war we mounted maps for people on Major Ansell's method, including some for the Prince of Wales. The Patent number was 12844/105 and was dated June 21st 1905....

Winterbotham pushed Stanford a little further on this: on 7 October 1931:

I presume that all patent rights in that connection have now expired. Is that not so?

Stanford's reply to this is unrecorded, but Winterbotham seems to have been satisfied on the copyright question since he determined to go ahead and use this mounting on the three-sheet topographical ten-mile map as well as on his new road map. He issued instructions on 2 October 1931, including details of the indexes of related series he wished to see inside the covers. Their absence from the first issues caused considerable recriminations later. As a first step, Winterbotham decided on 30 May 1932 to put on sale thirty copies of the topographical map in Ansell-fold at 12/6d per copy, and he followed this by a production run at the same price, in brown board covers (H.8.3c2). But a price even at this level scarcely seems to have covered the costs of preparing this map, much of which must have been made up by hand. The generous overlaps of the three-sheet map had to be removed, as were the margins. This in itself resulted in enormous wastage. Then nearly half an inch more was trimmed from the left hand side, inside the border. The Shetland Islands inset was cut from its original position and placed nearer the Scottish mainland. A patch of sea covers the hole! That the Shetlands survived at all is the more inexplicable because the Orkney Islands above the word "Hoy" also succumbed to the guillotine. But at 12/6d few copies were in fact sold (148 in 1935 figures), and these depressing sales figures eventually persuaded Winterbotham to cancel it.

The Road Map, on the other hand, was designed with the requirements of Ansell-fold fully in mind. It was thus on two sheets, with no overlap. With the exclusion of the Western Isles, considered unnecessary on a road map, the sheet width (at 620,000 yards) was some 80mm narrower than the three-sheet map, and in Ansell-fold would reduce to 225mm W-E by 192mm S-N. The Orkney and Shetland Islands were included on separate inset maps. Ireland and France were ignored. Title, scale-bar and symbols panel were all brought together on a new title panel within the border, which would survive the loss of margins in the Ansell-fold version. Grid values were within the border for the same reason. In an endeavour to keep the map clear by avoiding overcrowding detail, many minor roads and topographical details were omitted. Prominence was naturally given to features that would interest the motorist. Both "A" and "B" class roads were shown, distinguished by colour. The principal "A" roads were drawn in thicker lines, with the "A1" being honoured with the widest gauge of all. Pecked lines were used for roads being built (not listed in the legend), and the information on the map was regularly updated. Vehicle ferries and telephone call boxes were recorded, and there was a mileage measurement system.

Perhaps the most surprising aspect of this Road Map was the number of features common with the topographical map that were given new treatment, even to the extent of inventing a new symbol for the national boundary. Hills were shown by relief rather than layers, and marine contours were deleted. The sea colour joined the water plate, first by ruling, and from 1937 by stipple. But the inland part of the water plate was not repaired where overbridges for minor roads and railways had been removed. Roads were shown by uncased coloured lines.

From the start the Road Map was issued flat, mounted and folded in covers (H.21, designed by Ellis Martin), and in Ansell-fold with turquoise boards (H.8.3c1). Winterbotham's attempts to reduce the costs involved in this last were to a large extent negated by the necessity of maintaining duplicate impressions of Sheet 2, especially printed without the title panel, which of course was an essential feature of sheets selling separately. This problem was overcome on 30 October 1934 by the simple expedient of allowing title panels to appear twice on future Ansell-fold versions.

The map sold well in Ansell-fold (570 in 1935), and it remained in production even after the sheet size was widened to the west to 680,000 yards in 1937 to permit the inclusion of the Western and Scilly Isles. This increase had also been determined on 30 October 1934, obviously as a direct result of the decision to recast the topographical map also on the new National Projection (see Appendix 1). With its road information updated to 1 April 1935, the new Road Map was noted as nearly ready for publication in OSR 1934-35. What caused the two-year delay until eventual publication in 1937 (with 1935 remaining the printed publication date throughout its life) is unclear. The making of its new plates seems to have run parallel with those of the topographical map. Certainly features in the extension west can be found common to both, like the smaller figure used at 7°W in the lower border of Sheet 1. Possibly the recognition that the earlier policy of deleting unnecessary features from the Road Map had been carried too far may have had something to do with it. This resulted in the restitution of spot heights and certain water features that had been omitted, like Loch Venachar. Further physical features were named. A category of unclassified roads, coloured black, and associated placenames, was added. Isle of Man roads were declassified. The unique gauge of the "A1" was retained, though somewhat reduced.

2. Topographical editions

Winterbotham drew on the lessons of the Road Map when in September 1934 decisions had to be taken on the future of the topographical ten-mile map. He recognised that three overlapping sheets and the additional special Scotland map caused enormous wastage in the duplication of resources. He made his decision on 3 October 1934 to recast the map on two sheets. The joining line would be the same as on the 1932 Road Map, but since the Western Isles naturally had to be included, a wider sheet than that became necessary. The Orkney and Shetland Islands would be inset. A model was produced, and approved on 26 October, though, with the sheet size now so large, it was considered imperative that all printed matter should be within the bounding lines drawn. Thus a title panel, including symbols, within the borders was necessary, as on the Road Map. By positioning this bottom right on Sheet 1, it proved possible to combine the Orkney and Shetland Islands in one inset rather than the two of the 1932 Road Map, which in this respect was not changed in its larger sheet version. On 30 October 1934, Winterbotham confirmed that both road and layered map should be produced in two sheets, with sheet lines and general arrangement the same for both maps.

Other developments followed. A revision of road classification was anyway enforced with the OS decision to employ MOT numbers for all small-scale maps. The new Director General, M.N.MacLeod, approved their use on the two-sheet ten-mile map, with only the Class "A" roads coloured. He also decided that the full co-ordinates of the National Yard Grid, which was to be used on the new map, should appear only at the corners of the sheet. It thus proved possible to enter the figures inside the border rather than on the map face, as on the Road Map. He also insisted on the reinstatement of the alpha-numerical system concurrently with the grid co-ordinates.

Proof stage was reached by the summer of 1936, and a note reveals that the layers on Sheet 2 were made by the enamel plate process. MacLeod asked to see second proofs omitting contours, and it was this version that he approved on 10 August. Consideration was also given to associated two-sheet editions: a new outline map was published, and a new Physical Features map in an endeavour to reap profit from the recent schools drive. But the fifteen year's supply of the three-sheet water and contours edition still in stock persuaded MacLeod not to make a new one. It was about to make its debut as a period map, **Monastic Britain**, when war broke out.

The topographical version proved instantly successful, and reprints was regularly required. Railway and reservoir revision in particular were kept up to date. Redesdale was named in 1940. It was the only civilian coloured map to be reprinted complete in 1941. As late as 17 August 1943 (long after the publication of the 1:625,000 Planning Series map) it was OS post-war policy for the imperial map that it: *"will be retained and published with the metric grid on it. C.S.Ch. will verify that the negatives of the gridded edition are in existence"*,[10] and for some years after the war, reviving it remained OS policy (see p.145). The Land Utilisation Survey of Britain (LUS) recognised in it an excellent base for national summary information, but with the intervention of war, the only maps to approach publication dealt with Land Utilisation, Land Fertility and Types of Farming. South sheets of two of these were printed by Bacon for publication by LUS. But they were victims of the bombing and few copies survive of either.

3. Military editions

The War Office quickly espoused the two-sheet map, and GSGS 3993 superseded the three-sheet GSGS 3955 in 1938. It made considerable use of the map throughout the war, overprinting it in several different ways for army, RAF and home forces. The War Office would remain faithful to the 1:633,600 scale until at least the mid-1950s, when even the new National Grid military system is to be found on it rather than on the equivalent metric map which had long superseded it for civilian and administrative maps. The War Office also requested reprints of the Road Map throughout the war, and it was a military rather than a civilian requirement that led to a reprint as late as 1947, complete, *mirabile dictu*, with the National Yard Grid - a year after the new National Grid metric **Roads** map had been issued!

---

3. Ministry of Town and Country Planning Series

During the Second World War, as in the first, the OS suspended its normal civilian programme in favour of military and Government necessities. Military mapping at the ten-mile scale was briefly mentioned in the last section. Government requirements for it were not determined until 1941, by which time the need for recording a survey of the national life and resources of Great Britain on a series of planning maps, in order to support both the war effort, and the reconstruction of the country afterwards, had become all too apparent.[11]

Such a concept had in fact originated before the war. A committee appointed in 1938 by the Council of the Town Planning Institute under the chairmanship of Sir Leslie Scott recommended the setting up of a Commission to carry out a national survey of the country's natural and economic resources. And at its Cambridge meeting the same year, the British Association for the Advancement of Science set up its own National Atlas Committee under the chairmanship of Eva Taylor. MacLeod served as a member. Taylor set out her committee's objectives in her report to the British Association:

The proposed Atlas aims at a strictly objective and scientific presentation of the natural conditions, natural resources and economic development of the land and the adjacent sea, of the history and prehistory of our country, and of the distribution, occupations, movement and social conditions of the population...It would...be of service to administra-

---

10. PRO OS 1/789 26A: Extract from Technical Post War Planning No 3.
11. Unless otherwise cited, documents supporting this account of events until July 1942 may be found in PRO OS 1/155.

tors, public men, educationists and research workers in many fields, since it would present in convenient form the data upon which many conclusions, and decisions of national importance, must be based. The National Atlas would present...on a uniform cartographic plan the results of the various surveys, returns and censuses made by Government departments.

Taylor's subsequent endeavours to achieve the publication of a National Atlas were discussed by Christopher Board in the E.G.R.Taylor Lecture at the RGS in 1989. Space does not permit a reprise of these events here. Her ambition was anyway thwarted by the war.

Nonetheless, a few salient points need to be made. The National Atlas Committee considered the 1:1,000,000 OS physical maps the most suitable base maps, perhaps because certain thematic maps on agricultural and population subjects had already appeared from Southampton at that scale. No-one seems to have suggested the ten-mile, which, as we have noted, had already been taken up by the LUS. MacLeod was an enthusiastic supporter of the idea of an atlas, though not without reservations. He did not wish to see OS involvement offering a concealed subsidy, as had occurred when they undertook to publish maps for the LUS. He was concerned about the already pressing commitments of his department. He thought that commercial map publishers would view the project as their natural educational and scientific preserve, though in his view none of them would have the resources to fund so extensive a project. MacLeod also drew attention to the unifying influence the use of the coming National Grid could have on the atlas.

Professor Taylor refused to allow the war to kill this project, and the pressing needs she could see all about her - for reconstruction and, even in 1940, for planning a post-war Great Britain - moved her from fulfilling the needs of scientists and educationists as her objective towards those of administrators and those whose task it would be to rebuild Great Britain. She approached Lord Reith, then Minister of Works, who on 26 February 1941 had announced the establishment of a Consultative Panel for Physical Reconstruction. Early in March, Taylor met H.G.Vincent, an undersecretary who, though initially sceptical of the value of cartographic representation and even more of how quickly anything of use could be achieved, was impressed by her views. On 8 April after further consultation, Vincent minuted Reith:

We asked you to defer consideration of Professor Taylor until we had examined more closely the working of research. Professor Holford is now sure that we must use her National Atlas scheme, and we should like you to appoint her to the Panel.....We can then have a group to work on mapping, including Dudley Stamp and Professor Taylor from the Panel, and Mr Jellicoe and Professor Fawcett from outside.

She was invited the following day.

After meeting Taylor, Vincent seems to have spent time writing a proposal for a National Survey, which was going to be of fundamental importance to the Ministry of Works (MOW)[12], as the central planning authority, in its work of reconstruction. Included was a list of subjects which could best be expressed in cartographic form. He completed this task on 7 May 1941. It fell to his colleague H.L.Davis to prepare the covering application to the Treasury for funding for the maps. Davis also sent a memorandum to A.R.Manktelow of MAF on 13 May 1941, no doubt to keep the OS's ruling department abreast of events. He included what must be a transcript of Vincent's proposal as to the purpose of a National Survey:

It has been agreed that a central planning Authority must, as its first function, be responsible for the compilation of a comprehensive National Survey. It is desirable that much of the information should be represented and available in maps, produced to similar scale and format for comparison.

Use can be made of the material and expert services of the National Atlas Committee of the British Association, who have already agreed scales, projection, grid and general format for the National Atlas. The final choice of planning subjects should be decided in consultation with experts....eg Stamp and Taylor - but the attached note shows subjects which will definitely be required and on which work can begin at once (see next page). The subjects in the First Stage are those on which information is readily available. The cost of preparing roughs would not average more than £40-50 per map. The subjects in the Second Stage are those on which information must be surveyed and collected. The cost of preparing roughs will be about £100 per map. It may be possible for the Ordnance Survey to handle the printing.

The expenditure on roughs would be mainly in the form of salaries to Dr Willetts (sic), the Organising Secretary of the LUS, and his staff....When a central planning Authority is formed, it will need staff of this character (geographical draughtsmen).

It is interesting to note how in this memorandum the "National Atlas" has already been subsumed into maps for a "National Survey". Commenting on the list in a letter to MAF on 20 May 1941, MacLeod pointed out that it:

...could readily be expanded into the "National Atlas" which the British Association had been advocating....If the OS produces the maps for the Central Planning Authority...it might be well to undertake also the production of the National Atlas as part of its normal programme.

He thought the OS should undertake the production of all the maps required.

With MAF support, OS participation in the National Survey evolved quickly. A meeting between MacLeod and

---

12. Ministry of Works and Buildings (MOW) became Ministry of Town and Country Planning (MTCP) in 1943, and in 1951 was renamed Ministry of Local Government and Planning, then Ministry of Housing and Local Government (MHLG).

List, probably by H.G.Vincent, of National Survey subjects to be expressed in cartographic form

First Stage (Information already mapped or readily available)	Second Stage (information to be collected and roughs prepared within 12 months)
1. Physical	
2. Geological (stratigraphical)	3. Geological (drift)
4. Mineral deposits and workings: (coal, iron and other metals)	
5. Ditto: (quarries, brick, cement etc.)	6. Water supply: (sources, hardness etc.)
8. Afforestation (including some vegetation)	7. Soil
9. Land Utilisation	
10. Types of Farming (see Agricultural Atlas)	
11. Transport: (road, rail, waterways, ports, civil aerodromes, airways)	
12. Electricity and gas	
13. Pre-war industrial location	14. Changes in industrial location due to war (trends only)
15. Population: 1931	
16. Rate of population change: 1921-31	17. Population forecast
18. Pre-war unemployment distribution	
19. Administrative areas	
20. Town Planning areas	
21. Regions: Civil Defence	
22. Regions: Ministry of Labour including Special Areas	
23. Regions: Ministry of Transport	
24. Regions: Ministry of Agriculture Inspectorate	
25. Regions: Home Office	
26. Regions: Post Office and BBC	
27. National Trust Properties (and Youth Hostels)	28. Common Lands
	29. State-owned lands (including Ancient Monuments)

MOW officials was arranged for 9 June 1941. From the Ministry were Professor Holford, Vincent and a man whose name MacLeod did not catch, but who must have been John Dower[13]. MacLeod thought the OS might be responsible for the fair drawing of material supplied by the Central Planning Authority, advice on publication, printing base maps, and possibly compiling some maps, such as population, from statistics. In a long minute written on 12 June, MacLeod summarised the various opinions on base map scale. He reported that Holford inclined to the ten-mile as being the largest scale which would allow of the whole of Great Britain being shown on two sheets. MacLeod also listed the map subjects to be considered, divided into groups concerning the base map, physical and natural features, industrial and economic features, climatic, cultural, administrative and miscellaneous subjects. The climatic group (Rainfall & Snowfall, Temperature, Water Supply, Light, Wind & Pressure) was the only significant addition to Vincent's list. MacLeod visualised the base map as consisting only of permanent features, requiring very little revision, and the physical features essentially showing a clear division between mountain and plain. He requested cost estimates to produce at his next meeting, arranged for 23 June.

MacLeod chaired the 23 June meeting, which Stamp and Taylor also attended. Much of the discussion concerned the base map scale. The recommendation of the Davidson Committee, that OS maps on scales smaller than one-inch should remain mathematically related to this parent, was recalled.[14] MacLeod pointed out that this recommendation had received only the narrowest of majorities. Therefore, now that a new base map had anyway to be created, he thought their choice should be free as to scale and size. Eva Taylor wanted the smallest possible, with National Atlas publication still her ultimate objective. But Holford and Dower, who had produced a paper which included Vincent's list of map subjects as the basis for discussion, had as their objective not educational but office use. They had been using the OS imperial ten-mile map, which they liked. The official minute records their opinion that most of the maps for planning use should be at a scale not less than ten miles to an inch, and recommended either 1:500,000 or 1:625,000. The remainder, and the bulk of the National Atlas, if and when this followed, should be on a related smaller scale, respectively 1:1,000,000 or 1:1,250,000. MacLeod agreed

---

13. Dower was a civil servant who is now chiefly remembered for his report on *National Parks in England and Wales* (BPP(HC) 1945 Cmd 6628, V, 283-339).
14. *Final Report.....* (1938) 10,11.

to create pilot maps at the 1:1,000,000 and 1:500,000 scales, the latter by enlargement of the existing 1:1,000,000, in grey, in black outline with blue water features, and in a generalised physical format. He also requested further production cost estimates.

The official minute credits MacLeod with the suggestion that a Maps Office, with a paid officer and assistants, should be formed within the Planning Branch of the MOW, with a remit to prepare for publication maps considered essential to the Central Planning Authority and its derivatives. There would be a standing committee, which quickly came to be known as the Maps Advisory Committee, to monitor its affairs. On 10 July, Lord Reith himself invited MacLeod to join this committee, once the Central Planning Authority had been established. Taylor and Stamp from the British Association's National Atlas Committee were also invited, as was Sir Edward Bailey, Director of the Geological Survey (GS), to represent the Department of Scientific and Industrial Research. MacLeod saw the role of Taylor and Stamp on the committee as promoting the cause of the National Atlas, of which maps for the Central Planning Authority and post-war reconstruction would serve as a nucleus. Dr Edward

further. Holford's own preference remained the imperial ten-mile, but on 11 July 1941 Dower completed a second paper discussing the comparative merits of linked metric scales, either 1:500,000, 1:1,000,000 and 1:2,000,000, or 1:625,000, 1:1,250,000 and 1:2,500,000. He chose to recommend not a two-sheet, but a three-sheet map at the 1:625,000 scale which could be enlarged from the imperial ten-mile, had sufficient detail and could be accommodated in a double elephant plan chest (40ins by 26½ins), wherein the 1:500,000 would be uncomfortably tight. The 1:1,250,000 scale could be used for condensed summaries, or as England and Wales, or Scotland, sheets of a National Atlas, and the 1:2,500,000 scale to show the British Isles as a whole. But for comparative purposes, the 1:500,000 family should yet be considered. Dower provided a diagram of how he envisaged his three-sheet map would look. England and Wales would be covered in two landscape sheets, which would fit together to make a wall map. Scotland would be on a portrait sheet overlapping and offset to the west.

The second half of Dower's paper concerned map subjects, and he put forward a list of ten categories of map that he thought should be produced first:

No	Map subject and sources
1.	General OS, made up of (a) physical (water and relief), and (b) topographical (places with names, roads, railways etc), from existing 1/633,600 map
2.	Geology (Solid), from existing 25 and 4 miles to the inch maps
3.	Land Utilization, from existing L.U. Survey 1/633,600 map (not yet available for Scotland?)
4.	Agricultural Land Fertility, from Dr Stamp's 1/633,600 mss maps (not yet available for Scotland?)
5.	Minerals (coal, iron and other metals) from existing map, with some amplification (other minerals - stone, clay, gravel etc - on a separate later sheet)
6.	Population Density (1931), from existing 1/1,000,000 map
7.	Population Changes (1921-31), from Census tables - not yet mapped
8.	Roads, from existing 1/633,600 map (with some amplification?)
9.	Railways, from existing 1/633,600 map (with some amplification - information from Railway Companies)
10.	Administrative Areas, from existing 2 and 4 miles per inch County diagram maps

Christie Willatts, until then the Organising Secretary of the LUS, was invited to become Maps Research Officer, initially part-time. For some years the MTCP's Maps Office was responsible for examining data from sources such as Annual Surveys of the Board of Trade, Census Returns, Lists of Quarries and Mines, Lloyds Register, besides much unpublished information, then charting the results on large-scale manuscript maps. These were maintained in a state of revision and were always available for official consultation. Whenever appropriate these maps were summarised at the ten-mile scale, and it became an OS duty to print some of these, and publish those considered worthy of wider distribution. Later a similar Maps Office under A.S.Butler was created in the Department of Health for Scotland (DHS) in Edinburgh.

But this anticipates events. After the June meeting, Holford and Dower investigated the scale problem

He suggested print runs of one thousand, including separate printings of the two components of the base map.

On 22 July, MacLeod sent Vincent some much needed cost estimates. About £800 should get the first ten maps to proof stage, and printing them would cost about £5 each for the first hundred copies. MacLeod also agreed to provide pilot maps at the 1:625,000 and 1:500,000 scales, as suggested by Dower, for the first meeting of the Maps Advisory Committee. Davis, for MOW, was able to present the OS costings to the Treasury in his updated submission of 8 August. He explained the intention to publish some thirty titles overall, with ten in the first instalment. And he asked whether this, if approved, should be regarded as a repayment service or part of a National Atlas and thus the cost borne by the OS vote. The Treasury replied to Davis's letter on 4 September, giving approval to the first ten titles.

Further consultation would be required before any more were produced. Treasury advice was that MOW should not look upon this project as an OS repayment service, but if the OS insisted on this, the Treasury should be advised. It was later (29 December 1941) confirmed that the OS would not do so.[15]

It was also on 4 September 1941 that the Maps Advisory Committee first met, at MOW under MacLeod's chairmanship. Willatts attended and acted as secretary. The meeting gave especial consideration to matters of the format and scales of the base maps, and the grid to be superimposed upon them. It was agreed that the National Grid, at the time referred to as the Ordnance Survey Grid, must be an integral feature in order to provide a convenient system of reference, and an index to larger scale maps. Although Dower was absent through illness, his opinions still dominated the will of the meeting. His options on base map scales were examined and his final recommendations, for a three-sheet map, two portrait and one landscape, at 1:625,000 with a sheet size of 32ins by 24½ins, and one-sheet maps at 1:1,250,000 and 1:2,500,000, were accepted.

Dower's list of map subjects, slightly revised from his 11 July prototype, was set down in his "Notes of matters for consideration and decision". He also reordered them on the basis of their importance for planning purposes.[16] Solid Geology he omitted, but he added a new group at the end: Commons, National Trust Lands, Youth Hostels, Forestry Commission lands, Water Catchment areas, Access to mountains. Also before the Committee was an extended list prepared by Holford and Dower of thirty titles, divided into three stages of map production. The first section, essentially the same as Dower's revised list, detailed maps to be published as soon as possible, the second, maps on which preparation should be begun immediately, and the third maps for post-war publication. This was probably the same list as that submitted by Davis to the Treasury on 8 August. Ten titles were in each group (see the table opposite). Increased print runs of 2500 were recommended. The Committee examined the situation of surviving pre-war maps and available source materials, and some caution was expressed on the map titles suggested. MacLeod in particular thought that republishing the subject matter of maps already in existence was hardly a priority at such a time. Nonetheless the Committee approved an immediate start on the first group.

MacLeod wrote up his own minute of this meeting the following day, and immediately got OS staff at work on the project. By 11 September work was in hand on clearing the pre-war 1:633,600 map of information not required on the new 1:625,000 scale base map. But he already foresaw problems, and on the same day wrote to Willatts: "*I am not quite certain whether we can get Great Britain on to three sheets of the 1/625,000 scale.*" Willatts replied four days later: "*I note that you have not verified that Britain will go into three sheets, but I imagine you will find it will, but with the disadvantage not shared by your excellent ten mile map, that the sheets will not fit to form a rectangle.*" At this point this somewhat mysterious episode of the three-sheet 1:625,000 map disappears from the files. Whether or not its demise was speeded on by Willatts's comments cannot be confirmed. We can at this distance anyway only hazard guesses as to what influenced Dower in recommending it in the first place. His preferred sheet size was even smaller than that of the 1926-1937 OS three-sheet map, and a portrait format Scotland could have pleased only the Scots. How much notice we should take of his desire that the map should fit a double elephant plan chest is also debatable, but it does suggest that he did not even know of the 1937 map, and was still himself using its predecessor. But the Committee's acceptance of Dower's recommendation is remarkable, and MacLeod's is nothing short of incredible, since he must have understood the relative simplicity of the direct conversion of his 1937 map into a two-sheet metric map, as opposed to the complete restructuring of sheet lines required by Dower's proposal. But the files reveal no concern on his part until his letter of 11 September.

But whatever the chain of events, the next entries in the file refer to the two-sheet map as a *fait accompli*. In mid-October, cost estimates were requested, and on 27 October, comparative estimates for the drawing, proving, photo and printing 1000 or 2500 copies of each map in the first instalment were provided. Specific reference was made in this document to two-sheet maps. On why and when the conversion from a three-sheet to a two-sheet map was made, the files are silent. There was no Maps Advisory Committee meeting, no further written comment from Willatts or Dower, not even an internal memorandum giving instructions. But the change in sheet size may account for the need to return yet again to the Treasury for confirmation, as Vincent advised MacLeod on 5 December 1941.

Willatts meanwhile was in dire need of base maps on which to plot data and begin drawing. On 6 January 1942 MacLeod was able to notify him of the dispatch of some twenty proof copies, both black and grey, and on 13 January 1942 he ordered that the full MOW issue of 200 black and 200 grey outline copies of each sheet be printed before any further alterations were made. This was completed by 17 March 1942. The files reveal many subsequent requests by Willatts for further stocks, but they have been omitted from the cartobibliography below. He needed copies on tracing paper, but usually they were printed on heavy paper, and sometimes in blue on Whatman's paper. These were used for final drawing in black ink: on photographing, the blue disappears, leaving only the drawing from which overprint plates could be made.

---

15. The Treasury file on this subject is lost, but copies are in PRO OS 1/162 18.
16. No 1 on this list corresponds to No 1 on Dower's 11 July 1941 paper; 2=10; 3=6; 4=3; 5=8 and 9; 6=4; 7=5; 8=7.

The Holford-Dower list of Planning Map titles

No	Phase 1 titles	Author's notes
1.	Topographical (Base Map)	The outline base map: later lists also put Topography here
2.	Physical (Base Map)	Reached proof stage only
3.	General OS (1 and 2 combined)	Originally named Physical Background Map, published as Topography. Later lists put this in No 1: No 3 becomes the 1:25,000 Index
4.	Administrative Areas: (based on 3)	This never had physical attributes
5.	Population Density, 1931 (based on 1)	To be an enlarged version of the 1:1,000,000 map (1934) with fewer categories. Later also encompassed Population of Urban Areas
6.	Land Utilisation	Also encompassed Types of Farming
7.	Land Fertility	Became Land Classification: also encompassed Vegetation
8.	Communications and Traffic (based on 1)	Divided between Roads and Railways: others (Waterways [Canals] and Ports) never reached publication
9.	Minerals: coal, iron etc	Divided into Coal and Iron, Iron and Steel
10.	Population changes, 1921-1931 (based on 1)	Encompassed Population Total Changes 1931-39 as well as Population Change by Migration, 1921-31, 1931-39
	Phase 2 titles	
11.	Rainfall: fog: sunshine	Two rainfall maps only were published, in 1949 and 1967
12.	Forestry	Not published, but sundry Forestry Commission Area maps were overprinted by the Commission, mainly on OS Roads maps
13.	Minerals: gravel, cement, brick etc	The Economic Minerals group included Limestone, also Gravel including Associated Sands. Others, like Brickworks, Cement, Sandstone and Igneous and Metamorphic Rocks were never published, though the last reached proof stage
14.	Industrial Location, pre-war	Drawn but not published
15.	Incidence of Unemployment: pre-war average	Drawn but not published
16.	Electrical Supply and Distribution	There were to be two maps, one detailing the supply areas, and the other the transmission lines. The latter was never published
17.	Land Drainage	Not published
18.	Geological: solid	Published by GS
19.	Geological: drift	Long in preparation, but not published
20.	Ancient Monuments: National Trust: Youth Hostels: Commons (with Gazetteer)	Not published as such, though partly covered by period maps
	Phase 3 titles	
21.	Water Supply and Catchment	Not published, but called Water Distribution in 1945. *NB*: Rivers and their Catchment Basins (of Scotland) was reprinted in 1946
22.	Soils	Not published
23.	Recreation: centres and areas	Not published
24.	State Properties	Not published
25.	Communications: post-war (including re-classified roads)	New road maps were published in 1956 and 1964
26.	Population Density: post-war	Became Population Density, 1951
27.	Population Changes: 1931-post-war	Two population change titles for 1939-47 were published in 1954
28.	Industrial Location: post-war	Not published
29.	Industrial Localisation: post-war	Not published
30.	Administrative and Planning Areas: post-war	Later administrative area maps have been permanently available

Publication

Davis made his final approach to the Treasury on 9 January 1942. Print runs of 2500 were required (though in practice little notice was taken of this). OS participation was approved by MAF, and it was fully expected that production costs could be recovered from map sales, so the OS would not be pressing for repayment. The Treasury gave its approval for the preparation and publication of the first instalment of ten on 17 January 1942. A further application should be made when work started on any more.

Publication was put in hand on 23 March, when MacLeod announced his intention to print a sales edition of the base map. He took advice on 25 March, and on 28 March approved print runs of one thousand copies of each sheet selling at 2/-. On 2 April instructions were made to print all copies on the same heavy paper as the pre-war outline edition, and to have the price added to the sheets at the appropriate places. The printer received these instructions too late, and on 3 April 1942 reported: "*The copies have been printed prior to the above instructions. I suggest typing price on the copies.*"[17] He had also used both medium and heavy quality papers. In the event copies of this printing seem not to have been on public sale, but were sent mainly to Willatts and others responsible for production of other maps in the series. The October 1942 printing would be recorded in OSPR, with a fuller title panel describing to users how to employ the new National Grid, known during the war as the Ordnance Survey Grid. This was in fact the first map to be published carrying the National Grid, apparently without objection from the War Office.

It proved possible to bring one further title, **Land Utilisation**, to publication very speedily. Sheet 2 had, of course, been published before the war by George Bacon, only for the stocks of the map to be destroyed in an air raid on 10 May 1941. Through Stamp and Willatts the OS obtained from Bacon's the surviving production materials of this pre-war map. Thus Sheet 2 was created by photographically enlarging Bacon's eight printing plates to the new metric scale. Sheet 1 had survived the fire at Bacon's, somewhat charred, as a compiler's model, not a proof as is stated in PRO files. Willatts handed this over to MacLeod on 17 September 1941, and from it printing plates were created. Publication of both sheets was achieved early in 1943.

All other titles had to be newly drawn. For some, the material was available, even in map form,[18] albeit at the wrong scale and sometimes out of date, but in most cases a wholly new map was required, and in each case Willatts was confronted with different problems of how best to display the information.[19] Most production difficulties were resolved by correspondence and meetings between Willatts and OS representatives. The progress of each map through the planning, drawing and proof stages was also reviewed at the meetings of the Maps Advisory Committee, and suggestions for improvements made. The committee met for the second time on 22 July 1942. After its review of current map production, it considered further titles, and agreed that work should start on six more - Water, Gas and Electricity, Land Drainage, other minerals, Forestry, and Recreation, Commons and National Trust Areas.

It might prove tedious to continue with detailed accounts of these meetings, most of which pertain to the natural course of events towards published maps. Interesting highlights revealed are given in the notes to the maps concerned. A few maps caused particular problems, and of these the **Physical Map** and those at the 1:1,250,000 scale are dealt with separately in this account. For those interested, I append a note giving references to where minutes of these meetings may be found.[20] They continued regularly until 1947, though after the war they developed into bipartite conferences between representatives of the OS and MTCP. Meetings were held again more frequently in the early 1950s. The DHS also held similar meetings, beginning on 20 March 1944. At the second of these on 11 October 1944, they too produced a list of possible map subjects, amounting to 95 different titles in 28 sections. The OS also necessarily made use of the 1:625,000 base map until such time as they were able to publish their intended new one. They used it for period maps, indexes and sundry internal purposes, and also made it available to other Government Departments beyond MTCP and DHS. Some of these were not to achieve completion: one that was cancelled at the DG's conference on 17 December 1948 was a ten-mile map showing Parliamentary Boundaries.

One further event from these early years requires notice. This was the occasion of the public launching of the Planning Series, undertaken by MTCP and DHS, with the full support of the OS. Simultaneous press conferences to advertise the growing number of maps in the Planning Series were held on 6 September 1942 at the MTCP in London and DHS in Edinburgh. For the occasion Willatts wrote a six-page foolscap pamphlet.[21] But publicity of its maps was usually the concern of the OS itself, and thereafter the series was afforded generous space in the post-war editions of the DOSSSM booklet.

---

17. The only copies known to survive unpriced are in the RGS.
18. Administrative Areas, Population Density, 1931, Roads, Railways.
19. Willatts (1963) addresses this matter.
20. 4.9.1941 in PRO OS 1/155; 22.7.1942, 22.2.1943 in PRO OS 1/156; 10.5.1943, 5.11.1943 in PRO OS 1/162; 17.7.1944, 16.10.1944, 23.10.1944, 22.2.1945, 13.4.1945, 6.7.1945 in PRO OS 1/556; 5.2.1946, 8.5.1946, 30.7.1946 in PRO OS 1/557; 11.6.1947, 18.6.1949, 17.1.1950 in PRO OS 1/558; 23.11.1950, 19.3.1951, 9.4.1952, 12.5.1953 in PRO OS 1/433.
21. Copies are in several of the PRO files relevant to this subject, eg PRO OS 1/999.

Problems with the Treasury[22]

A new Planning Series chapter opened in 1949 with Willatts's proposal to prepare a map entitled Spheres of Influence, altered later to **Local Accessibility**. It is significant because this was the first title proposed which had not been one of the original thirty listed in 1941.[23] This prompted a meeting at the OS on 18 June 1949, which in turn led to an internal examination of the series.[24] Among other things, some serious shortcomings were revealed in the official arrangements with the Treasury over the funding of these maps. Treasury approval was of course theoretically required in order to acquire finance for any new map. Normally this was an OS responsibility, via MAF, their ruling government department. With the Planning Series the situation was more complex, with MTCP involved in the making of the maps, and the OS in the production and selling of them. We noted above that approval from the Treasury for the first ten maps was finalised on 17 January 1942, with the instruction that application for future groups of maps should be made when the occasion arose. But in the event even this was not as straightforward as it sounds, since the list then submitted, and approved, was more in the nature of map categories than map titles, with the result that something like fifteen maps were published under this first authorisation. Thus when the subject came up again in 1949, the OS found they had to initiate a full investigation of this matter. They discovered that there were several maps outside the first instalment in course of preparation (some were even in print), and decisions had to be made concerning the titles to be put into a second instalment. By July 1949 they had categorised the titles as they understood the situation,[25] and passed the whole matter over to Willatts to deal with, suggesting that he obtain approval for those apparently published or in preparation without Treasury sanction.

MTCP replied on 17 August admitting that they had no written record of authorisation for the second instalment. But they claimed that the Treasury had authorised it verbally, apparently in a conversation held on 30 July 1942. They also wrote to the Treasury in order to set the record straight:

...In July 1942, our records show that the Treasury verbally agreed to our proceeding with the production of the second instalment.

Since then we do not appear to have kept the Treasury informed of the progress that has been made with this series of maps and I am now writing....to ask for Treasury approval for the maps already produced (which differ slightly from those originally conceived) and for the continuation of the series. It was never intended that the original list discussed with the Treasury should be firm as regards the actual content of the maps and this was understood on both sides.

Sixteen maps have been published and six maps are in the Press. In addition a further six are being prepared. Further maps are also contemplated.[26]

A perturbed Treasury replied on 23 February 1950:

We are not happy about proposed continuation. We find no record of the verbal agreement allowing continuation after the first ten & I am rather concerned to see that the number of maps now published or in the press has reached 22. However, continue with those in the press since the OS will recover its costs, and there will be no net cost on the Exchequer.[27] *Do not proceed with the preparation of the other six which are unlikely to recover costs.*[28]

However, on appeal, even these were allowed, in a letter dated 10 May 1950, on condition that any further titles would be submitted to the Treasury first. At the same time it was agreed between the OS and MTCP that no reprints of maps in the Planning Series would occur without reference to Willatts, who naturally wished to retain some control over the possibility of updating information at such times. This was ignored when, two years later, a reprint of the **Administrative Areas** map was needed. Late in 1950, there was discussion within the OS about this map, and how, since the OS was the main Government authority on boundaries, it would seem appropriate that they should take over from MTCP responsibility for the map and its revision. By 1952 the map was out of date, and anyway there had been errors in the drawing of the boundaries in the first place. But by this time, decisions on revising and replacing 1:625,000 maps were taken against the background that within two years a new base map was expected to be ready. Administrative Areas was certainly one title that the OS were considering taking to their own - the decision may by then even have been taken - and it was therefore decided to reprint the map without correction or revision in an edition to last three years, until a new edition could be made on the new base.

By late 1954, the OS were finalising their policy with regard to their new 1:625,000 map, and what titles should constitute the OS 1:625,000 Series, now distinct from the Planning Series (see p.147). Involved in this decision was a fundamental review of the Planning

---

22. This section is mainly supported by documents in PRO OS 1/432 (from November 1950, PRO OS 1/433).
23. The list appears on p.105.
24. PRO OS 1/162 167: since 1 June 1948 the OS had changed the name from the MTCP Series to the Planning Series.
25. PRO OS 1/432 5A. See the reference list on p.110, column A.
26. PRO OS 1/432 47B. See the reference list on p.110, column B.
27. PRO OS 1/432 114. See the reference list on p.110, "c" and "d" entries in column C.
28. PRO OS 1/432 114. See the reference list on p.110, "e" entries in column C.

Series, its maps, their sales and costs. On 30 August 1954, details of present stock, issues over the previous eighteen months, date and quantity of last issue for each title were demanded. By the end of the year, the review was even more searching, in requiring dates and print runs of first printings as well. At the DG's conference on 12 November 1954, it was decided that:

When further statistics are available, approach is to be made to MHLG drawing attention (a) to expenditure incurred by the OS in producing and publishing certain of the maps and the comparatively small financial return, and (b) suggesting that consideration be given with the Treasury to a change in the financial arrangements.

The results of the review were incomplete and not encouraging. No production figures were available before 1947, but in the seven years from April 1947 to March 1954, total production costs on the Planning Series amounted to £21,500. Net returns for about half the sheets did not yet cover the costs of printing them, never mind their preparation. It was recommended to Director General J.C.T.Willis on 22 January 1955 that the OS Series proper should be restricted to the base map in coloured and outline versions, the **Administrative Areas** map, the **Roads** map and a new Physical Map - titles which the OS were confident would sell well and more than cover their costs. It was suggested that the other titles would then be produced for MHLG as a repayment service, similar to the method adopted for GS maps. But it was impossible to alter these arrangements unilaterally and without the approval both of the Treasury and MHLG. Meanwhile on 7 March 1955 instructions were issued that:

No map [sponsored by other Government Departments, (*de facto* MHLG)] *will be made in future until its costing and selling procedure has been fully agreed on the appropriate files with the department concerned, and the necessary authority to proceed received from Chessington.*

Work on the two maps currently in production, **Limestone** and **Igneous and Metamorphic Rocks** could proceed under the existing arrangements until the question of their completion had been determined at Chessington.[29]

On 23 May 1955, Willatts was put fully in the picture. He was reminded that Treasury authorisation for Planning Series maps had been a matter for his department. But since it had been treated as a normal OS series, a record of costs to the OS had been kept only from the printing plate stage, and that since only 1945. Figures for preparatory work were unavailable. However, printing costs alone revealed that the OS would never recover its costs, and it followed that the OS vote was covering expenditure for which the Director General was accountable, but which was being incurred on specialised maps produced at the request of MHLG and DHS. Consideration should therefore be given to the proposition that they would continue to print and publish future projects as the agent of MHLG, leaving it as Willatts's responsibility to find the funding necessary.[30]

This started a correspondence which went on for nine months, and eventually involved also the new Physical Map, which had been brought to proof stage without consulting Willatts in contradiction of standing arrangements. At this stage Willatts wrote to Willis, who, being unaware of earlier events, had then to be brought up to date. His reply on 22 March 1956 appeased Willatts on all fronts, and resolved the issue of funding the Planning Series with his decision, since the intention was now to publish very few new titles, to leave the arrangements as they were. But he asked Willatts to seek OS co-operation in his plans at an earlier stage. In his note of thanks Willatts pointed out that maps presently in production may be few, but there were plenty more subjects in preparation. There the matter rested. Every now and then[31], the files bear witness to another attempt to have the matter re-examined, and the financial arrangements altered, but the fact is that for the next thirteen years, no action was taken. At such times also, moves were made to alter the price of the maps to reflect the individual costs of producing them. But again, it was viewed as advantageous to maintain a uniform price for all Planning Series sheets, irrespective of their popularity or production costs. General price increases occurred in 1962 and 1964.[32]

The demise of the Planning Series

As the 1960s passed, concern grew about the increasingly out of date information being carried by the earlier of the planning maps still in print. Gas and Coke, Coal and Iron, Iron and Steel, Electricity Statutory Supply Areas and Local Accessibility had all been criticised. It was Willatts's view that, although sometimes made more than twenty years earlier, subjects such as these altered only very slowly, whereas others, such as population maps, were reviewed and updated as the opportunity occurred. On 25 April 1966 the OS sent Willatts a list of planning maps in stock, wanting his opinion on the future of out of print sheets, and those running out soon. The list he supplied six months later (19 October) is important, because it provided the basis of decisions taken three years later when OS responsibility for the maps terminated. Coal and Iron, Iron and Steel, Types of Farming and Vegetation should lapse, with no-one available to revise the last two. Gas and Coke was obsolete, but it was not yet time for its successor. Population Density, 1951 and Population of Urban Areas

---

29. PRO OS 1/705 11A.
30. PRO OS 1/433 177A, draft letter only.
31. When it came to making Gravel including Associated Sands, Rainfall Annual Average 1916-1950, etc.
32. PRO OS 1/999 108,121.

1951 should be maintained and revised every 20 years, so next in 1971. The other population maps were increasingly historical, but MHLG would keep a small stock of them for the future. A successor to Land Utilisation would be welcome at the end of the second Land Utilisation Survey. Land Classification, Local Accessibility, Gravel including Associated Sands and Electricity Statutory Supply Areas should be kept in stock. MHLG was investigating a revision of quarry data for the Limestone map, and a new rainfall map was imminent.

Instructions for storage and cancellation of surviving Planning Series map materials were issued following a MHLG policy schedule dated 4 July 1967.[33] After this date only a handful of reprints were made, from disaster store material.[34] Early in 1968 calls were revived to make MHLG work operate on a repayment basis. But there was now a stronger momentum, with so many obsolete titles by then selling so slowly, simply to delete much of the range from the catalogue. This resulted in another approach to Willatts, on 25 June 1968, listing those titles the OS wished to delete.[35] He was reminded of his 1966 offer, made in reference to population maps, to hold small stocks of obsolete maps, for distribution to newly founded universities and the like, and was asked if this idea could be seriously pursued. Willatts did not reply in detail until 15 October, too late for the matter to be dealt with in OSC 1969. He summarised his recommendations on the future of all the surviving titles: of those the OS wished to retain, they should reprint some as necessary, and consult with MHLG and SDD before reprinting others.[36] The OS followed these recommendations. Willatts agreed that MHLG would accept small stocks - usually one hundred, though smaller numbers of some probably represent the entire residue - of the titles that the OS wished to discontinue, provided it was without charge to MHLG.[37] His suggestion was that MHLG would pay an appropriate royalty on maps sold. A reference to these MHLG stocks should appear in OSC. A list was drawn up in September 1969,[38] giving the total numbers of each sheet, and annual sales of each. In all there were still a massive 19,800 copies in stock, and less than 2½% that number had been sold the previous year. Transfer of stock to MHLG took place on 15 December 1969,[39] but even then 17,707 maps still remained, destined for cancellation.[40]

The OS continued to sell their titles until they announced in 1973 that all remaining ten-mile Planning Series map titles, save the base map, Administrative Areas, Physical Map and Route Planning Map, would in future only be available from the DOE.[41] Stocks still held by the OS were then passed over to DOE. They remained on sale until 1986 when the DOE Maps Office was put under pressure to curtail its work and reduce its staff. Few copies survived the consequent destruction.

Specification

Sheet lines were identical to the pre-war two-sheet National Yard Grid 1:633,600 map, photographically enlarged to the metric 1:625,000 scale, with the National Grid somewhat uneasily superimposed. A one-colour map was required and clarity was essential, thus minor roads and spot heights were deleted, also submarine contours, while other elements of the blue plate were combined with the black. Usually the differences are slight, as at South Hylton near Sunderland where the river and the railway virtually merge. Not all the points of change were found, and on most editions Ailsa Craig retained its 10-fathom contour and spot height. There is also an uncorrected fault in that River Seph (SE 5794) is present in name only on Sheet 2. The deletion of the submarine contours left many water feature names, especially in the north (eg Pentland Skerries, Am Balg), applying to no visible feature. Others, like the Flannan Isles, were left with only a lighthouse. At Willatts's suggestion, the curious reference at Carreg Gwastad Point (SM 9340), that the French landed in 1797, was deleted. This was an anomaly first added to the map in 1926 and survived at larger scales until the introduction of the 1:50,000 Series. Other errors beset the border detail. A figure "4" in the bottom border, overlooked from the pre-war alpha-numerical system, survived on issues of Sheet 1 until deleted in 1944. Many other inconsistencies affect latitude and longitude values: some are missing, or lack the degree or minute symbol.

The base map was revised again in 1944, it appears for two principal reasons. The first was to incorporate the revised form of grid reference with the 100km reference separated out. The explanatory note on the grid was rewritten, eliminating the phrase Ordnance Survey Grid in favour of National Grid, and grid values were added around the title panel. This was done in March. Then in June and July the commas following the grid values were deleted, and the numerals themselves redrawn where possible between the border lines. These corrections had to be made to the many maps already in preparation individually, which explains why so many errors,

---

33. This information from the job files.
34. The last was of Land Classification Sheet 2 in November 1969.
35. See the reference list on p.110, column D.
36. See the reference list on p.110, column F.
37. See the reference list on p.110, column E3.
38. PRO OS 1/999 180A.
39. Notice of the transfer was recorded in OSC 1970.
40. See the reference list on p.110, columns E1, E2 and E4.
41. OSPR 4/73; see reference list on p.110, column F. See also PRO OS 1/999 159A, 164A.

Reference list of Planning Series titles

No	Map title	A	B	C	D	E1	E2	E3	E4	F	G
1.	Great Britain base map	1	a	a							+
2.	Great Britain: Topography	1	a	a							
3.	Administrative Areas	2	a	a							+
4.	Coal and Iron	9	a	b	+	0,500	0,35	0,100	0,368		
5.	Electricity: Statutory Supply Areas	a	a	c	+	2800,2270	11,14	100,100	2667,2172		
6.	Gas and Coke	b	b	d	+	850,1700	15,17	100,100	704,1572		
7.	Gravel: including Associated Sands	c		1961						a	
8.	Iron and Steel	9	a	b	+	0,85	0,64	0,69	0,0		
9.	Land Classification	7	a	a						b	
10.	Land Utilisation	6	a	a						a	
11.	Limestone	c	b	c						b	
12.	Local Accessibility	c	c	e	+					a	
13.	Population: Changes by Migration 1921-1931	10	b	c		500,280	21,9	100,100	389,175		
14.	Population: Changes by Migration 1931-39	10	b	c		0,475	0,15	0,100	0,370		
15.	Population: Changes by Migration 1939-47		c	e		100,350	16,12	78,100	8,243		
16.	Population: Total Changes 1921-1931	10	a	a		1300,1100	10,33	100,100	1196,1003		
17.	Population: Total Changes 1931-39	10	a	a		1160,950	9,11	100,100	1059,836		
18.	Population: Total Changes 1939-47		c	e		0,100	0,18	0,80	0,0		
19.	Population Change 1951-1961			1965						a	
20.	Population Density, 1931	5	a	a							
21.	Population Density, 1951			?						a	
22.	Population of Urban Areas	5	a	b							
23.	Population of Urban Areas 1951			?		30,0	5,0	26,0	0,0		
24.	Railways	8	a	b							
25.	Rainfall: Annual Average 1881-1915	b	b	c		2240,2580	17,17	100,100	2137,2476		
26.	Rainfall: Annual Average 1916-1950			1962						a	
27.	Roads	8	a	b							+
28.	Types of Farming	6	a	b							
29.	Vegetation (Sheet 2)	7	a	b		430	93	100	332		
29.	Vegetation (Sheet 1)	7		b						a	
30.	Igneous and Metamorphic Rocks	c	c	e							
31.	Physical Map	2	b	a,d							+
32.	Population Changes, 1921-31	10									
33.	Population Changes, 1931-38	10		a							
34.	Geology (solid)	a		c							
35.	Geology (drift)	c									
36.	Ports	8									
37.	Electricity Transmission Lines	c									
38.	Brickworks		c	e							
39.	Sandstone	c	c	e							
40.	1:25,000 Index	3									

A. OS list to MTCP 7.1949 detailing their assessment of the titles approved by the Treasury: those numbered 1-10 correspond with the first ten entries on Dower's list of thirty on p.105. The others were not part of the first instalment: a = those published, b = those at proof stage, c = those in preparation
B. MTCP view of publication situation at 9.1949: a = those published, b = those in the press, c = those in preparation
C. Treasury approval: a = granted 17.1.1942, b = assumed from 17.1.1942 provisions, confirmed 23.2.1950, c = apparently given verbally, confirmed 23.2.1950, d = granted 23.2.1950, e = refused 23.2.1950, granted 10.5.1950 on appeal
D. Titles OS wished to delete 25.6.1968
E. Details only of maps (Sheets 1,2) transferred from OS to MHLG 15.12.1969: E1 = Stock of each sheet, including Stanford's, E2 = Last year's issues, E3 = Number of each sheet sent to MHLG, E4 = Stock after 1.1970, to be destroyed
F. Titles retained by the OS 15.12.1969, to DOE 4.1973: a = reprint as necessary, b = consult MHLG before reprinting
G. Titles retained by the OS 15.12.1969 and 4.1973

both commas and outsize numerals, survived. It seems futile to make a more detailed analysis of these, since almost every title then in preparation will throw up some inaccuracies of this kind. Many versions of Sheet 1 have an oddly rounded figure "6" in the lower border. An area originally left blank, presumably intended for overprint titles, was eliminated. In the title panel, two distinctive versions of the figure "1" may be found, one with a pointed top and the other with a flat top. As to Sheet 2, National Grid values were omitted around the title panels of a few maps. These were later placed below the 380km line, and later still the water plate was also limited by this line. The base map publication date usually given was 1942: this was sometimes altered on maps in preparation between 1944 and 1946. These dates have no relevance to the overprint, and, as with the sheet price, were sometimes deleted. Legends exist in many different forms, and these are listed below.

The second reason for the 1944 revision was to adjust certain features on the map itself, a process that was begun in August. Some physical names were altered or moved: The Minch became North Minch, Outer and Inner Hebrides were individually named and the Sea of the Hebrides name moved. On Sheet 2, the spacing of the words Bristol Channel and The Solent was improved. A second railway revision also appeared in 1944, but curiously this did not displace the original since both thereafter were in use. Neither was particularly up to date, but, so far as can be seen, neither was ever changed. These alterations were reported to the Planning Meeting on 17 July 1944, when the negatives were ready for the new printing of the base map (1244/Ch). At least two of the overprinted maps, Sheet 2 of **Index to the 1:25,000 Series** and **Roads**, were in preparation too early to receive the revision of the earlier cramped letter spacing, though the new grid information was applied to them.

Graticule intersections were added in 1945. Later topographical revision was limited almost entirely to the depiction of reservoirs, and even this was very unevenly achieved in that many were not depicted at all. The building of Ladybower Reservoir west of Sheffield caused the mapmakers most problems even at this scale. All reservoirs references (save only Alwen Reservoir in SH 95, which was evidently missed) were deleted following a security directive in 1957.

The physical map

In spite of repeated attempts over 12 years, the 1942 physical map was never published, though there were many versions printed, either as proof states or for specific official purposes. The first version of the map was but a copy of the physical elements of the prewar ten-mile topographical map to which were added further names from the 1928 1:1,000,000 physical maps and metric contours from the 1:500,000 aviation map of 1937. First proofs (16.2.1943) had six layers derived from two colours. A 50-metre contour line was added to second proofs (15.10.1943), and the sea was sprayed blue.

The most contentious problem that permanently delayed this map was that of nomenclature: it had never been an OS official task to standardise the correct designation of physical features. Immediately he perceived the difficulty, Willatts discussed it with MacLeod, and accepted the latter's invitation to investigate it more fully. He convened a conference at the RGS on 18 September 1943 at which geographers were invited to suggest just what names should appear on a physical map. In informing the OS of the meeting's views, Willatts specifically disclaimed all responsibility for the making of the map, though he would give his opinion when asked.

Third proofs (7.9.1946) were printed with some additional names, and, at least on Sheet 1, an extract from the otherwise unused 1:2,500,000 Planning Series map, overwritten with major regional names. References to the Ordnance Survey Grid gave way to National Grid. One hundred copies of the map, presumably in this form, were printed, with title deleted, on 1 November 1946 for LM&S Railway Co, School of Transport, Derby. Following this, it was decided to replace metric contours with imperial ones, derived from the quarter-inch map, using three colours to provide eight layers. Much of the lettering was altered, to standardise alphabets drawn from the earlier 1:1,000,000 and ten-mile maps, to increase the size of many names in order to have them cover the relevant areas more satisfactorily, and to add more names. The title panel was redrawn, and, for the only time on a map of this generation, a Chessington address and 1948 publication date used. Sheet 2 was proved (20.10.1948) but apparently not even examined, and new proofs of both sheets were printed early in 1949, now with the title altered to **Physical Features Great Britain**.

The map in this form was printed especially for the Ministry of Health (MOH) in December 1949. There were two variants, one with black plate and water features only, the other also with contours. Willatts wrote another long memorandum on the deficiencies of the map on 5 December 1949. Criticisms concerned the grid, names, writing and orthography, layers and spot heights. Agreement on these matters was reached at a meeting between MTCP, DHS and OS on 17 January 1950, and the OS then embarked on another extended period of experimental work. The Snowdonia and East Anglia portions of Sheet 2 were subject to layer experiments, now using four layer colours. At least four different versions were printed. Quite independent experiments were undertaken in the use of monotype for lettering. In the midst of this, yet another commission arrived from the railway company at Derby, now British Railways, who requested one hundred copies, layered and uncontoured, and "Physical Features" deleted from the title. They were printed on 5 January 1951 using plates unaffected by the latest experiments.

Both layer and lettering experiments were to be brought together in a job started late in September 1951. National Grid figures would be redrawn upright on all sides. Work continued until about April 1952, when it seems to have been interrupted and never resumed, probably because of further criticisms from Willatts who was sent some unfinished early pulls. These preceded the application of the monotype lettering, and it seems unlikely that any including that feature were ever printed. Little else was done on the map, though it remained officially in preparation for some time in parallel with the new "Ten Mile" map to be published in 1955. The physical map was seen as a thing apart. Since so much of it would have to be independently drawn anyway, with additional rivers and a quite different name plate, it was considered unimportant to integrate it into the new plans. This policy was only reversed late in 1954. A full printing history of the 1942 Physical Map, but for a few pulls for internal purposes, is given in Hellyer (1992a), where this subject is more fully discussed. The list in this book, but for one exception, gives details of only those copies which are known to survive: for an unpublished map these are surprisingly abundant and delightfully varied.

4. Smaller scales

Both smaller scales, 1:1,250,000 and 1:2,500,000, appeared in Dower's final recommendations on base map scales. Both were accepted as partners of the 1:625,000 at the first Maps Advisory Committee meeting on 4 September 1941. But with all available energies concentrated on the 1:625,000 map, the gestation period of the smaller scale maps proved to be much longer.

Work was already in hand on both projects when Willatts visited the OS on 28 September 1942.[42] Actual map making began on 20 November 1942, with the issue of outline negatives of the 1:625,000 sheets. Blues were photographed from them, and the drawing of the much sparser detail required for the 1:1,250,000 map was made upon the blues. Once this was completed, the two sheets were pinned together before reduction. By the end of June 1943, a negative at 1:1,250,000 had been produced, and on 10 July first proofs were printed and a copy sent to Willatts. He replied in detail on 25 April 1944 with sweeping criticisms.[43] Willatts's suggested corrections, concerning the grid, the selection of towns, roads and railways, names and physical features, were considered so drastic that all the material was forwarded on 2 May 1944 to Director General G.Cheetham for his opinion. Even more damaging criticisms were received from Butler on behalf of DHS. On 3 May, Cheetham permitted work to go ahead to clear up the problems south of the border. But he delayed further consideration of Scotland until he had had the opportunity of discussing Butler's criticisms with Willatts.

This was the situation reported to the Maps Advisory Committee at its meeting on 15 July 1944. Correspondence leading to a resolution of the difficulties continued during the summer, and from the discussion we learn more of the intended specification of the map. The grid was in the colour of the base map. Willatts thought this too strong when in black, and pushed for a rouletted grid. But he gave way when it was pointed out that overprinted versions of the map, when this may have become a problem, would not be in black anyway. Important physical features were to be named, and the usual problems of their selection and positioning reappeared. Butler's request for the inclusion of smaller streams was considered disastrous. County names and boundaries, and the names of important towns only were to be shown, linked by the MOT trunk road system. Those selected seem originally to have been chosen at random from the 1:625,000 map, and further inconsistency would have resulted from the desire of Butler and others who wanted additional smaller places in Scotland and Wales, linked by non-trunk roads. The railways depicted were drawn from a wartime 1:1,000,000 **Essential Railways Map.** The selection was not necessarily a good one from a civilian point of view, and it was generally possible to accommodate Butler's changes here, which were mostly deletions.

Further proof copies were printed on 28 July 1944 for Willatts's use, even though amendments in Scotland were still under review. It took until an MTCP-OS meeting on 13 April 1945 to sort these out. It was agreed to add further names, and to adjust the names of some of the Scottish shire counties. The number of rivers shown was increased slightly. But Butler's plea for further roads was rejected. Work on these corrections began on 27 August 1945, further proofs were printed on 17 April 1946, and again Willatts's comments were awaited before proceeding further. The base map was finally published at the start of 1947. By then Willatts was already at work on the "National Parks" and "Wild Life" maps, for publication in two Royal Commission Reports on conservation which were the responsibility of MTCP, so Willatts was kept supplied with adequate stocks of pre-publication base maps.

But the 1:1,250,000 was destined to become a map for which nobody ever found a worthwhile use. Dower probably recommended it in the first place in order to dispose of any further resistance from the National Atlas lobby, but while the war continued there was no possibility of specific plans for its use from that quarter. Once the map was in print, the National Atlas project was still dormant, and the correspondence reveals that neither the OS nor MTCP had any strong views on its use. When on 19 October 1951 he wrote a policy document for the map,[44] the current DMP was forced to admit: "*As far as I can make out the present policy for this map is to do nothing.*" It seems, in fact, to have been used only three times for its conceived purpose, as a base map for the

---

42. Much of the information in this section is drawn from PRO OS 1/156.
43. Much of the remainder is based on PRO OS 1/157.
44. PRO OS 11/5 2.

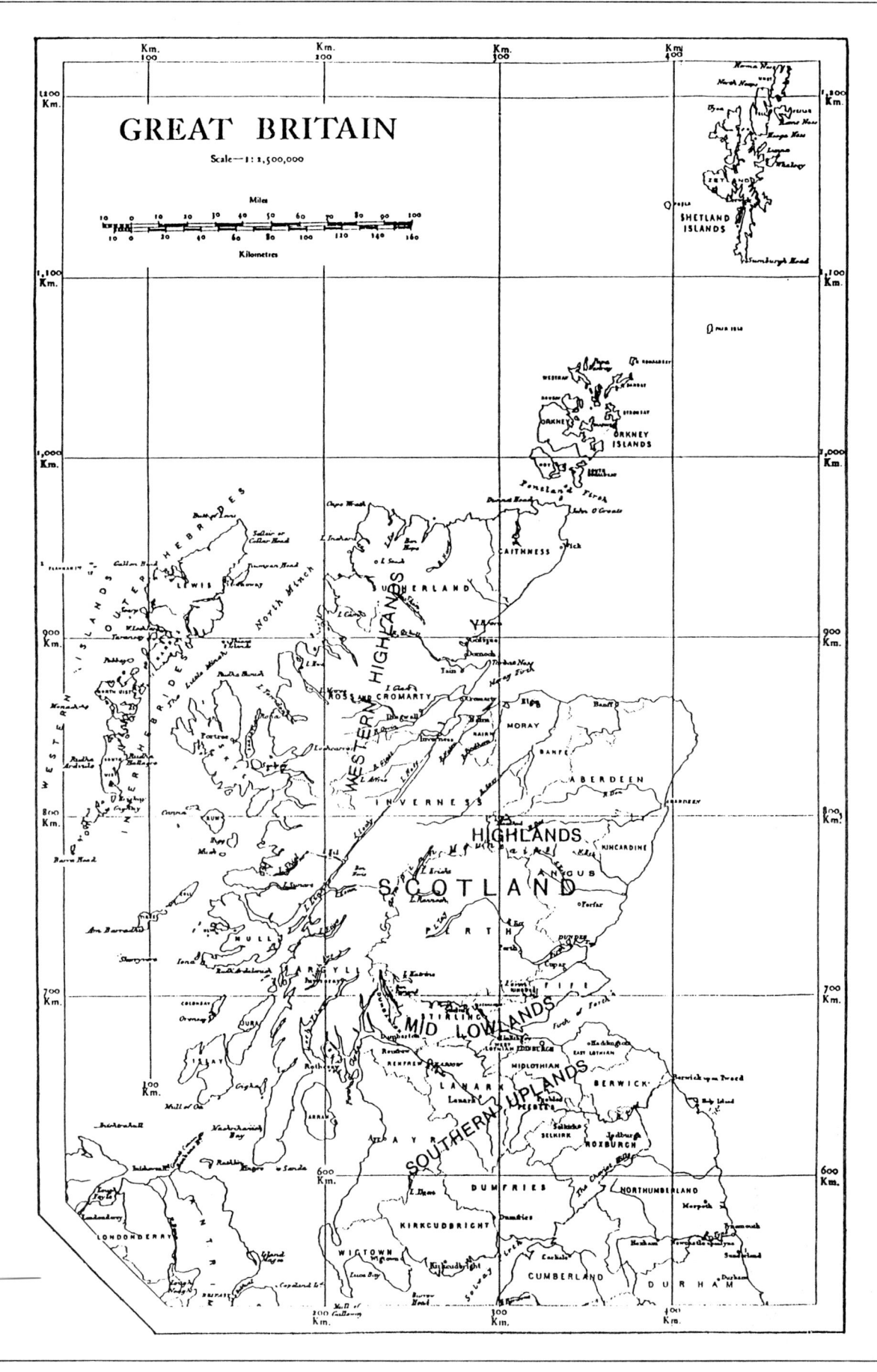

Plate 12. The northern half of the unpublished 1:2,500,000 map, 1942.

Planning Series. Of the three, one failed at proof stage (unique number 3252), one is but an outline diagram (3851), and only the third became a fully coloured map (3846). The base has been in most regular use for index purposes, in particular to the 1:50,000, 1:25,000 and Six-Inch Series, for professional and trade use, and for a handful of comparative indexes for internal OS use. Many of these have proved remarkably rare, and of the six-inch index in particular, which seems to have been updated quarterly for nearly six years, many states have not been discovered at all as yet, since copies were presumably normally thrown away when superseded. Late issues are noticeably more common.

But, in spite of this lack of purpose, the map was constantly being revised, and there are several remarkable aspects to its history. The state of revision of the map in its pre-"B" and "D" states will reveal aberrations that will fascinate inquisitive students. It went to five differently lettered reprints, but only "D" reached an OSPR, and one wonders why this one, and not "B", which was as heavily revised, achieved this privilege. It is inexplicable with the "E" edition, because this was a new map derived from the RPM "M" edition, with new alphabets, more detail, and redrawn in accordance with the true National Grid co-ordinates used on 1:625,000 maps since 1955. But for all this activity, the map sold only very slowly, and was probably maintained in print more for its value as an index than for any other reason. It was last recorded in OSC 1981-82.

The smallest of the scales, the 1:2,500,000, proved even less successful. Work was in progress on 28 September 1942, and a first proof copy was supplied to Willatts on 18 February 1943, five months earlier than the 1:1,250,000. This had not been returned over a year later. But part of it does survive - just. In 1946 the north half was inset on the Sheet 1 third proof of the doomed 1:625,000 Physical Map. Should Sheet 2 ever turn up, we may even find the rest of the 1:2,500,000 map.

Standard attributes. *NB*: Many of these are omitted or altered on military states

IV.1. 1:633,600 Great Britain (three sheets)

Map dimensions: 38 inches (965mm) W-E by 26 inches (660mm) S-N (OS figures)
Mapped area per sheet: 98,800 square miles, with a six-inch (sixty-mile) overlap between the sheets

1. Map border showing:
   a. Latitude and longitude values at 30' intervals, related to a diced graduation outside them at 5' intervals
   b. Alpha-numeric system originating in NW corner at two-inch (20-mile) intervals divided by transverse lines
   c. The start or end of names situated near the edge of the map
2. In the top margin:
   a. Top left corner: ORDNANCE SURVEY
   b. Top centre: map title: TEN MILE MAP OF GREAT BRITAIN
   c. Top right corner: SHEET 1 (or 2 or 3) (*Northern* (or *Central* or *Southern*) *Section*)
3. In the bottom margin:
   a. Bottom left: legend ("br","bs","bt","cr","cs","ct" types), boxed either 4 by 4 or 2 by 8 as convenient
   b. Next right: diagram of adjacent sheets (sometimes right of the scale-bars)
   c. Bottom centre: Scale of Ten Statute Miles to One Inch = 1/633600, with separate imperial (10+50 miles) and metric (10+100km) scale-bars. Crown Copyright Reserved is below (sometimes above within the border)
   d. Next right, usually in same column: *Altitudes* statement; *Submarine Contours* statement; on c.1930-34 issues *Price* statement, with *Catalogue* note (sometimes separated). The price statement may include no price
   e. Bottom right: revision and publication details and dates:
   1. *Drawn, Heliozincographed and Printed at the Ordnance Survey Office, Southampton* (line dropped after 360/32)
   2. *Published by* Colonel Commandant E.M.Jack,C.M.G.,D.S.O., Director General, *19xx*. Later wording: from 1933: *Printed and Published by the* Director General. *at the Ordnance Survey Office, Southampton, 19xx*; from ?1935: *First published by the* Director General *at the Ordnance Survey Office, Southampton,* 19xx (sometimes a new date), *with periodical corrected reprints* (...ORDNANCE SURVEY OFFICE, SOUTHAMPTON...on Sheets 2,3)
   3. *Minor corrections 19xx* (occasional, and sometimes within the border)
   4. Print or reprint code (sometimes bottom left)
   5. Standard *right of way* warning
4. A reference grid across the map at two-inch (twenty-mile) intervals, continued into the border
5. Symbols: as in legends illustrated on p.80, though not all features are present there. See col."K" on p.9
6. Layer box in lower right margin, with HEIGHT IN FEET AND METRES (Sheet 1 with 13 sections, Sheet 2 with 11, Sheet 3 with 10). With empty lower sections on aviation versions, sometimes on the map face for Michelin-fold aviation maps
7. Framed inset maps of Shetland Islands and St Kilda are on Sheet 1. The Irish and French coasts are shown in outline, with some coastal placenames and appropriate symbols

IV.1   115
___

Colour plates:
1. Black for everything except:
2. Blue: coastline, inland water, sand, submarine contours, water feature names, layer mix from 3600ft
3. Blue spray: sea and inland open water
4. Brown: contours and contour values
5. Brown: 3 layer mixes from 200ft (400ft from c.1934 - air and military editions only)
6. Screened green: layer to 200ft (topographical editions only); solid green: woodland (air and military editions only)
7. Yellow: layer mix from 200ft (topographical editions only); main roads (air editions only)
8. Burnt sienna: main roads (topographical and military editions only)
9. Burnt sienna: 3 layer mixes from 800ft
Outline edition: plates 1,2. Water and contours edition: plates 2,4. Physical features edition: plates 2,4,5,6,7,9

Lettering:
   The lettering styles of placenames are illustrated in the legend, though there is no indication as to how the distinction between sizes of community was defined. See col."K" on p.11 for more detailed analysis of the use of lettering

Cover types recorded: H.8.3c2, H.17, H.19.1a, H.19.1b, H.77, H.78

___

   IV.2. 1:633,600 Great Britain (two sheets).   *NB*: References to the **Road Map** are classified IV.2R
___

National Yard Grid co-ordinates: 680,000yds E - 1,300,000yds E (1932); 1,120,000yds N - 1,650,000yds N - 2,180,000yds N
                                 620,000yds E - 1,300,000yds E (1937); 1,120,000yds N - 1,650,000yds N - 2,180,000yds N
Mapped area per sheet: 620,000 yards E-W by 530,000 yards S-N (1932) = 106,082.128,1 square miles
                       680,000 yards E-W by 530,000 yards S-N (1937) = 116,348.140,5 square miles

1. Map border showing:
     a. Latitude and longitude values at 30' intervals, related to a diced graduation outside them at 5' intervals
     b. Alpha-numeric system originating in the NW corner of Sheet 1 (Sheet 2 letters continue at "M") in accordance
   with the grid at 50,000 yard intervals (not IV.2R)
     c. The values at 50,000 yard intervals of the National Yard Grid (IV.2R inside the neat line on the map face)
     d. The start or end of names situated near the edge of the map
     e. The place of destination of main roads
     f. SHEET 1 (or 2) (top right corner)
2. The title panel: bottom right (IV.2 Sheet 1), top right (IV.2R Sheet 1, IV.2, IV.2R Sheet 2)
     a. Title: ORDNANCE SURVEY TEN MILE MAP OF GREAT BRITAIN (for IV.2R title see list)
     b. Scale: Ten Miles to One Inch = 1/633600, with separate imperial (10+50 miles above, 10,000+90,000 yards
   below) and metric (10+100km) scale-bars
     c. Insets of Town Areas are on the Scale of Five Miles to One Inch = 1/316800 (IV.2R, 1937 only)
     d. Legend (headed REFERENCE): "du" type (IV.2), in four boxes; "qa","qb" types (IV.2R)
     e. *Transverse Mercator Projection, origin 49°N, 2°W*, and description of the *Co-ordinate lines* (rewritten 1937)
     f. Copyright notice: *are necessary* is altered to *is necessary* in 1937
     g. *Altitudes, Submarine Contours, right of way* statements (IV.2 only)
     h. *Printed and Published by the* Director General *at the* ORDNANCE SURVEY OFFICE. *Southampton 1932* (IV.2R 1932),
   altered to: *First published by the* Director General *at the* ORDNANCE SURVEY OFFICE, *Southampton 193x, with periodical
   corrected reprints* (IV.2 and IV.2R from 1937)
     j. Reference to *glossary of Welsh words* (IV.2 Sheet 2 only)
     k. Price (Flat & unmounted) 1/- Net (IV.2R). *NB*: No closing parenthesis, and 2/- Net (IV.2)
     l. References to *Prices* in *Lists & Catalogue*, and to other *Editions* of the ten-mile map (IV.2 only)
     m. Sheet index (IV.2 only)
     n. Print or reprint code (sometimes off the panel)
3. The National Yard Grid across the map at 50,000 yard intervals, ceasing at the neat line
4. Symbols: IV.2 as IV.1. IV.2R: see col."L" on p.9 and "qa","qb" legend types illustrated on p.80
5. Layer box, with HEIGHTS IN FEET AND METRES (Sheet 1 with 13 sections, Sheet 2 with 10): on sea areas (not IV.2R)
6. Framed inset maps of the Skerry, Rona and St Kilda groups (with a location diagram), the Orkney and Shetland Islands
   with Fair Isle are on Sheet 1. The Isles of Scilly and Man extrude into the border. The Irish and French coasts
   are shown in outline, with some coastal placenames and appropriate symbols

Colour plates (IV.2R):
1. Black for everything, including unclassified roads (uncased) from 1937, except:
2. Blue: coastline, inland water, sands, sea and inland open water (screened: stippled from 1937), water feature names
3. Brown: stippled for ground tint, more concentrated for hill relief
4. Red: "A" class roads (uncased) and numbers
5. Purple: "B" class roads (uncased) and numbers

Colour plates (IV.2): IV.1 plates 1,2,3,5,6,7,8,9, ie without the contour plate. Layer colours are similar, with sienna for brown. After the first issues, MacLeod found that "*the green is too anaemic & should be strengthened*". It was. Outline edition: plates 1,2. Physical Features: plates 2,4,5,6,7,9, ie with the contour plate.
NB: See notes for exceptional uses of the blue water plate

Lettering: IV.2 as IV.1. IV.2R: see col."L" on p.11 for a detailed analysis

Cover types recorded: H.8.3c1, H.17, H.21, H.175 (original hand-lettered version - unused)

IV.3. 1:625,000 Great Britain (two sheets)

Co-ordinates: as IV.2 (1937), but converted to National Grid values. *NB*: See Appendix 1, p.178 for formulae.

1. Map border showing:
    a. Latitude and longitude values at 30' intervals, related to a diced graduation outside them at 5' intervals. From 1945 usually with associated graticule intersections on the face of the map
    b. The values at 10km intervals of the National Grid (Ordnance Survey Grid)
    c. The start or end of names situated near the edge of the map
    d. The place of destination of main roads
2. The title panel, usually top centre (Sheet 1) and top right (Sheet 2). *NB*: "X" type has elements "a" and "b" only
    a. Title: GREAT BRITAIN, followed by SHEET 1 (or 2)
    b. Scale: 1/625,000 or about Ten Miles to One Inch, with imperial (10+50 miles) and metric (10+70km) scale-bar
    c. Legend (headed REFERENCE): "ev","ew","fv","fw","fx","fy","fz" types, in four boxes
    d. EXPLANATORY NOTE ON THE ORDNANCE SURVEY GRID (rewritten for NATIONAL GRID)
    e. Sheet index showing 100km lines (100km squares numbered on revised version), with SHEET 1 (or 2) label
    f. Transverse Mercator Projection, origin 49°N, 2°W
    g. Copyright notice
    h. Publication statement: ORDNANCE SURVEY, Southampton, 19xx. (except **Physical Features**), sometimes deleted
    j. PRICE (without overprint) flat and unmounted 2/- net (**Topography** 5/-), sometimes deleted
3. In the bottom margin: bottom left: print or reprint code (occasionally on title panel)
4. The National Grid across the map at 10km intervals, ceasing at the neat line, emphasised at 100km intervals
5. Symbols: as IV.1
6. Four section layer box, with HEIGHTS IN METRES: on the sea area of the map (on some versions with layers only)
7. Framed inset maps of the Skerry, Rona and St Kilda groups (with a location diagram), the Orkney and Shetland Islands with Fair Isle are on Sheet 1. The Isles of Scilly and Man extrude into the border. The Irish and French coasts are shown in outline, with some coastal placenames and appropriate symbols
8. Overprint title panel, with price, usually bottom right for Sheet 1 and in various positions on the left for Sheet 2. Planning Series prices: 5/-, 7/6d (from 1 January 1962), 12/6d (shown without price, from 1 December 1964). Different price structures obtained for period or parliamentary maps and indexes. Original Planning Series publication statement: Published in colour by the DIRECTOR GENERAL, ORDNANCE SURVEY, 19xx, changed in 1950 to: Printed and published (or Published) by the Director General of the ORDNANCE SURVEY, CHESSINGTON(,) SURREY, 19xx, and in 1955 to *Printed and Published by* The DIRECTOR GENERAL of the ORDNANCE SURVEY, CHESSINGTON, SURREY, 19xx. Exceptions are noted individually. These title panels may cause interruption or alteration to topographical detail

Colours: Outline maps in black, grey or brown. Surviving water features were added to this plate. Some have the sea sprayed blue, or six coloured layers, produced from two strengths of brown. Otherwise, only overprints are coloured

Lettering: as IV.2

Cover types recorded: H.96.3, H.101.1, H.101.2, H.175, H.176, H.177, H.179.1, H.179.2, H.180, H.194.2

IV.4

IV.4. 1:1,250,000 Great Britain, to "D" edition

Co-ordinates: as the combined 1942 1:625,000 sheets (IV.3), which had been drawn to pre-war National Yard Grid values

1. Map border showing:
     a. Latitude and longitude values at 30' intervals (no 30' values in the south border), related to a diced graduation outside them at 10' intervals
     b. The values at 10km intervals of the National Grid
2. The title panel, top centre, initially similar to the IV.3 1:625,000 title, redrawn as necessary
     a. Title: GREAT BRITAIN (for later wording see "t" type legends on p.118)
     b. Scale: 1/1,250,000 or about Twenty Miles to One Inch, with imperial (10+60 miles) and metric (10+80km) scale-bar. "D" edition has Scale 1:1,250,000, with a redesigned scale-bar
     c. EXPLANATORY NOTE ON THE NATIONAL GRID (reduced to NATIONAL GRID REFERENCE SYSTEM example from "D" edition)
     d. Gridded sheet index with numerical (literal from "C" edition) values: the SHEET 1 (or 2) label is deleted
     e. Transverse Mercator Projection, origin 49°N, 2°W (dropped from "D" edition)
     f. Legend (headed SYMBOLS), "tc","td" types ("D" edition only). Earlier "sa","sb" types relate to entire panel
3. In the bottom margin:
     a. Bottom left: print code, sometimes [combined with] overprint code
     b. Bottom centre, initially: Crown Copyright Reserved; PRICE (without overprint) flat and unmounted 2/- net
     c. Bottom right: Published by the ORDNANCE SURVEY, 1946, changed in 1962 to: Made and Published by the Director General of the ORDNANCE SURVEY, CHESSINGTON, SURREY, 19xx (SOUTHAMPTON from 1969)
4. The National Grid across the map at 10km intervals, ceasing at the neat line, emphasised at 100km intervals
5. Symbols: see col."N" on p.9 for a detailed analysis. A legend did not appear before 1969
6. Framed inset maps of the Skerry, Rona and St Kilda groups (with a location diagram), the Orkney and Shetland Islands with Fair Isle. The Isles of Scilly extrude into the border. The Irish and French coasts are shown in outline, with some coastal placenames (including Kingstown) and appropriate symbols

Colours: black or grey outline maps: only overprints to the map are coloured

Lettering: see col."N" on p.11 for a detailed analysis

Variable attributes (IV.1,2,3,4)

1. Sheet number or name. *NB*: North of England strip maps (usually to 710km N) appear as **n**
2. Print code
3. Base map publication date. Legend. Price. *NB*: For publication statement variants, see standard attributes above. Legend types: (*NB*: Some symbols may be deleted by overprinting) (see illustration on p.80)
     "b" types: for three-sheet 1:633,600 maps (IV.1). Changes from "a" types (III.1): "Roads; First Class" become "Main Roads", "Roads; Second Class" becomes "Other Roads". "Roads in course of construction" added. "Rivers and streams" becomes "Rivers, Streams & Canals". "Contours" (200[ft]) and "Submarine Contours" (5 [fathoms]) added. There are five categories of community classification. The lighthouse symbol is alight, "Aerodrome" is added
          br With "Aerodrome" (16 symbols)
         [**bs** With "Aerodrome or Seaplane Station" (16 symbols). *NB*: This type may not exist, but see "cs")]
          bt "Aerodrome" category deleted (15 symbols)
     "c" types: "cr" introduced c.1932, "cs" c.1934, "ct" c.1935
          cr,cs,ct As "br","bs","bt" with a "National" boundary symbol added to the "County" boundary entry
     "d" type: Title panel of two-sheet 1:633,600 National Yard Grid maps (IV.2. ■IV.2 detail described in col.13)
          du 15 symbols. Changes from "c" type: no contour symbol, national boundary symbol is on a separate line, "Main Roads" superseded by "Ministry of Transport Roads, Class 1", there are two grades of "Other Roads"
     "e" types: Title panel of two-sheet 1:625,000 maps annotated Ordnance Survey Grid (IV.3)
          ev As "du" type legend lacking "Submarine Contour", "Canal" written between the canal and the river
          ew Title panel lacking legend
     "f" types: Title panel of two-sheet 1:625,000 maps annotated National Grid (IV.3)
          fv,fw As "ev","ew"
          fx As "fv" type legend with "Other Roads" reference, but lacking symbol
          fy As "fx" type legend lacking "Other Roads" reference
          fz As "fy" type legend lacking also "Roads in course of construction"

"q" types: Title panel of two-sheet 1:633,600 National Yard Grid Road Maps (IV.2R)
- **qa** Legend with "Class 1(A) Numbers 1 to 99", "Class 1(A) Other Roads", "Class 2(B)", "International" and "County" "Boundaries", "Royal Automobile Club" and "Automobile Association" "Telephone Call Boxes", "Counties", five different sized communities, and mileage measurement indicators
- **qb** As "qa" type, with "Roads under construction" and "Unclassified Roads". Minor adjustments elsewhere

"s" types: Title panel of 1:1,250,000 map, to "C" edition (IV.4)
- **sa** Original type, with National Grid (100km square numbered) sample "Newmarket", no legend
- **sb** As "sa" type, with text rewritten and respaced left of new National Grid literal index

"t" types: Title panel of 1:1,250,000 map, "D" edition (IV.4)
- **tc** With "Ordnance Survey of GREAT BRITAIN" sanserif title on redesigned panel with legend, listing "Motorways", "Trunk Roads", "Railways", "Large Towns", "Other Towns or Villages", National Boundaries", "County Boundaries". With redesigned scale-bar
- **td** With "Ordnance Survey GREAT BRITAIN" title, "and other selected roads" added to "Trunk roads", community symbols redrawn and placed flush right, more lower case lettering, National Grid sample "Dover"

"X" type (IV.3). The title with IV.3 standard attribute elements 2a, 2b only. The bottom margin has: Crown Copyright Reserved / Price Flat & unmounted 2/- Net / Ordnance Survey 1942

4. Railway revision, from O-riginal to post-war states 1 and 2 (sample changes only). *NB*: Sheet numbers as at 1937. IV.3 coupled with reservoir revision (see col.5); + or - graticule intersections (IV.3)

Sheet 1: **O**-riginal state defined by the absence of the changes noted below
- **Ou** Pateley Bridge-Lofthouse closed, Plumpton East-North Jcns (near Ulverston) opened. (no "Ov","Ow","Ox","Oy")
- **Oz** "Disused" Solway Viaduct closed
- **1** Catterick Camp Branch closed
- **2** Campbeltown-Machrihanish, Brayton-Abbey Holme, Drumburgh-Port Carlisle, Dolphinton-Broomlee, Tow Law-Burnhill, Distington-Rowrah, Brigham-Bulgill, Elvanfoot-Wanlockhead closed; mineral lines shown: eg Newton Aycliffe-Stillington, Carlton-Stockton, Burnhill-Stanhope, Barton Branch, North Monkland Loop, Dykehead Branch
- **2a** Newton Aycliffe-Stillington, Carlton-Stockton, Burnhill-Stanhope closed

Sheet 2: **O**-riginal state defined by the absence of the changes noted below
- **Os** Eastry-Sandwich Road, Totton-Fawley, Torrington-Halwill opened
- **Ot** Kenton-Debenham closed; Dungeness-Hythe, Clay Cross-Ashover opened
- **Ou** Halesworth-Southwold, Gosport-Stokes Bay, Wrangbrook-Wath closed
- **Ov** Gosport-Stokes Bay, Wrangbrook-Wath reinstated; the erroneous LM&S & GWR label on ex-L&M deleted; Stoke Jcn-Allhallows opened; mineral lines shown: eg Carn Brea-Portreath, Carn Brea-Point Quay (Devoran), Burngullow-St Denis, Ludlow-Clee Hill, Scalford-Waltham-on-the-Wolds Branches, Cromford-Parsley Hay, Mouldsworth-Helsby, Buckley-Connahs Quay, Bawtrey-Haxey, Dinnington-Kirk Sandall, Braithwell-Aire, Wrangbrook-Conisborough
- **Ow** WH&FR relabelled WHR, deviation round tunnel on Cheadle Branch opened
- **Ox** Mineral lines listed in "Ov" state deleted, the erroneous LM&S & GWR label on ex-L&M reinstated
- **Oy** Linton-Barnstaple, Chichester-Selsea closed, Lydd-New Romsey deviation opened
- **Oz** Fort Brockhurst-Lee-on-the-Solent, Quainton Road-Brill, Savernake-Marlborough (GW) closed
- **1** Pateley Bridge-Ramsgill opened
- **1z** Original Lydd-New Romsey, Pateley Bridge-Ramsgill, Welsh Highland Line, Sand Hutton Ry, Weston-Portishead closed
- **2** Sleaford-Cranwell closed, Ulceby-Immingham Dock opened, ex-L&M correctly labelled LM&SR

5. Reservoir revision. *NB*: Sheet numbers as at 1937. IV.3 reservoir information appears in col.4

Sheet 1. **a**: Blackwater (NN 3060), Ballo (NO 2405), St Andrews (NO 4712), Earlsburn (NS 7089), Kilmannan (NS 4878), Hillend (NS 8467), Threipmuir (NT 1564), Crosswood (NT 0961), Cobbinshaw (NT 0258), Gladhouse (NT 3054), Talla (NT 1122), Catcleugh (NT 7403), Waskerley (NY 0244), Tunstall (NY 0741). (no "b"). **c**: Loch Rannoch (name in upper and lower case, NN 5557). **d**: Hawes Water (NY 4713). **e**: Loch Rannoch (in upper case, NN 5557), Dunalastair Water (NN 7058), Loch Treig (NN 3370), Loch Ericht (NN 5370), Loch Laggan (NN 4585), Katrine (NN 4509). **f**: Clatteringshaws Loch (NX 5476). **g**: Carron Valley (shown as complete, NS 7084). **h**: Tongland (NX 7050), Earlstoun Loch (NX 6183), Carsfad Loch (NX 6186), Kendoon Loch (NX 6091), Carron Valley "under cons[truction]" (NS 7084). **i**: Carron Valley (completed, NS 7084). **j**: Reservoirs references deleted, but Res^r at Loch Treig remains (NN 3370), Catcleugh name remains. **k**: Deletion completed with removal of Res^r at Loch Treig (NN 3370) and Catcleugh name (NT 7403)

Sheet 2. **a**: Killington (SD 6991), Derwent (SK 1791), Alwen (SH 9454), Lake Vyrnwy (SJ 0021), Caban Coch, Craig Goch, Garreg-ddu, Pen-y-garreg (Elan Valley Reservoirs, SN 9069). **b**: Trawsfynydd placename moved (SH 6937). **c**: Trawsfynydd (SH 6937). **d**: Stocks (SD 7356), Gorple and Widdop (SD 9331), Broomhead and Moor Hall (SK 2696), Queen Mary TQ 0769. **e**: Abberton TL 9717. **f**: Eyebrook (SP 8695). **g**: Derwent placename deleted and water area added (SK 1887). **h**: "Reservoir (Lady Bower)" (SK 1887). **i**: Lady Bower water area increased (SK 1887). **j**: Lady Bower marked Res^r (SK 1887). **k**: Reservoir references deleted except Alwen (SH 9454)

6. Minor correction date (IV.1)
    Map revision (IV.3): **A**: 1942, **B**: 1944, **C**: 1948 (Physical Map only). For details see p.109 and p.111
    Map revision (IV.4): codes refer to pre-"B" edition states. Two negatives (A,B) were in use from 3363 issue onwards:
    A1: Tunbridge Wells (original state, 1947). **A2**: Grassholm Island, The Smalls, Haskeir I. named (by 11.1952).
    A3: Flintshire (det) set within the Welsh Border (23.7.1954), and St Aldhelm's Head name redrawn, the apostrophe cutting the grid (23.8.1954). **A4**: Price 2/3. The longitude value in bottom right corner of the Orkney and Shetland Isles inset panel corrected to 0°30' (15.8.1960)
    B1: Royal Tunbridge Wells (by 25.4.1951). **B2**: Flintshire (det) set within the Welsh Border (23.7.1954), and St Aldhelm's Head name redrawn, the apostrophe below the grid (23.8.1954)
8. Copyright statement or date (IV.4).   9. Graticule or grid details.   10. Magnetic variation date
11. Colour of base map. Some greys are marked **g***. These are named grey in the job files when I would have thought them brown or purple: this is perhaps explained by one reference to a mixture of 95% krystal black and 5% concentrated scarlet. There are many different shades. Krystal black looks grey as well, but those listed here are confirmed by the job files. IV.1,2: By default the water features plate is present in blue. Exceptions: + in base map colour, **x** lacking marine features, - absent. IV.3: The water plate is lacking, except: + when present on the blue plate
12. Number of coloured layers. Colour of sea. ■Number of colours in overprint. IV.1 maps by default have contours. IV.2,3 maps by default have no contours. Exceptional cases are preceded by - or +
13■ Overprint panel publication date (IV.3 published in Chessington, unless preceded by **S**-outhampton), followed by **c** if a Crown Copyright date. Price. Prefaced > if on base plate. NB: ■IV.2: This information is in col.3
14■ Date of overprinted information.   15. Print run. Date of printing.   16. Cover type
17. OSPR or GSPR (suffixed **g**).   18. Location of copies (PRO maps in ZOS 4 unless quoted otherwise).   19. Notes

---

IV.1. 1:633,600 Great Britain (three sheets), 1926

1	2	3	4	5	6	9	10	11	12	■14	16	17	18	19

Ten Mile Map of Great Britain

1	5000/26	1926br:np	0	a	-	AN	-	j	13b0	-	17	1926/4	CPT	1
2	7000-26	1926br:np	0	a	-	AN	-	j	11b0	-	17	1926/3	CPT	1,2
2	7000-26	1926br:np	0	a	-	AN	-	j	11b0	-	17	-	RH,PC	2
2	7000/26,2750/29	1926br:np	0t	a	1928	AN	-	j	11b0	-	-	-	book	3
2	1935	1926cs:np	0w	d	-	AN	-	j	11b0	-	17	-	RH	
3	no code	1925br:np	0	a	-	AN	-	j	10b0	-	17	1926/1	CPT	
3	3950/30	1925br:4/-	0u	c	1930	AN	-	j	10b0	-	17	-	Og,RH	1,4
3	1035	1933ct:np	0w	c	-	AN	-	j	10b0	-	17	-	RH	5

Sheet price 4/-,5/-,7/6d, later 2/-,3/-,5/- (DOSSSM [1935]). Extracts: of Sheet 2 in DOSSSM 1927,[1930],[1935],[1937], Sheet 3 in DOSSSM 1925, another in Cox (1932) et seq (7000/31,8000/33,8000/36). Superseded by the 1937 map
    1. Also in H.8.3c2 covers (price 12/6d)
    2. There were two printings with this code: the "7" is under the "P", then to the right of it, with larger numerals
    3. In *Report on War Office Exercise No. 3 (1929). Harrogate 1st-3rd May*, (WOP 4506, 3500 7/29 - 25% with maps). There is a copy at PRO WO 279/68
    4. Contour symbol missing
    5. Contour symbol 300[ft]. With an inset map at 1:253,440 of London. This inset, first introduced on aviation issues in 1934, already shows revision in built-up areas. The woodland is uncoloured, though the casing survives

Ten Mile Map of Great Britain: Outline Edition

1	5000/26	1926br:np	0	a	-	AN	-	jh	-0w0	-	-	1926/4	CPT	
2	7000-26	1926br:np	0	a	-	AN	-	jh	-0w0	-	-	1926/3	CPT	
3	no code	1925br:np	0	a	-	AN	-	jh	-0w0	-	-	1926/1	CPT	

Sheet price 2/6d. Superseded by the 1937 map

[Ten Mile Map of Great Britain: water and contours alone]

[1]	no code	-	-	-	-	-	-	-h	-w0	-	-	-	BFq,Eg,Lrgs	
[2]	no code	-	-	-	-	-	-	-h	-w0	-	-	-	BFq,Eg,Lrgs,NTg	
[3]	no code	-	-	-	-	-	-	-h	-w0	-	-	-	BFq,Eg,Lrgs	

Without borders or marginalia. Sheet price 1/6d (DOSSSM 1927). Breaks for overbridges have not been repaired on the blue plate. 1936 sales figures (Sheets 1,2,3: annual sales 12,16,22; stocks 385,344,319) suggest print runs of 500

120   IV.1

1	2	3	4	5	6	9	10	11	12	■14	16	17	18	19

[Ten Mile Map of Great Britain: physical features alone]
[1] no code   -   - - - -   -h 13w0 -   -   1931/4 CPT
[2] no code   -   - - - -   -h 11w0 -   -   1931/4 CPT
[3] no code   -   - - - -   -h 10w0 -   -   1931/4 CPT   1
   Without borders or marginalia. Publication date not given, but estimates for printing fifty copies of each sheet on stout (143lb) paper were made on 5 January 1932 (PRO OS 1/144 23A). Sheet price was calculated as 3/9, but, according to DOSSSM [1935], fixed at 2/- per sheet. 1936 sales figures (Sheets 1,2,3: annual sales 14,16,20; stocks 120,113,66) suggest print runs of 200. Breaks for overbridges have not been repaired on the blue plate. Superseded by the 1937 map
   1. Contour symbol missing, which would suggest the 3950/30 printing as the parent

   Ten Mile Map of Great Britain: ■War Office Staff Exercise Nº.2, 1927
3 no code   1925br:np   Os a -   AN -   j 10b2 1927   -   -   book
   In *Report on War Office Exercise No.2 (1927)*. Winchester, 9th-12th May, (WOP 3176, 5000 8/27 - 25% with maps). Copy at PRO WO 279/59

   Ten Mile Map of Great Britain: ■Map A issued with War Office Exercise Nº.2 (1929)
3 1000/29   1925br:np   Ot a 1929 AN -   j 10b3 1929   -   -   book
   In *Report on War Office Exercise No.2 (1929)*. The War Office, 22nd-24th April, (WOP 4472, 2250 7/29). Copies at PRO WO 279/67, RH (map only)

   Ten Mile Map of Great Britain: ■Map A [issued with War Office Exercise Nos 1 and 2 (1931)]
3 7250/31   1925br:4/-   Ou c 1930 AN -   j 10b2 1927   -   -   book
   In *Report on War Office Exercises Nos 1 and 2 (1931)*, (WOP 5697, 3750 9/31). Copy at PRO WO 279/73

   Ten Mile Map of Great Britain, overprinted with air information
      A1. ■**Special Air Edition (Provisional)** [title in red]
1 1100/29   1926bt:np   O   a 1928 ■gt ■1928 j 12b2 3.29   19.1a   1929/2 ABn,Lpro,PC
1 1100/29:750/29   1926bt:np   O   a 1928 ■gt ■1928 j 12b2 3.29   19.1a   -   CPT,RH
2 1100/28   1926bt:5/-   O   a 1928 ■gt ■1928 j 10b2 10.28   19.1a   1929/2 Lraf
3 1100/28   1925bt:5/-   O   a 1928 ■gt ■1928 j 9b2   10.28   19.1a   1929/2 Lraf,PC
      A2. ■**Air Edition** [title in blue]
2 1100/28:1650/30   1926bt:5/-   Ou c 1930 ■gt ■1928 j 10b2 5.30   19.1a   -   CPT,Lpro,Lrgs
2 1210/31   1926bt:np   Ou b 1931 ■gt ■1928 j 10b2 11.31   19.1a   -   RH   3
3 1650/30   1925bt:5/-   Ou c 1930 ■gt ■1928 j 9b2   5.30   19.1a   -   CPT,Lpro,Lrgs
3 1000/31   1925br:np   Ou c 1930 ■gt ■1928 j 9b2   1.31   -   PC
      A3. ■**Civil Air Edition** [title in blue]
3 360/32   1925br:5/-   Ou c 1931 ■gt ■1932 j 9b2   4.32   -   ABn
      A4. ■**Civil Air Edition** [title in purple]
1 2220/34   1926cr:np   O   a 1933 ■gt ■1934 j -11b2 30.6.34 19.1b   1934/3 CPT
1 2220/34   1926cr:■3/6   O   a 1933 ■gt ■1934 j -11b2 25.4.35 19.1b   -   Lrgs,LDg
1   O   a   ■gt ■1934 j 11b2 1.5.36   19.1b   -   PC   5
1 8037   1926cr:■3/6   O   c -   ■gt ■1934 j 11b2 29.4.37 19.1b   -   Lmd,Lu,Rg   1c
1 WO 8200/37   1926cr:■3/6   O   c -   ■gt ■1934 j 11b2 13.11.37 19.1b   -   Lraf,RH   1f
2 3095/34   1926cs:np   Ou c 1933 ■gt ■1933 j -9b2 30.6.34 19.1b   1934/3 CPT   11
2 3095/34   1926cs:np   Ou c 1933 ■gt ■1933 j -9b2 15.5.35 19.1b   -   Lrgs   11
2   Ow d ?1933 ■gt ■1933 j 9b2   12.6.36 19.1b   -   PC   5,11
2   Ow d ?1933 ■gt ■1933 j 9b2   3.12.36 19.1b   -   PC   5,11
2 8037   1926cs:■3/6   Ow d 1933 ■gt ■1933 j 9b2   10.8.37 19.1b   -   Lmd,Lu,Rg   1e,11
3 ?1900/34   Ov c ?1933 ■gt ■1933 j 8b2   19.1b   1934/3 PC   5,6,?11
3 1900/34:1950/34   1933ct:np   Ov c 1933 ■gt ■1934 j -8b2 30.6.34 19.1b   -   CPT   11
3 1900/34:1950/34   1933ct:np   Ov c 1933 ■gt ■1934 j -8b2 18.6.35 19.1b   -   Lrgs   11
3 1836   -   not found
3 1836:2036   1933ct:np   Ow c -   ■gt ■1934 j 8b2   12.6.36 19.1b   -   RH   1a,5,11
3   Ow d   ■gt ■1934 j 8b2   19.7.37 19.1b   -   PC   1d,5,11
3 9750/37   1933ct:■3/6   Ow d -   ■gt ■1934 j 8b2   7.12.37 19.1b   -   Lraf   1g,11
3 no code   1933ct:■3/6   Ow d -   ■gt ■1934 j 8b2   7.12.37 19.1b   -   Cg,Lmd,Rg,RH   1g,11

IV.1   121

1	2	3	4	5	6	9	10	11	12	■14	16	17	18	19
	B1.	■Royal Air Force Edition												
1													not found	2
2													not found	2
3													not found	2
	B2.	■Royal Air Force Edition												
1						CG							not found	4
2						CG							not found	4
3						CG							not found	4
	B3.	■Royal Air Force Edition												
1	?2220/34												not found	
1	?1035									14.11.35			not found	8
1	436	1926cr:np	O	c	-	-	-	j	11b0	-	-	-	BSg	10
1	436	1926cr:np	O	c	-	■CG	■1934	j	11b2	12.6.36	-	-	Lraf	1a,12
1	8037	1926cr:np	O	c	-	■CG	■1934	j	11b2	1.3.37	-	-	Lmd,PC	1b,12
2	3095/34	1926cs:np	Ou	c	1933	■CG	■1933	j	9b2	30.6.34	-	-	Lraf	
2	?1935									14.12.35			not found	8
2	1935:1036	1926cs:np	Ow	d	1933	■CG	■1933	j	9b2	12.6.36	-	-	Lraf,PC	1a,12
2	8037	1926cs:np	Ow	d	1933	■CG	■1933	j	9b2	1.3.37	-	-	Lmd,RH	1b,12
3	1950/34:1500/34	1933ct:np	Ov	c	1933	■CG	■1934	j	8b2	24.8.34	-	-	RH,PC	7,12
3	1035	1933ct:np	Ow	c	-	■CG	■1934	j	8b2	14.12.35	-	-	Lraf	
3	1836:2036	1933ct:np	Ow	c	-	■CG	■1934	j	8b2	12.6.36	-	-	Lraf	1a,12
3	8037	1933ct:np	Ow	c	-	■CG	■1934	j	8b2	1.3.37	-	-	Lmd	1b,12
	C.	■Military Edition												
1	?2220/34										-	-	not found	9
2	?3095/34										-	-	not found	9
3	1900/34:1950/34	1933ct:np	Ov	c	1933	■gt	■1934	j	8b2	30.6.34	-	-	PC	9,12

The print code inter-relationship of these editions will be apparent, and it is certain that several states have yet to be located. Sheet price 5/-,6/- (DOSSSM 1930). The 1934 editions were advertised as on sale flat (3/6d), mounted and folded (6/-), and flat and unmounted on Place's waterproof paper (4/6d). No Place's copies have yet been located. From 1934, Sheet 3 carries an inset map at 1:253,440 of London. The contour sample seems to have been deleted from the legend from 1934, and not restored even on those issues with contours. Superseded in 1938 by the 1:500,000 **Aeronautical Map** (RAF and Civil Air Editions), and GSGS 3955 by GSGS 3993.

Four experimental specimen portions of a map with revised characteristics are in PRO OS 1/55 22a-d, of South East England. All have relief colouring, with no contours. They carry red danger areas and a dark blue coastline. In black are railways, roads (on three), placenames, spot heights, a symbols key and compass rose, with MV 1933. 22a shows roads, one, two and four track railways, and green uncased woods. 22b has no roads or woods, 22c has woods but no roads, 22d has woods and grey roads. Their fate was sealed at the 12.12.1933 conference.

    1a 1936(NA 81), 1b 1937(NA 34), 1c 1937(NA 71), 1d 1937(NA 149), 1e 1937(NA 165), 1f 1937(NA 245), 1g 1937(NA 257)
    2. Pre-dating 1927, when this edition was modified for civilian use (PRO OS 1/55)
    3. The alpha-numerical system has been retained in the border, perhaps in error
    4. With updated air information, and the War Office Cassini Grid in its British Modified Grid form (OSR 1931-32)
    5. Only trimmed copies in 1934 covers so far recorded, lacking much marginalia
    6. A defective copy, certainly of the 1934 type, since it has only eight layers, but odd in that it has contours, supposedly lacking in 1934. Its air information predates 30.6.1934. It could well be the missing 1900/34 state
    7. This edition also carries the 1900/34 print code
    8. It is unlikely that these printings (Sheet 1 unknown as a civilian map) would have existed only on the map below
    9. GSGS 3955. Published by the War Office 1934. Listed in War Office (1936). The main roads are burnt sienna
    10. This is the base map, without overprint, to the next. A Civil Air Edition version is also possible
    11. With a note, bottom right: ■Subsequent corrections will be issued in AIR MINISTRY NOTICES TO AIRMEN
    12. With a note, bottom right: ■Subsequent corrections will be issued in AIR MINISTRY ORDERS

Ten Mile Map of Great Britain: ■(Showing Main Overhead Electrical Grid Lines)

1	1035	1926cr:np	O	a	-	-	■1934	j	-Ow2	14.11.35	-	-	Lraf
2	1935	1926cs:np	Ow	d	-	-	■1933	j	-Ow2	14.12.35	-	-	Lraf
3	1035	1933ct:np	Ow	c	-	-	■1934	j	-Ow2	14.12.35	-	-	Lmd,Lraf

The air information overprint is blue, the main overhead electrical grid lines red. See note 8 above

Supplement. Special sheets

1 2   3   4 5 6   9   10   11 12 ■14   16   17   18   19

Ordnance Survey Map of the Solar Eclipse 29th June, 1927.
- 5000/27   1927-:2/6   0 a -   AN -   j 10b1 29.6.27   77   1927/1   CPT

"*This map has been prepared in co-operation with the Joint Permanent Eclipse Committee of the Royal Society and the Royal Astronomical Society to show the path of the shadow of the Moon during the total eclipse, the magnitude of the partial eclipse visible outside the limits of totality, the Greenwich Mean Time of middle of eclipse, and the altitude of the Sun at that time. <u>Summer Time is one hour later</u> The curves have been calculated for the Committee by Dr L.J. Comrie, of the Nautical Almanac Office.*"   The cover title is: "ECLIPSE MAP: THIS *Map, on the Scale of* Ten Miles to One Inch, *has been prepared in conjunction with the Joint Permanent Eclipse Committee of the* Royal Society *and the* Royal Astronomical Society. *It shows* Times *and* Places *where the* TOTAL ECLIPSE OF THE SUN *can be seen in* England *on* JUNE 29th, 1927". In spite of this wealth of expertise, the track of the moon was still shown a mile south east of its true position, and four seconds too early! The OS printed a pamphlet to explain the errors (OSO 3000 12.5.1927). A proof copy of the cover has the erroneous wording "*Joint Permanent Committee*" (PRO OS 1/15/3). The map size is 23ins W-E by 32ins S-N, the grid sections 1 to 12½ and A to Q concurring with 5 to 16½ and C (Sheet 2) to H (Sheet 3) of the three-sheet map. The Pembrokeshire coast extrudes into the margin. A single line frame surrounds the border

Map of Scotland: 10 Miles to the Inch
- 3000/27   1927br:4/-   0 a -   AN -   j 13b0 -   78   1927/3   CPT

The NW corner is as on Sheet 1 of the three-sheet map. The map size is just over 27ins W-E by 33ins S-N. With inset maps of St Kilda and the Shetland Islands: the latter has its own scale-bar, and a title in the hatched lettering of the map title. Sheet prices 4/-,5/-,7/6d, reduced to 3/-,4/-,6/6d (DOSSSM [1935]). Copies exist with a price reduction to 3/- overprinted in red. Superseded by the 1937 map. The general notes refer also to the outline map below

Map of Scotland: 10 Miles to the Inch: Outline Edition
- 3000/27   1927br:2/6   0 a -   AN -   jh -0w0 -   -   1927/3   CPT

Airship Raid, 2-3 May, 1916
- 3084/30   1930-:np   0 a -   grid -   j 0w1   2-3.5.1916 -   -   book

Map No 31 in H.A.Jones *The War in the Air* Volume 3, Maps (OUP, 1931). Part of Sheet 1

■London and South-East England: Anti-Aircraft Defence Scheme. January, 1918
- 3100/35   1935-:np   0 a -   -   -   g 0b5   1.1918   -   -   book

In H.A.Jones *The War in the Air* Volume 5 (OUP, 1935). Part of Sheet 3

■Airship Raid maps
- 3100/35   1935-:np   0 a -   -   -   g 0b2   1917-1918 -   -   book

Map Nos 2,5,9 in H.A.Jones *The War in the Air* Volume 5, Maps (OUP, 1935). Various parts of Sheets 2,3

■Aeroplane Raid, 7th-8th March, 1918
- 3100/35   1935-:np   0 a -   grid -   g 0b2   7-8.3.1918 -   -   book

Map No 27 in H.A.Jones *The War in the Air* Volume 5, Maps (OUP, 1935). Part of Sheet 3

Indexes to the three-sheet map

Index to the Ordnance Survey Map of Great Britain on the scale of Ten Miles to One Inch (Scale 85 miles to 1 inch, in DOSSSM 1925, showing the three-sheet layout with hatched overlaps)

Index to the Ordnance Survey Map of Great Britain on the scale of Ten Miles to One Inch Published 1927 (Scale 85 miles to 1 inch, in DOSSSM 1927 and 1930, showing the three-sheet layout with hatched overlaps, and **Map of Scotland**)

With a relettered title, the first of these was was reduced in scale for use on the back cards of covered maps. The updated index was not used. The reduced version also appears with a new title: **Index to the Ordnance Survey Aviation Map of Great Britain, on the scale of Ten Miles to One Inch,** and was used on pre-1934 aviation map covers.

Index to the Sheets of the Ordnance Survey "Ten-mile" Map of Great Britain. Civil Air Edition (a newly drawn index showing towns rather than rivers and county boundaries, for use with the Michelin style aviation map cover. No overlaps are shown, though there is a note *Approximate overlap of 10 miles is given between Sheets*)

IV.2. 1:633,600 Great Britain (two sheets), 1932 and 1937

1	2	3	4	5	9	10	11	12	∎14	16	17	18	19
\multicolumn{14}{l}{Road Map of Great Britain: showing the Numbers and Classification of the Ministry of Transport}
1	5000/32	1932qa:1/-	-	-	YG	-	jx	Rb2	-	21	1932/2	CPT	2
1	2600/34	1932qa:1/-	-	-	YG	-	jx	Rb2	-	21	-	PC	2
1	2600/34:800/35	1932qa:1/-	-	-	YG	-	jx	Rb2	-	21	-	SOrm	2
1	1536	1932qa:1/-	-	-	YG	-	jx	Rb2	-	21	-	PC	2
1	15/37:1537 M37 R37	1935qb:1/-	-	-	YG	-	jx	Rb3	-	21	-	RH,PC	2
1	15/37:1838 M38 R38	1935qb:1/-	-	-	YG	-	jx	Rb3	-	21	-	RH,PC	2
1	1539 M39 R39	1935qb:1/-	-	-	YG	-	jx	Rb3	-	21	-	PC	
1	1040 M39 R39	1935qb:1/-	-	-	YG	-	jx	Rb3	-	21	-	Lrgs,RH	
1	2000/10/43 Ch	1935qb:1/-	-	-	YG	-	jx	Rb3	-	-	-	BSg,Og	
1	1544/Ch	1935qb:1/-	-	-	YG	-	jx	Rb3	-	-	-	En,Lrgs,SOrm	
2	5000/32	1932qa:1/-	-	-	YG	-	jx	Rb2	-	21	1932/2	CPT	
2	5000/32	-	-	-	YG	-	jx	Rb2	-	8.3c1	-	RH	1
2	5000/32:2000/33	1932qa:1/-	-	-	YG	-	jx	Rb2	-	21	-	RH,PC	
2	5000/32:2000/33	-	-	-	YG	-	jx	Rb2	-	8.3c1	-	PC	1
2	5600/34	1932qa:1/-	-	-	YG	-	jx	Rb2	-	21	-	Lrgs,Ng	
2	5600/34	-	-	-	YG	-	jx	Rb2	-	8.3c1	-	Lpro ZOS 4/93	1
2	2536	1932qa:1/-	-	-	YG	-	jx	Rb2	-	21	-	PC	2
2	25/37 M37 R37	1935qb:1/-	-	-	YG	-	jx	Rb3	-	21	-	RH	2
2	4038 M38 R38	1935qb:1/-	-	-	YG	-	jx	Rb3	-	21	-	Lrgs,RH	2
2	2539 M39 R39	1935qb:1/-	-	-	YG	-	jx	Rb3	-	21	-	Ob	
2	1540 M39 R39	1935qb:1/-	-	-	YG	-	jx	Rb3	-	21	-	Lrgs	
2	1500/10/43 Ch	1935qb:1/-	-	-	YG	-	jx	Rb3	-	-	-	RH	
2	2044/Ch	1935qb:1/-	-	-	YG	-	jx	Rb3	-	-	-	Lrgs,SOrm	
2	1000/12/47	1935qb:np	-	-	YG	-	jx	Ow3	-	-	-	Lrgs,RH	3

With inset maps at 1:316,800 of the Glasgow, Manchester, West Riding towns (Keighley was added in 1937), West Midlands and London areas. The scale of these was not noted until 1937. The "overprint" colours here refer to road colours. Covered sheet prices 2/3d,2/9d,5/6d, Ansell-fold 6/6d (DOSSSM [1935]) and 7/6d (DOSSSM [1937]). Later versions of the Sheet 1 cover include the word "Scotland". The cover exists also in Benderfold. An extract of Sheet 2 appears in DOSSSM [1935],[1937].

1. For use in Ansell-fold and wall mounted versions, therefore lacking title panel. Being flat and still retaining its margins, the PRO copy is complete, and so far the only one recorded in this condition. It has the print code and CCR in the bottom border. These details would have been removed during mounting
2. These issues can also be found in H.8.3c1 covers: later Ansell-fold issues have yet to be confirmed
3. The town stipple on the outline negatives was replaced with more widely spaced dots in October 1944. Proofs were sent to the War Office in November 1947, and these 1000 copies printed in December following order WOPD 4184/OS, one would hope, the final printing of a map showing the National Yard Grid!

Ordnance Survey Ten Mile Map of Great Britain

1	6037	1936du:2/-	O	e	YGan	-	j	13b0	-	17	1937/2	CPT	1
1	1538 M37 R36	1936du:2/-	Oz	f	YGan	-	j	13b0	-	17	-	PC	
1	1539 M39 R39:1541	1936du:2/-	Oz	g	YGan	-	j	13b0	-	17	-	Lpro,RH	2
2	6037	1936du:2/-	Ox	d	YGan	-	j	10b0	-	17	1937/2	CPT	
2	M37 R36 4038	1936du:2/-	Oy	d	YGan	-	j	10b0	-	17	-	Lrgs	
2	M39 R39 3039	1936du:2/-	Oz	e	YGan	-	j	10b0	-	17	-	RH	
2	2041	1936du:2/-	Oz	e	YGan	-	j	10b0	-	17	-	BSg,Lpro,SOrm	

Proof copies with and without contours were printed, but these have not been located. Covered sheet prices 3/-,5/- (DOSSSM [1937]): the map was apparently not sold unmounted in covers. Later copies of the Sheet 1 cover include the word "Scotland". The cover exists also in Benderfold. A new Characteristic Sheet appeared to accompany the publication of this map (DOSSSM [1937] 12 - not found). Not replaced until the 1955 map. There is an extract of Sheet 2 in *Final Report...*, (Specimen 6, 2000/38, lacking grid)

1. 6037 changed, probably from 5036; the sea around the northern isles inset on the sheet diagram is uncoloured
2. No 1539 M39 R39 version has been found, and with work apparently ceasing before the issue of the paper, the evidence of the job file is that it was not printed. It was probably resumed in 1941

1	2	3	4	5	9	10	11	12	■14	16	17	18	19

Ordnance Survey Ten Mile Map of Great Britain (Outline Edition)
1	6037	1936du:2/-	0	e	YGan	-	jh	Ow0	-	-	1937/2	CPT	1
1	?15/38								-	-	-	not found	
1	15/38:1540 M39 R39	1936du:2/-	Oz	h	YGan	-	jh	Ow0	-	-	-	Lrgs,CCS	2
1	1000/7/41	1936du:2/-	Oz	h	YGan	-	jh	Ow0	-	-	-	En,Lpro,SOrm	
2	6037	1936du:2/-	Ox	d	YGan	-	jh	Ow0	-	-	1937/2	CPT	
2	M39 R39 3039	1936du:2/-	Oz	e	YGan	-	jh	Ow0	-	-	-	Lrgs	
2	1000/7/41	1936du:2/-	Oz	e	YGan	-	jh	Ow0	-	-	-	Lpro,SOrm	

1. 6037 changed, probably from 5036.   2. 1569 copies printed on 11.1.1940 (printer's tally on CCS copy)

Ordnance Survey Ten Mile Map of Great Britain: Physical Features alone
| 1 | 337 | 1936-:2/- | - | e | - | - | -h | +13w0 | - | - | 1937/2 | CPT | |
| 2 | 337 | 1936-:2/- | - | d | - | - | -h | +10w0 | - | - | 1937/2 | CPT | |

No border. No base map: the title panel was added to the contour plate which was unused on the topographical map

Ordnance Survey Ten Mile Map of Great Britain (Outline Edition) / [Valuation Regions]
| 1 | 500/39 | 1936du:np | Oz | f | YGan | - | j | Ow5 | nd | - | 1939/1 | CPT,Lrgs | |
| 2 | 500/39 | 1936du:np | Oy | d | YGan | - | j | Ow6 | nd | - | 1939/1 | CPT,Lrgs | |

Prepared for the Central Valuation Board, Mines Department, under the provisions of the Coal Act, 1938. Counties in Wales and in England north of the Shropshire-Suffolk line are overprinted, as are all county boundaries

■Monastic Britain (South Sheet)
| 2 | proof | ■1939:np | Oy | - | YGan | - | g- | Ow1 | nd | 175 | - | SOrc | |

Compiled by R.N.Hadcock. O.G.S.Crawford was informed of Hadcock's portrayal of the monastic tradition in map form in 1935, and Hadcock quickly agreed to publication by the OS. Early moves to publish the map at 1:1,000,000 were overruled by MacLeod, who insisted on the new two-sheet ten-mile map. This south sheet proof was printed on 6.1.1940, whereupon instructions for a print run of 3000 were made, but there is no evidence that this ever took place. The water plate is lacking in its entirety, though it would presumably have appeared on any published map. The map was finally published in 1950 on the 1:625,000 base. Hadcock's model at 1:633,600 of the north sheet (showing the 50,000 yard lines of the National Yard Grid) is in SOrc. The hand lettering on Jerrard's cover design (see illustration on p.94), designed for this map, was replaced in 1950 by letterpress. Lit: Hellyer (1987),(1988),(1989a),(1989d)

■Retriangulation of Great Britain: Primary Observations
| 1 | | | | | | | | | | | | not found | |
| 2 | | | | | | | | | | | | not found | |

■Retriangulation of Great Britain: Secondary Blocks
| 1 | | | | | | | | | | | | not found | |
| 2 | | | | | | | | | | | | not found | |

There is scanty documentary evidence that these maps were printed in November 1945 (fiuo). Subsequent work in 1951 (at 1:625,000) suggests that they were at the imperial ten-mile scale. See R5050 file and control cards (CCS), and IV.3

Supplement 1. Editions published by the Land Utilisation Survey of Britain

---

■Land Utilisation Map of Great Britain: South Sheet
| 2 | no code | ■nd:np | Oz | e | YGan | - | j | Ow6 | 1931-39 | x | - | CPT,Lse | |
| 2 | no code | ■nd:np | Oz | e | YGan | - | j | Ow0 | - | - | - | Lse | 1 |

Sheet 2 only, reduced from one-inch LUS maps by Phyllis M.Boyd and Marguerite V.Coldman. Printed on the OS outline and water base map in 1940 by G.W.Bacon & Co. Ltd, and published by LUS. Some were mounted and folded into green quarto covers. The British Library copy was acquired on 20.12.1940. The stock of 2500 copies was largely destroyed by enemy action on 10.5.1941. Bacon had made plates for Sheet 1, but these were destroyed in the air raid before any proof copies were run off. All that survived was a compiler's model, which Willatts handed over to the OS on 17.9.1941 (PRO OS 1/155 96). From it printing plates were created for the post-war map. It has not been located. See pp.106,132

1. Used, probably by Stamp, for a drawing in black of the area boundaries for the planned pre-war Land Fertility (Land Classification) map. Lundy and the Scilly Isles were not overdrawn, though the Isle of Man was. Lit: Stamp (1962) 352; see also note on **Land Classification**, p.132

1	2	3	4	5	9	10	11	12	■14	16	17	18	19

■Types of Farming Map of England and Wales

n	no code	-		Oz	h	YGan	-	j	Ow8	6.39	x	-	Lbl,Ob,PC
2	Copyright 0.70	■nd:np		Oz	e	YGan	-	j	Ow8	[6.39]	x	-	Lbl,Ob,PC

Prepared by MAF in June 1939. Printed in 1941 by G.W.Bacon & Co. Ltd on OS outline and water base maps, and published by LUS. Some copies were folded and bound into blue quarto covers with the Explanatory Bulletin *A Farming Type Map of England & Wales* (London, published for LUS by Geographical Publications Ltd, 1941). The conterminous edges were left without marginalia for mounting together. Titular areas not overprinted: Isles of Lundy, Caldy, Skomer, Skokholm, Holy Isle. One hundred advance copies were supplied to MAF and distributed by them. The British Library acquired a copy on 6.5.1941. The remainder of the stock, including the copyright copies, together with the printing plates, was destroyed by enemy action on 10.5.1941. The text reappeared in a second (temporary) edition in 1941 with a monochrome map at 1:1,267,200 printed by George Philip & Son, Ltd, and published for LUS by Geographical Publications Ltd.

A ten-mile coloured map of Scotland was prepared by LUS in conjunction with DAS, showing a simple division of Scotland into five predominant farming types, but it proved impossible to print it before the war. Two monochrome diagrams at approximately 40 miles to the inch were prepared instead, entitled **A Classification of predominant farming types in Scotland** and **Types of Farming in Scotland**, and published in Stamp (1941b). A coloured map at 1:625,000 appeared in 1944

Supplement 2. Military issues

Ordnance Survey Ten Mile Map of Great Britain: ■Military Edition

1	WO 11273/39	1936du:np	Oz	h	■CG	■1.38	j	13b1	nd	-	-	BSg	
1	■15000/6/42 LR	nd:du:np	Oz	g	■CG	■1.40	j	13b1	nd	-	-	Lmd,Lpro,Lrgs,RH	1
1	■14000/3/43 LR	nd:du:np	Oz	g	■CG	■1.40	j	13b3	nd	-	-	PC	
1	■20500/12/43 Wa	nd:du:np	Oz	g	■CG	■1.40	j	13b3	nd	-	-	Lmd,Lrgs	
1	■20500/12/43 Wa	nd:du:np	Oz	g	■CG	■1.40	j	13b3	nd	-	-	Lrgs,Ob	2
1	1000/48	nd:du:np	Oz	g	■CG	■1.40	j	13b3	nd	-	-	LPg,RH	
2	WO 20000/39	1936du:np	Oz	e	■CG	■1.38	j	10b1	nd	-	-	BSg,Mg,RH	
2	■20000/6/42 LR	nd:du:np	Oz	f	■CG	■1.40	j	10b1	nd	-	-	Lpro,Lrgs,Ob	1
2	■14000/3/43 LR	nd:du:np	Oz	f	■CG	■1.40	j	10b3	nd	-	-	Lmd	
2	■20500/12/43 Wa	nd:du:np	Oz	f	■CG	■1.40	j	10b3	nd	-	-	Lmd,Lrgs	

GSGS 3993. No publisher or publication date given, but War Office, 1938. First issues have no road destinations in borders. After 1939 brown hill tints replace sienna. The three-colour overprints distinguish the Irish Grid area (red, hence "GSGS 3993" on Sheet 1 is thereafter also in red), the North European Grid Zone III (blue: new grid letters after 3/43), the French Lambert Zone [I] (red) and the Nord de Guerre Zone (blue) from the War Office Cassini Grid area (purple). In order to incorporate the North European Grid Zone III, the title panel on Sheet 1 was moved to the top of the sheet. The only notes then supplied referred to projection (previously removed), altitude and submarine contours

1. The print code is added to the water plate. PRO copies are at ZOS 3/262,263.   2. Print code redrawn

Ordnance Survey Ten Mile Map of Great Britain (Outline Edition): ■Military Edition

?1		1936du:np	Oz	h	■CG	■1.38	j	0b1	nd	-	-	not found	
1	■7502/6/41	nd:du:np	Oz	g	■CG	■1.40	j	0b1	nd	-	-	Lpro Air 20/8543	
1	■20500/12/43 Wa	nd:du:np	Oz	g	■CG	■1.40	j	3b3	nd	-	-	BSg,Lmd,RH	1
2	no code	1936du:np	Oz	e	■CG	■1.38	j	0b1	nd	-	-	Mg	
2	■7502/6/41	nd:du:np	Oz	f	■CG	■1.40	j	0b1	nd	-	-	Lpro Air 20/8543	
2	■20500/12/43 Wa	nd:du:np	Oz	f	■CG	■1.40	j	0b3	nd	-	-	BSg,Lmd,RH	1

GSGS 3993. Publication and panel details as above. The main roads are coloured

1. This issue derived from the layered map, hence the three blue layers on Sheet 1 where the top three layer bands of the original required various permutations of blue, a need which was fulfilled from the water plate

Ordnance Survey Ten Mile Map of Great Britain (Outline Edition): Military Edition

1	7000/12/41 LR	nd:du:np	Oz	h	CG	1.40	g	+0w0	nd	-	-	Lpro Air 20/8543	
1	3500/2/48	nd:du:np	Oz	g	CG	1.40	g	+0w0	nd	-	-	Lmd	1
2	5000/4/42 LR	nd:du:np	Oz	f	CG	1.40	g	+0w0	nd	-	-	Lpro Air 20/8543	
2	3500/2/48	nd:du:np	Oz	f	CG	1.40	g	+0w0	nd	-	-	Lmd	2

GSGS 3993. Publication and panel details as above. First issues used for "Top Secret Defence Measures" KPID 1042

1. With an inset of Northern Ireland.   2. GSGS number lacking

1	2	3	4	5	9	10	11	12	■14	16	17	18	19

Ordnance Survey Ten Mile Map of Great Britain: ■Second Military Edition

| 1 | 15000/4/50 | ■1950du:np | 0z | i | ■NGm | ■1.50 | j | | 13b3 | nd | - | - | Lmd,Lrgs,Ob | |
| 2 | 20000/4/50 | ■1950du:np | 1z | g | ■NGm | ■1.50 | j | | 10b3 | nd | - | - | Lmd,Lrgs,Ob,RH | |

GSGS 3993. Published by War Office, 1938, this second edition in 1950. The hill tints are brown, not sienna

Ordnance Survey Ten Mile Map of Great Britain (Outline Edition): Second Military Edition

| 1 | 3000/4/50 | 1950du:np | 0z | i | NGm | 1.50 | g | | +0w0 | nd | - | - | Lmd | |
| 2 | 3000/4/50 | 1950du:np | 1z | g | NGm | 1.50 | g | | +0w0 | nd | - | - | Lmd | |

GSGS 3993. Published by War Office, 1938, this second edition in 1950

Commands, Areas and Administrative Area Divisions
Home Commands Boundary Map [from Fifth Edition]

GSGS 3993A: First Edition ?1939; Second Edition 10.1951; Third Edition "Restricted" 10.1955; reprinted "Restricted" 7.1957; Fifth Edition "Restricted" 1957; reprinted 9.1958; Seventh Edition 1961

Essential Traffic Routes Map / Ordnance Survey Ten Mile Map of Great Britain (Outline Edition): Military Edition

1	5036 M39 R39:15/38	1936du:np	0z	h	CG	1.38	j+	0w2	nd	-	-	RH	1
1	15/38:■4200/40	1936du:np	0z	h	CG	1.38	j+	0w2	8.40	-	-	RH	2
1										-	-	not found	3
1	7000/12/41 LR	nd:du:np	0z	h	CG	1.40	g+	0w2	12.41	-	-	PC	4
1	5040/4/43 LR	nd:du:np	0z	h	CG	1.40	g+	0w2	3.43	-	-	En	5
2										-	-	none found	

GSGS 3993B. The information is not given, but the map was presumably printed by the War Office, and in 1940
1. Overprinted "Secret"
2. Overprinted "Second Edition / For Official Use Only: Not to be published". This edition also carries the print code 5036 M39 R39
3. Third Edition
4. Overprinted "Fourth Edition December 1941 / For Official Use Only. Not to be published"
5. Overprinted "Fifth Edition / Security". With military road numbers

Control of Flying in Balloon Areas

GSGS 3993C (1941). This title was subsumed into GSGS (Misc.) 501 (qv) on 16.8.1949

Reserve Group Boundaries (RAF)

GSGS 3993/1D

Home Command Boundary Map

GSGS 3993E (1961). With an overprint on GSGS 3993A of Civil Defence Regional Boundaries. Corrected to 1.4.1961

Ordnance Survey Ten Mile Map of Great Britain: ■Civil Aviation Edition

GSGS 4395, c.1943. Not found and almost certainly not printed. A two-sheet map, listed in War Office (1944)

Geographical Organisation Static Engineer Force

HQ Engr Ser. SOS ETO USA, 1942. The ten-mile map with an American overprint. Sheet 2 in Lmd

---

Supplement 3. Royal Air Force issues

"*The Ten Mile base was used for a wide range of RAF products, usually classified. On some the OS title panel is replaced by an RAF legend. Some do not have the GSGS number blocked out. Some do not have the Misc. number printed on them. Some have AD Maps numbers. Some have GSGS "OR" numbers associated with the overprint. Most wartime versions have an Air Ministry referenced copy number eg SD229, CD1022, F06 etc. Almost any combination of these can be found*" (letter Ian Mumford to author, 9 December 1991). The following RAF versions are recorded:

Restricted Flying Areas Defended Areas and Balloon Areas

GSGS (Misc.) 501

1	2	3	4	5	9	10	11	12	■14	16	17	18	19

■Danger! Overhead Electricity Cables. Look - or you'll leap!
1	15000/7/44 Ch	■1944:np	Oz	h		CG■g	1.40	r+	Ow1	nd	-	-	Lrgs,Ob
2											-	-	not found	1

GSGS (Misc.) 502. Compiled and drawn at GSGS (AM), February 1944, from information supplied by DAS (AM). With an inset map entitled **Northern Ireland** showing the War Office Irish Grid. The map also insisted that it was "*Not to be taken into the air*". There was also a "*Warning. This sheet indicates the main overhead electricity transmission system and isolated groups of two or more wireless masts as far as can be ascertained. Individual masts and other high obstructions have not been shown.*"

1. A copy was in the MTCP Library (listed in the MTCP 1947 unpublished catalogue at U/E/7/4)

■Security-Released Airfields in the United Kingdom: Correct to 31ˢᵗ December 1944
1	10000/2/45 Ch	■1944:np	Oz	h		CG■g	1.40	r+	Ow1	31.12.44	-	-	Lrgs,RH
2	10000/2/45 Ch	■1944:np	Oz	f		CG■g	1.40	r+	Ow1	31.12.44	-	-	En,Lrgs

GSGS (Misc.) 505. [?Compiled and drawn at] GSGS (AM), 1944. With the same inset map of Northern Ireland

RAF Stations in the British Isles
GSGS (Misc.) 506

Armament Training Areas
GSGS (Misc.) 507

RAF Stations
GSGS (Misc.) 517 (1946)

■Aerodrome and Royal Air Force Non-Flying Unit Location Map
1	2000/10/54 OS	■1954du:np	2a	i		NGm■g	1955	g+	Ow2	1.1.54	-	-	Ob
2	2000/10/54 OS	■1954du:np	1z	g		NGm■g	1.55	g+	Ow2	1.1.54	-	-	Ob

GSGS (Misc.) 1636. Published by War Office, Edition 1 GSGS, 1954. Drawn and reproduced by OS. With inset map of Northern Ireland. The graticule is on the blue plate, together with aerodrome co-ordinates, the information supplied by DD Ops (Nav.) Air Ministry; the RAF non-flying unit information is on the green plate, this information supplied by Org. Stats. Air Ministry. The maps were classified "Restricted"

Supplement 4. County indexes. *NB*: Sheets 1-3 of the Scotland index are at 1:253,440

Map of the Counties of Caithness, Inverness, Orkney, Ross and Cromarty, Sutherland, Zetland, showing Burghs, Civil Parishes and Parliamentary Areas

4	950/8/42LR	1942:1/6	0	-	-	-		gh+	Ow0	-	-	1942/3	CPT,PC	1
4	950/8/42LR	1942:2/6	0	-	-	-		gh+	Ow1	1.8.42	-	1942/3	CPT	2
4	950/8/42LR	1942:3/6	0	-	-	-		gh+	Ow2	1.8.42	-	1942/3	CPT	3
4		1942:1/6	0	-	-	-		-t+	0-0	-	-	1942/3	not found	4

1. Index to sheet lines only.   2. Showing Civil Parish boundaries [red]
3. Showing Parliamentary and Municipal boundaries, etc.   4. On tracing paper

Map of the Counties of Caithness, Inverness, Orkney, Ross and Cromarty, Sutherland, Zetland,: ■showing Burghs, Districts and Civil Parishes: The Parliamentary Constituencies scheduled in Representation of the People Act 1948

4	L316	1947:3/-	0	-	-	-		sh+	Oy0	-	-	-	not found	1
4	L316:■ba	1947:3/-	0	-	■NG	-		sh+	Oy3	1.11.49	-	-	not found	2
4	L316:■baa	1947:4/6	0	-	■NG	-		sh+	Oy4	1.11.49	-	1/50	CPT	3

1. Combined index to sheet lines of the Old Six Inch Series.   2. Showing Local Government boundaries [red]
3. Showing Parliamentary [green] and Local Government [red] boundaries

Map of the Counties of Caithness, Inverness, Orkney, Ross and Cromarty, Sutherland, Zetland, showing The sheet lines of the Six Inch maps (County Series)

4	L536	1953:3/6	0	-	-	-		gh+	Ow0	-	-	4/53	DRg	

Publication date as given bottom left. Also surviving (bottom right): Ordnance Survey 1947

128   IV.3

IV.3. 1:625,000 Great Britain, 1942. *NB*: An alphabetical listing follows the base maps

1	2	3	4	6	11	12	■13	■14	15	16	17	18	19

NB: Titles marked "+" are members of the Ministry of Town & Country Planning Series

Great Britain: Sheet 1 (or Sheet 2): Scale: 1/625,000 or about Ten Miles to One Inch [outline base map]

1	no code	1942X:np	1h	A-	j	Ow0	-	-	1052:3.4.42	-	-	Lrgs	1
1	no code	1942X:2/-	1h	A-	j	Ow0	-	-	do do	-	-	Bg,Eu,Og	1
1	no code	1942X:2/-	1h	A-	g	Ow0	-	-	do do	-	-	BSg	1
1	no code	1942X:2/-	1h	A-	jh	Ow0	-	-	do do	-	-	Lrgs	1
1	500/10/42 LR	1942ev:2/-	1h	A-	g*	Ow0	-	-		-	6/43	CPT,Lrgs	
1	1244/Ch	1942fx:2/-	2h	B-	j	Ow0	-	-	1200C:10.8.44	-	-	Lrgs,RH	
1	1244/Ch	1942fx:2/-	2h	B-	k	Ow0	-	-	do do	-	-	SOrm	
1	1244/Ch	1942fx:2/-	2h	B-	s	Ow0	-	-	do do	-	-	Eu,Og,SOrm	
1	4045/Cr	1942fx:2/-	2h	B+	j	Ow0	-	-		-	-	Bg,Lrgs,RH	
1	4045/Cr	1942fx:2/-	2h	B+	kh	Ow0	-	-		-	-	Lpro,SOrm,RH	
1	3801	1942fx:2/-	2i	B+	j	Ow0	-	-	2900:17.8.48	-	-	Lraf,RH	
1	3801	1942fx:2/-	2i	B+	k	Ow0	-	-	do do	-	-	Cg,RH	
1	3801	1942fx:2/-	2i	B+	gh	Ow0	-	-	990:21.10.49	-	-	RH	
1	C	1942fx:2/6	2i	B+	j	Ow0	-	-	2047:11.7.53	-	-	Gg,Lse	
2	no code	1942X:np	1f	A-	j	Ow0	-	-	996:3.4.42	-	-	Lrgs	1
2	no code	1942X:2/-	1f	A-	j	Ow0	-	-	do do	-	-	Bg	1
2	no code	1942X:2/-	1f	A-	g	Ow0	-	-	do do	-	-	BSg	1
2	no code	1942X:2/-	1f	A-	jh	Ow0	-	-	do do	-	-		1
2	500/10/42 LR	1942ev:2/-	1f	A-	g*	Ow0	-	-		-	6/43	CPT,Lrgs	
2	1244/Ch	1942fx:2/-	2f	B-	j	Ow0	-	-	1200C:10.8.44	-	-	Lrgs,RH	
2	1244/Ch	1942fx:2/-	2f	B-	k	Ow0	-	-	do do	-	-	SOrm	
2	1244/Ch	1942fx:2/-	2f	B-	s	Ow0	-	-	do do	-	-	Og,SOrm	
2	4045/Cr	1942fx:2/-	2f	B+	j	Ow0	-	-		-	-	Bg,Lrgs	
2	4045/Cr	1942fx:2/-	2f	B+	kh	Ow0	-	-		-	-	Lpro,SOrm	
2	3802	1942fx:2/-	2j	B+	j	Ow0	-	-	3912:17.8.48	-	-	Cg,Lraf,RH	
2	3802	1942fx:2/-	2j	B+	k	Ow0	-	-	do do	-	-	PC	
2	?3802	1942fx:2/-	2j	B+		Ow0	-	-	1000C:25.1.49	-	-		
2	3802	1942fx:2/-	2j	B+	gh	Ow0	-	-	1010:21.10.49	-	-	RH	

Only published maps are listed here: reference to MTCP printings is made on p.104. Superseded by the "Ten Mile" Outline Map, 1955. See also **Newlyn Levelling** and **Retriangulation of Great Britain** below
1. Cumulative print run information from OS date stamps on the unpriced copies in the RGS

Great Britain: **Topography**: Sheet 1 (or Sheet 2)

1	proof	1944fx:5/-	2h	B-	g	3b0	-	-	6.11.44	-	-	BSg
1	2044 Ch	1944fx:5/-	2h	B-	g	3b0	-	-	2166:14.12.44	-	1944/4	CPT
2	2044 Ch	1944fx:5/-	2f	B-	g	3b0	-	-	2168:15.12.44	-	1944/4	CPT
2	3823	1942fx:5/-	2i	B+	g	3b0	-	-	1585:29.6.51	-	-	Ng,CCS

First known as the **Physical Background Map**. The decision to add the 50-metre contour line was taken at the planning meeting on 22.2.1943. The **Topography** title was adopted on 16.10.1944, and the wording of the title panel was compressed to include it (PRO OS 1/162). This base map, or its layer plates, was used for **Ancient Britain**, **Monastic Britain**, **Railways** and **Roads**. Unique numbers allocated: 3822,3823 (allocated in 1947, but not used until 1951. See p.191)

+Administrative Areas

1	no code	1944fx:np	1h	B-	g	Ow6	>1944:5/-	1.11.44	2143	-	1944/4	CPT,Lrgs	
1	3803	1944fx:np	1h	B+	g	Ow6	>1944:5/-	1.11.44	2000C:7.12.46	-	-	Cg,NTg,Sg,RH	
1	3856	1944fx:np	1i	B+	g	Ow6	>1944:5/-	1.4.51	1060:11.2.53	-	-	Lbl,Mg,RH	
2	proof	1942ev:np	1f	A-	g	Ow1	>nd:np	1.5.43	9.11.43	-	-	Lrgs	1
2	■2044 Ch	1944fx:np	1f	B-	g	Ow6	>1944:5/-	1.11.44	2105	-	1944/4	CPT,Lrgs	
2	3804	1944fx:np	1f	B+	g	Ow6	>1944:5/-	1.11.44	3000C:7.12.46	-	-	Og,Sg	
2	3804	1944fx:np	1f	B+	g	Ow6	>1944:5/-	1.11.44	1000C:26.4.50	-	-		
2	3804	1944fx:np	1f	B+	g	Ow6	>1944:5/-	1.2.50	1645:11.2.53	-	-	Lbl,Lu,RH	

IV.3   129

1   2   3   4   6   11 12 ■13   ■14   15   16   17   18   19

Compiled by OS, with MTCP and the Scottish Home Department and DHS. Colours used: purple for county borough, green for municipal borough, red for urban district: wash for infill, solid for names, and solid red for most boundaries. With inset map at 1:253,440 of London. Superseded by the 1956 map. An extract of Sheet 2 is in DOSSSM 1947,1951

1. The proof has a smaller title panel with title **Administrative Areas Map**. It lacks the green and purple plate of the final maps, but boxes have been prepared for them, and there is space in the lists. This copy was overprinted in green by the "spotted dog" technique by MTCP for **Bomb Damage: Towns with over 100 Dwellings destroyed or damaged beyond repair by 30th June 1943**. The overprint was derived from confidential information supplied by MOH

+**Administrative Areas** [outline edition]
1	no code	1942fx:np	1h	B-	g	Ow1	>1944:5/-	1.3.44	500C:24.10.44	-	-	Cg,PC	1
1	no code	1942fx:np	1h	B-	gh	Ow1	>1944:5/-	1.3.44	200C:24.10.44	-	-	[MTCP]	1
1	■3803	1944fx:np	1i	B+	gh	Ow1	>1944:5/-	1.11.44	200C:1.9.48	-	-	RH	
2	no code	1942fx:np	1f	B-	g	Ow1	>1944:5/-	1.3.44	300C:24.10.44	-	-	Cg,PC	1
2	no code	1942fx:np	1f	B-	gh	Ow1	>1944:5/-	1.3.44	200C:24.10.44	-	-	[MTCP]	1
2	■3804	1944fx:np	1i	B+	gh	Ow1	>1944:5/-	1.11.44	100C:1.9.48	-	-	Mg,RH	

Compiled as above. Boundaries marked, but no names. With inset map at 1:253,440 of London

1. An issue for MTCP: some of the medium weight paper copies at least were put on sale

■**Ancient Britain**: (North [or South] Sheet): [First Edition]: ("B" edition covers give "Second Edition")
1	■3624	nd:fx:np	2i	B+	g	3b1	1951:6/9	nd	6000C:15.3.51	179.1	4/51	CPT	
1	■3624	nd:fx:np	2i	B+	g	3b1	1951:7/-	nd	500C:4.7.55	-	-	Lse	1
1	■4224	nd:fx:np	2k	B+	g	3b1	1951:7/-	nd	600C:22.4.59	-	-	Ng,NTg	1
■N ■B		-	2k	B+	g	3b1	1964c:np	-1066	6000C:5.10.64	180	3/65	CPT	
■N ■B		-	2k	B+	g	3b1	1964c:np	-1066	13000C:12.3.68	180	-	RH	
2	■3625	nd:fx:np	2i	B+	g	3b1	1951:6/9	nd	10000C:15.3.51	179.1	4/51	CPT	
2	■3625	nd:fx:np	2i	B+	g	3b1	1951:7/-	nd	1000C:4.7.55	-	-	Lse	1
2	■4225	nd:fx:np	2k	B+	g	3b1	1951:7/-	nd	1200C:22.4.59	-	-	Ng,NTg	1
■S ■B		-	2k	B+	g	3b1	1964c:np	-1066	11000C:5.10.64	179.2	3/65	CPT	2
■S ■B		-	2k	B+	g	3b1	1964c:np	-1066	28000C:12.3.68	179.2	-		
■S ■B		-	2k	B+	g	3b1	S1964c:np	-1066	4500C:9.5.77	179.2	-	RH	3

Compiled by C.W.Phillips and his staff. Conceived as the **1951 Festival Tourist Map**. It was aimed at the popular market in that every site marked had to be visible on the ground and even significant, but not later than 1066. With letterpress. On the **Topography** base. Superseded by the "C" edition, 1982. Lit: Hellyer (1987),(1988),(1989a),(1989d)

1. These were paper flat issues: covered copies were still in stock. The paper flat price is 3/6d (PRO OS 1/440)
2. Leighterton long barrow from SO to ST 819913, Legis Tor (SX 569652) symbol changed from neolithic to bronze age
3. Probably, by some eight years, the last reprint on the 1942 base map

Boundary Commissions for Scotland & England (Sheet 1) **England & Wales** (Sheet 2): **House of Commons (Redistribution of Seats) Acts, 1944 and 1947**: ■**Map showing Parliamentary Counties, County Divisions and Parliamentary Boroughs and the Corresponding Administrative Areas**
1	L54:■aa	1946fx:np	1h	B+	g	Ow3	nd:np	31.10.47		-	-	BPP	1
1	L58:■aa	1946fx:np	1h	B+	gh	Ow3	nd:5/-	31.10.47		-	-	Eu	
2	L55:■aa	1946fx:np	1f	B+	g	Ow3	nd:np	1.6.47		-	-	BPP	1
2	L59:■aa	1946fx:np	1f	B+	gh	Ow3	nd:5/-	1.6.47		-	-	Eg	

With inset maps at 1:253,440 of the London area and at 1:126,720 of South Staffordshire. Both title panels of Sheet 2 are split, the remnants sharing a third panel on the left

1. Maps 1 and 2 in *Initial Report of the Boundary Commission for England* (BPP(HC) 1947-48 Cmd 7260, XV). Map 1 also appears in *Initial Report of the Boundary Commission for Scotland* (BPP(HC) 1947-48 Cmd 7270, XV), and map 2 in *Initial Report of the Boundary Commission for Wales* (BPP(HC) 1947-48 Cmd 7274, XV)

Boundary Commissions for Scotland & England (Sheet 1) **England & Wales** (Sheet 2): **House of Commons (Redistribution of Seats) Act, 1949**: ■**Map showing Parliamentary Constituencies and the Corresponding Administrative Areas**
1	L615:■aa	1946fx:np	1h	B+	g	Ow3	1954:np	1.6.54		-	-	BPP	
2	L615:■a	1946fx:np	1f	B+	g	Ow3	nd:np	1.6.54		-	-	BPP	

With inset maps at 1:63,360 of Edinburgh, and at 1:253,440 of Birmingham and South Staffordshire. They appear as maps 1 and 2 in *First Periodical Report of the Boundary Commission for England* (BPP(HC) 1953-54 Cmd 9311, IX). Map 2

1	2	3	4	6	11	12	■13	■14	15	16	17	18	19

also appears in *First Periodical Report of the Boundary Commission for Wales* (BPP(HC) 1953-54 Cmd 9313, IX). Both title panels of Sheet 2 are split, the remnants sharing a third panel on the left. The double use of the same unique number would appear to be an error: probably Sheet 2 should carry L616

Boundary Commission for ■Scotland>: House of Commons (Redistribution of Seats) Act, 1949: ■Map showing Parliamentary Constituencies as proposed by the Report on the Boundary Commission for Scotland, 1954, and the Corresponding Administrative Areas / ■Scotland

| 1 | L709:■aa | 1946fx:np | 1h | B+ | g | Ow3 | 1954:np | 1.6.54 | - | - | BPP | |

With an inset map at 1:63,360 of the Edinburgh area. Scotland only is overprinted. It appears as Map 1 in *First Periodical Report of the Boundary Commission for Scotland* (BPP(HC) 1953-54 Cmd 9312, IX)

+■Coal and Iron

1	proof 1	1944fz:2/-	2h	B+	g*	Ob4	1945:5/-	1940	31.8.45	-	-	BSg	
1	5046/Cr	1944fz:2/-	2h	B+	g*	Ob4	1945:5/-	1940	3741:8.2.46	-	2/46	CPT	
1	5046/Cr	1944fz:2/-	2h	B+	g*	Ob4	1945:5/-	1940	1200:8.2.46	-	2/46	RH	1
2	proof 1	1945fx:2/-	2f	B+	g*	Ob3	1945:5/-	1940	20.8.45	-	-	BSg,RH	
2	5046/Cr	1945fx:2/-	2f	B+	g*	Ob3	1945:5/-	1940	3809:14.2.46	-	2/46	CPT	
2	5046/Cr	1945fx:2/-	2f	B+	g*	Ob4	1945:5/-	1940	1200:14.2.46	-	2/46	Ng	1
2	■4354	1945fx:np	2i	B+	g*	Ob3	1945:7/6	1940	1036:28.5.62	-	-	Lrgs,RH	2

Compiled by the MTCP from GS maps and information from MFP and MOS. With "Generalised Sections through the Chief Coalfields". The title panel was redrawn after the proof state. Unique numbers allocated: 3820,3821

1. Overprinted by MFP: "*The Names of the Colliery Districts of the Ministry of Fuel and Power are shown in dark blue*". Not the responsibility of MTCP.   2. The base map price statement has not been completely deleted

+■Electricity: Statutory Supply Areas

1	proof 1	1944fz:2/-	2h	B+	g*	Ob6	1946:5/-	1945	4.1.46	-	-	RH	
1	4000/7/46 Ch	1944fz:2/-	2h	B+	g*	Ob6	1946:5/-	1.46	4192:29.7.46	-	9/46	CPT	
2	proof 1	1944fx:2/-	2g	B+	g*	Ob6	1946:5/-	1945	4.1.46	-	-	RH	
2	4000/7/46 Ch	1944fx:2/-	2g	B+	g*	Ob6	1946:5/-	1.46	4113:29.7.46	-	9/46	CPT	

Compiled by the MTCP and the DHS, from information in the possession of the Electricity Commissioners and the North of Scotland Hydro-Electric Board. With inset map at 1:253,440 of London. Areas left in outline and lacking overprint: Isles of Man, Lundy, Scilly. An error on the green plate meant that this one colour had to be reprinted, on 20.9.1946. Unique numbers allocated: 3828,3829

■Gas Act 1948 Areas of Gas Boards superimposed on Map of Administrative Areas: [3382,3383] Map A,B

1	3370	1944fx:np	1i	B+	g	Ow7	fouo:np	12.48	250E:12.11.48	-	-	not found	1
1	3370	1944fx:np	1i	B+	g	Ow7	>fouo:5/-	12.48	50:9.12.48	-	-	not found	1
1	3382	1944fx:np	1i	B+	g	Ow7	>nd:5/-	21.12.48	998:26.1.49	-	2/49	CPT	
1	3382	1944fx:np	1i	B+	g	Ow7	>nd:5/-	21.12.48	387E:13.8.56	-	-	Gg	2
2	3371	1944fx:np	1i	B+	g	Ow7	fouo:np	12.48	250E:12.11.48	-	-	not found	1
2	3371	1944fx:np	1i	B+	g	Ow7	>fouo:5/-	12.48	50:9.12.48	-	-	not found	1
2	3383	1944fx:np	1i	B+	g	Ow7	>nd:5/-	21.12.48	1173:26.1.49	-	2/49	CPT	
2	3383	1944fx:np	1i	B+	g	Ow7	>nd:5/-	21.12.48	571:31.5.50	-	-		
2	3383	1944fx:np	1i	B+	g	Ow7	>nd:7/6	21.12.48	458:22.8.62	-	-	Cg,RH	

An OS repayment job for MFP. The OS sold reprints of the map through their agents and accounted for the sales to MFP. With inset maps at 1:253,440 of the London area and Yorkshire (West Riding). Area left in outline and lacking overprint: Isle of Man. The administrative areas are as at 1.11.1944

1. Not seen, so details uncertain. First issues for MFP only, so not published. The reprints have corrected gas area boundaries.   2. This version has the lower half of the unique number obliterated

+■Gas and Coke

| 1 | ■3854 | 1942fx:2/- | 2i | B+ | g* | Ob4 | 1951:5/- | 5.49 | 1850:25.10.51 | - | 11/51 | CPT | 1 |
| 2 | ■3855 | 1942fx:2/- | 2j | B+ | g* | Ob4 | 1951:5/- | 5.49 | 2893:25.10.51 | - | 11/51 | CPT | 1 |

Compiled by MTCP and DHS from information supplied by MFP. Area left in outline and lacking overprint: Isle of Man. Unique numbers allocated: 3840,3841, but for some reason these were in the event not used after proofs carrying them had been printed on 20.4.1947 (Sheet 1) and 13.5.1947 (Sheet 2) (not found)

IV.3   131

1	2	3	4	6	11	12	■13	■14	15	16	17	18	19

■Geological Map: of Scotland and the North of England (Sheet 1) of England & Wales (Sheet 2)

1	proof	1945fx:np	1h	B-	gh	Ow12	1946:np	nd	undated	-	-	BSg	1
1	■2500/12/47	1945fx:np	1h	B-	gh	Ow12	1948:12/6	nd		194.2	1948/1g	CPT	
1	■3000/49	1945fx:np	1h	B-	gh	Ow12	1948:12/6	nd		194.2	-	CPT	
1	■4000/51	1945fx:np	1h	B-	g	Ow12	1951:12/6	nd	1829:29.9.51	194.2	-		
1	■4000/51	1945fx:np	1h	B-	gh	Ow12	1951:12/6	nd	1250:11.5.51	-	-	Ng,RH	
1	■550/56	1945fx:np	1h	B-	gh	Ow12	1948:12/6	nd		-	-	DRg,Og	
2	■2500/48	1942fx:np	2f	B+	gh	Ow12	1948:12/6	nd		194.2	1948/3g	CPT	
2	■3000/49	1942fx:np	2f	B+	gh	Ow12	1948:12/6	nd		194.2	-	CPT	
2	■4000/51	1942fx:np	2f	B+	g	Ow12	1951:12/6	nd		194.2	-		
2	■4000/51	1942fx:np	2f	B+	gh	Ow12	1951:12/6	nd		-	-	Ng,RH	
2	■3200/54	1942fx:np	2f	B+	g	Ow12	1951:12/6	nd	1820:28.6.54	194.2	-		
2	■3200/54	1942fx:np	2f	B+	gh	Ow12	1951:12/6	nd	1221:28.6.54	-	-	Lpro,Mg	
2	■500/56	1942fx:np	2f	B+	gh	Ow12	1951:12/6	nd		-	-	DRg,Og	

Prepared by the Geological Survey. Part of Ireland also overprinted on Sheet 1. Some island names were moved offshore - Coll, Tiree, Rhum, Eigg, Barra, Mull. Islay's name was also deleted, but apparently not replaced. NG values are missing around the title panel of Sheet 2. Superseded by the 1957 map. *NB*: Number of overprint colours uncertain

1. Undated proof, with eleven colour dots in the left hand border: yellow, deep yellow, orange, scarlet, blue, deep blue, pale pink, purple, sepia, grey and gold

■Geological Map: of Scotland and the North of England (Sheet 1) of England & Wales (Sheet 2) [outline edition]

| 1 | ■500/51 | 1945fx:np | 1h | B- | gh | Ow1 | 1951:3/6 | nd | 509:27.11.51 | - | 12/51 | CPT | |
| 2 | ■500/51 | 1942fx:np | 2f | B+ | gh | Ow1 | 1951:3/6 | nd | 525:27.11.51 | - | 12/51 | CPT | |

Prepared by the Geological Survey. General note as above. Also recorded in GSPR 1951/4

+■Igneous and Metamorphic Rocks

| 1 | proof | 1942fx:2/- | 2i | B+ | g | Ob6 | 1955:5/- | 1950 | 6.2.56 | - | - | CCS,RH | |
| 2 | proof 1 | 1942fx:2/- | 2j | B+ | g | Ob6 | 1955:5/- | 1950 | 10.11.55 | - | - | RH | |

Compiled by MHLG and DHS in collaboration with GS and with the assistance of the Granite and Whinstone Federation. Data relating to quarry output supplied by the MFP. Unpublished (but listed in DOSSSM 1957), eighteen proof copies only of each sheet being printed. MTCP manuscript version dates from 1946. The model for this map by S.H.Beaver is recorded on 6.12.1948 (PRO OS 1/162). Proofs for colour models were made 1.1951, then little else occurred until these proofs were printed: what the problem with them was is not recorded. The roadstone trade groups of rocks were divided into granite, gabbro, porphyry, basalt, gritstone, quartzite, schist, hornfels and unclassified categories, with quarries and mines distinguished by their 1950 output into five tonnage categories. Area left in outline and lacking overprint: Isle of Man. Explanatory Text 7 would have accompanied this map (listed in Explanatory Texts 4,8)

Index to 6-inch Air Photo Mosaics

| 1 | | | | | | | | | | - | - | not found | |
| 2 | | | | | | | | | | - | - | not found | |

In *Price List of Ordnance Survey Publications* for 1.7.1952 at 2/- per sheet

Index to the Ordnance Survey Maps at Scale of 1:25,000 on National Grid Sheet Lines - see National Grid Index

■Index to the Ordnance Survey Maps at Scale of Six Inches to the Mile on National Grid Sheet Lines

1	■42	1944fz:2/-	2i	B+	s	Ow1	nd:2/-	1.3.50		-	-	Dtf	1
1	no code	nd:fz:2/-	2i	B+	k	Ow1	1954:2/-	nd		-	7/54	CPT	2
2	■42	1942fx:np	1i	A+	s	Ow1	nd:2/-	1.3.50		-	-	Dtf	1
2	no code	nd:fx:np	1i	A+	k	Ow1	1954:2/-	nd		-	7/54	CPT	2

With inset maps at 1:253,440 of Glasgow and London. Superseded on the 1955 base

1. With National Grid numerical references. The base map title panel is crossed through with the blue overprint
2. With National Grid literal references. With an enlarged overprint title panel. The base map NG note is deleted

+■Iron and Steel

| 1 | proof 1 | 1944fz:2/- | 2h | B+ | g* | Ob7 | 1945:5/- | 1944 | 30.8.45 | - | - | BSg,RH | |
| 1 | 3046 Cr | 1944fz:2/- | 2i | B+ | g* | Ob7 | 1945:5/- | 1944 | 3054:6.2.46 | - | 2/46 | CPT | |

132   IV.3

1	2	3	4	6	11	12	■13	■14	15	16	17	18	19
2	proof	1945fz:2/-	2f	B+	g*	Ob4	1945:5/-	1944		-	-	RH	1
2	proof 1	1945fz:2/-	2f	B+	g*	Ob7	1945:5/-	1944	30.8.45	-	-	BSg	
2	3046 Cr	1945fz:2/-	2f	B+	g*	Ob7	1945:5/-	1944	3075:16.2.46	-	2/46	CPT	
2	■4355	1945fz:np	2i	B+	g*	Ob7	1945:7/6	1944	803:28.5.62	-	-	Lrgs,RH	

Compiled by MTCP from maps of the GS and from data supplied by MFP and MWT. With inset maps at 1:253,440 of the Glasgow to Motherwell and Swansea areas, and at 1:63,360 of the Middlesborough, Sheffield and Scunthorpe areas. Unique numbers allocated: 3826,3827

 1. Undated proof overprinted in the black, red, maroon and brown plates only

+■Land Classification

1	proof 1	1944fz:2/-	2h	B+	g	Ob8	1945:5/-	1938-42	4.2.46	-	-	BSg,RH	
1	5046 Cr	1944fz:2/-	2h	B+	g	Ob8	1945:5/-	1938-42	5170D:25.4.46	-	4/46	CPT	
2	proof 1	1942fv:2/-	2f	A-	g	Ow8	1944:5/-	1938-42	31.5.44	-	-	BSg	1
2	2044 Ch	1944fx:np	2f	B-	g	Ob8	1944:5/-	1938-42	1955:1.12.44	-	1945/1	CPT	
2	50/46 Cr	1944fx:np	2g	B+	g	Ob8	1944:5/-	1938-42	4750:22.2.46	-	-	Bg,Lrgs,RH	
2	5046/Cr	1944fx:np	2i	B+	g	Ob8	1944:np	1938-42	750:20.11.69	-	-	Eu,Mg,CCS	2

A Tentative Land Fertility Map of England and Wales was drawn at 1:1,000,000 in 1938 by Stamp and Willatts at the request of Sir Montague Barlow for the use of his Royal Commission on the Geographical Location of the Industrial Population. It was reduced and reproduced in *Nature* CXLIII, No 3620, 18.3.1939, 456-459. Stamp and others continued work on the subject and completed a ten-mile model which was shown at the Oxford Conference organised by the Town and Country Planning Association, 28-31.3.1941. This may be the uncoloured copy of the pre-war **Land Utilisation Map**, now at LSE (qv). Hopes of early publication were dashed by the bombing of the LUS headquarters, so three temporary monochrome versions entitled **Fertility Map of England and Wales** were printed by George Philip & Son, Ltd, and published at 1:1,267,200 in Stamp (1941a), with the subtitles **Land of Good Quality**, **Land of Intermediate Quality** and **Land of Poor Quality**. Work on a revised map began in 1943, incorporating new data supplied by MAF and MTCP. Stamp and Willatts drew land classification categories on LUS one-inch maps: these were reduced to quarter-inch and finally ten-mile scale. Stamp himself prepared the data for Scotland, which was developed into a map by DHS and DAS. Willatts and his MTCP Maps Office brought the map to publication. Areas lacking overprint: Isles of Sule Skerry, Rona, St Kilda. With Explanatory Text 1. Unique numbers allocated: 3818,3819. Lit: Stamp (1962) 351ff

 1. There are no NG values around the title areas.   2. The final reprint of a Planning Series map on this base

+■Land Utilisation

1	proof 1	nd:X:np	1h	A-	j	Ob5	nd:np	1931-39	17.7.42	-	-	Lpro	
1	2000/1/43 LR	1942ev:np	1h	A-	j	Ob5	nd:5/-	1931-39	1895:9.3.43	-	6/43	CPT	1
1	4045/Cr	1942fx:np	2h	B+	j	Ob5	1944:5/-	1931-39	4065:19.11.45	-	-	Bg,Og	
1	4353	1942fv:np	2i	B+	j	Ob5	1944:7/6	1931-39	1046:18.5.62	-	-	Lrgs	
1	4353	1942fv:np	2i	B+	j	Ob5	1944:np	1931-39	2550:4.10.67	-	-	Cg,Mg,Ng	
2	proof 1	1942ev:2/-	1f	A-	j	Ob6	nd:np	1931-39	26.10.42	-	-	Lpro	
2	2000/1/43 LR	1942ev:np	1f	A-	j	Ob6	nd:5/-	1931-39	1975:9.3.43	-	6/43	CPT	1
2	4045/Cr	1942fx:np	2f	B+	j	Ob6	1944:5/-	1931-39	4044:19.11.45	-	-	Bg,Og	
2	4063	1942fx:np	2i	B+	j	Ob6	1944:5/-	1931-39	1660:9.7.56	-	-	Lkg,Lrgs	
2	4063	1942fx:np	2i	B+	j	Ob6	1944:7/6	1931-39	1450:18.5.62	-	-	Lrgs	
2	4063	1942fx:np	2i	B+	j	Ob6	1944:np	1931-39	3000:4.10.67	-	-	Cg,Mg,Ng	

See also the pre-war map (p.124), and p.106. Willatts and his MTCP Maps Office brought the map to publication. Blue stipple is uniquely used for the sea. The reprints have an improved overprint title panel. The 1944 revisions were made to the reprints of this map, but not by the OS: they too were evidently added by LUS or MTCP (PRO OS 1/156, which also contains the proof maps). An extract of Sheet 2 is in DOSSSM 1947,1951. Unique numbers allocated: 3807,3808

 1. Of these LUS took one thousand copies of each sheet in payment for the production materials provided by Stamp

Large Scale Field Programme

1	3802				k	Ow4	fiuo	10.52	90:20.2.53	-	-	not found	1
1	■3965				k	Ow4	fiuo	1.55	200:21.6.55	-	-	not found	
2	■3781				k	Ow4	fiuo	10.52	93:20.2.53	-	-	not found	1
2	■3966				k	Ow4	fiuo	1.55	200:21.6.55	-	-	not found	

 1. Proof copies carrying the unique numbers were printed on 3.2.1953 (Sheet 1) and 20.12.1952 (Sheet 2), which explains why they are not consecutive. On the proof, 3802 was on the black overprint plate. Superseded in 1958

IV.3   133

1	2	3	4	6	11	12	■13	■14	15	16	17	18	19

+■Limestone (Sheet 1); ■Limestone: including Chalk (Sheet 2)
```
1   3977        1942fx:2/-   2i   B+   g   Ob5   1955:5/-    1949    585:17.8.55     -     2/57   CPT
1   3977        1942fx:2/-   2i   B+   g   Ob5   1955:5/-    1949    1033:18.6.57    -     -
1   no code     1942fx:np    2i   B+   g   Ob5   1955:np     1949    700:2.2.68      -     -      Mg            1
2   3978        1942fx:2/-   2j   B+   g   Ob5   1955:5/-    1949    865:17.8.55     -     2/57   CPT
2   4125        1942fx:2/-   2j   B+   g   Ob5   1955:5/-    1949    1807:18.6.57    -     -      Cg,Lrgs,Mg    2
```
Compiled by MHLG and DHS in collaboration with GS. Data relating to quarry output supplied by MFP, MOT, MOS, MOW. The symbols for quarries are in four sizes according to output. The choice of set sizes, rather than symbols directly proportional to output, was necessitated by an undertaking given to those who supplied the information that the map would not reveal the output of individual quarries. MTCP manuscript versions dated from 1943. Listed as in preparation in the 1945 press release. The map was delayed for a further eighteen months after printing in order to resolve some unfortunate misunderstandings with quarry owners, and to add "including Chalk" to the Sheet 2 title. This was overprinted on 6.12.1956. With Explanatory Text 4. Unique numbers allocated: 3844,3845
   1. The base map price reference has been poorly erased.   2. With an additional site at TQ 4657

+■Local Accessibility: The Hinterlands of Towns and Other Centres as determined by an Analysis of Bus Services
```
1   ■3944       1942fx:2/-   2i   B+   k   Ob4   1955:5/-    1947-51  1114:8.2.55    -     2/55   CPT
1   ■3944       1942fx:2/-   2i   B+   k   Ob4   1955:5/-    1947-51  2535:28.3.56   -     -
2   ■3945       1942fx:2/-   2j   B+   k   Ob4   1955:5/-    1947-51  1539:8.2.55    -     2/55   CPT
2   ■3945       1942fx:2/-   2j   B+   k   Ob4   1955:5/-    1947-51  3000:28.3.56   -     -
```
Compiled by MHLG and DHS from information supplied by the bus operators and the Ministry of Transport and Civil Aviation. The map was proposed as Spheres of Influence on 18.6.1949 (PRO OS 1/432 1). Areas lacking overprint: London Metropolitan area, Isles of Scilly. See also p.107. With Explanatory Text 6

+■Ministry of Labour local office areas
```
1   3849        1942fx:np    2h   B+   g   Ow1   fouo        11.46    200:13.5.47    -     -      RH            1
1   3849        1942fy:np    2i   B+   g   Ow1   fouo        11.46    ?do            -     -      NTg           2
1   ?3849       1942fy:np    2i   B+   g   Ow2   fouo        11.46    ?do            -     -      [MTCP]        3
2   3850        1946fy:np    2i   B+   g   Ow1   fouo        11.46    200C:8.3.47    -     -      Bg,NTg,RH     2
2   ?3850       1946fy:np    2i   B+   g   Ow2   fouo        11.46    ?do            -     -      [MTCP]        3
```
Not published, though printed for Government use. Plates entitled **Employment Exchange Areas** survived until 1966 when the OS asked Willatts if they were still required (PRO OS 1/999 138). Renamed when the Ministry of Labour changed the name of its offices. The need of MTCP for the map, with town insets, was discussed at the 30.7.1946 meeting (PRO OS 1/557). With inset maps at 1:253,440 of the East Lancashire and West Midlands conurbations and London, and at 1:126,720 (with National Grid!) of Glasgow and the Tyneside and Wearside area. Areas lacking overprint: Isles of Shetland, Man, Scilly, Fair Isle. The information was compiled by MTCP from data supplied by Ministry of Labour and National Service. Superseded by (?unpublished) maps on the 1955 "Ten Mile" base
   1. Print run information from a printer's tally. This version lacks area names. No similar Sheet 2 has been found
   2. Prepared MTCP 1948, printed OS 1948. Copies in NTg are inscribed: "*This map is not normally published. Presented to the Department by the Regional Office, Ministry of Town and Country Planning, Jan. 1950*". This legend state is otherwise unrecorded. Listed in the unpublished MTCP Catalogue: Maps Accession List 10 (Nov.-Dec.1948)
   3. Prepared MTCP 1948, printed OS 1948. With names in red (MTCP Catalogue: Maps Accession List 10 (Nov.-Dec.1948))

+■Ministry of Labour local office areas / Administrative Areas [as at 1.11.1944]
```
1   ■3803       1944fx:np    1i   B+   gh  Ow1   >1944:5/-   11.46    25C:24.9.48    -     -      Bg
1   ■3803       1944fx:np    1i   B+   gh  Ow1   >1944:5/-   11.46    30C:19.5.49    -     -      not found
2   ■3804       1944fx:np    1i   B+   gh  Ow1   >1944:5/-   11.46    25C:24.9.48    -     -      Bg
2   ■3804       1944fx:np    1i   B+   gh  Ow1   >1944:5/-   11.46    20C:19.5.49    -     -      not found
```
For official use only, despite the survival of the **Administrative Areas** publication details

■Monastic Britain: (North Sheet) or (South Sheet) [First Edition]
  ■Monastic Britain: North (or South) Sheet: Second Edition [from 3980,3927 editions]
```
1   3395        nd:fx:np     2i   B+   g   3b1   1950:2/6    nd       2187:27.9.50   175   11/50  CPT           3
1   3618        nd:fx:np     2i   B+   g   3b1   1950:2/6    nd       2150:16.3.51   175   -      RH            3
1   3980        nd:fx:np     2i   B+   g   3b2   1955:3/-    nd       6212:2.9.55    175   11/55  CPT
1   3980        nd:fx:np     2i   B+   g   3b2   1955:np     nd       1305:20.4.65   175   5/65                 5
```

134 IV.3

1	2	3	4	6	11	12	■13	■14	15	16	17	18	19
1	A	nd:fx:np	2i	B+	g	3b2	1955:np	nd	2090:5.7.65	176	-	RH	
1	A	nd:fx:np	2i	B+	g	3b2	1955:np	nd	9000:14.2.68	176	-	RH,CCS	
2	proof 2	1942fx:2/-	2i	B+	g	Ow1	1948:np	nd	1.12.47	-	-	SOrc	1
2	proof	1942fx:2/-	2i	B+	g	Ow0	1948:np	-	undated	-	-	SOrc	2
2	3394	nd:fx:np	2i	B+	g	3b1	1950:2/6	nd	3258:27.9.50	175	11/50	CPT	
2	3394	nd:fx:np	2i	B+	g	3b1	1950:2/6	nd	2838:16.3.51	175	-	RH	4
2	■3927	nd:fx:np	2i	B+	g	3b2	1954:3/-	nd	7583:10.11.54	175	11/54	CPT	
2	3927	nd:fx:np	2i	B+	g	3b2	1954:3/-	nd	5812:28.11.55	175	-	RH	
2	A	nd:fx:np	2i	B+	g	3b2	1954:np	nd	5143:17.2.65	177	5/65	RH	
2	A	nd:fx:np	2i	B+	g	3b2	1954:np	nd	13000:13.2.68	177	-	RH	

Compiled by R.N.Hadcock. With an inset map at c.1:22,588 entitled **Religious Houses in London.** On base maps 3801 and 4045/Cr. With letterpress. Superseded by the "B" edition, 1976. Lit: Hellyer (1987),(1988),(1989a),(1989d)

1. Second proof for the unpublished 1948 version. Discontinued once the decision was made to remove the Ogilby road system, to use modern names, and to revise the diocesan peculiar boundaries. Lacking the inset map
2. A proof copy, without overprint, used as the model for the first published version, 3394. Lacking the inset map
3. All known copies of 3395 have a correction sticker in the symbols panel: the correction is made on 3618
4. With an additional site at Rhedynog-Felen (SH 4657), and a deletion at SH 4150
5. The price was "blocked out" on these copies of the previous issue: it seems to be this announced in OSPR 5/65

■**National Grid Index to 1:25,000 Sheets** [50046 only]
■**Index to the Ordnance Survey Maps at Scale of 1:25,000 (about 2½ Inches to the Mile) on National Grid Sheet Lines**

1	■50046	1944fz:np	2h	B+	g	Ow1	nd:2/-	nd	11.45	-	5/46	CPT	
1	■768	1944fz:np	2i	B+	k	Ow2	1954:2/3	nd		-	10/54	CPT	
1	■768	1944fz:np	2k	B+	k	Ow2	1954:np	nd		-	-	Lkg	
1	no code	1944fz:np	2k	B+	k	Ow2	1954:np	nd		-	-	Ob	
2	■50046	1942fx:np	1f	A+	g	Ow1	nd:2/-	nd	11.45	-	5/46	CPT	
2	■769	1942fx:np	1i	A+	k	Ow2	1954:2/3	nd		-	10/54	CPT	
2	■769	1942fx:np	1k	A+	k	Ow2	1954:np	nd		-	-	Ob,RH	

Printed 11.1945. The overprint colours were blue in 1946, red and purple in 1954. Both versions seem to have been maintained in print in order to provide indexes with both numerical and literal National Grid references. The note on the National Grid in the base map title panel was deleted. With inset maps at 1:253,440 of Glasgow and London

National Grid Index to 1:25,000 Sheets / Administrative Areas

| 1 | | | | | | | | | | - | - | not found | |
| 2 | | | | | | | | | | - | - | not found | |

A special printing for MTCP was reported to the Maps Advisory Committee meeting on 30.7.1946 (PRO OS 1/557)

■**Newlyn Levelling: Diagram of Levelling Blocks**

| 1 | 3801 | 1942fx:2/- | 2i | B+ | gh | Ow1 | fiuo | nd | 56 | - | - | RH | |
| 2 | 3802 | 1942fx:2/- | 2j | B+ | gh | Ow1 | fiuo | nd | | - | - | RH | |

Print run information from a printer's tally. There is reference to block numbers, secondary and tertiary levelling in hand and completed, geodetic and secondary lines actual and proposed, fundamental bench marks and tidal stations

Physical Map Great Britain: Sheet 1 (or Sheet 2)
Physical Features Great Britain: Sheet 1 (or Sheet 2) [from 1949]

1	proof 2	1942ew:2/-	-h	A-	j+	6b1	-	-	15.10.43	-	-	RH	
1	proof 3	1942fw:np	-i	A-	j+	6b1	-	-	6.9.46	-	-	SOos	1
1		1942fw:np	-i	A-	jh+	6b1	-	-	154:1.11.46	-	-	not found	2
1	proof	C1948fw:np	--	C-	j+	8b1	-	-	22.3.49	-	-	CCS	3
1	no code	C1948fw:np	--	C-	j+	Ow1	-	-	116:12.12.49	-	-	Ob	4
1	no code	C1948fw:np	--	C-	j+	+Ow1	-	-	16:14.12.49	-	-	RH,CCS	4
1		C1948fw:np	--	C-	j+	8b1	-	-	147:5.1.51	-	-	not found	5
2	proof 2	1942ew:2/-	-f	A-	j+	6b1	-	-	15.10.43	-	-	Lkg,RH	
2	proof 3				j+	6b1	-	-	7.9.46	-	-	not found	1
2					jh+	6b1	-	-	113:1.11.46	-	-	not found	2
2	proof	C1948fw:np	--	C-	j+	8b1	-	-	20.10.48	-	-	CCS	

IV.3   135

1	2	3	4	6	11	12	■13	■14	15	16	17	18	19
	2	no code	C1948fw:np	--	C-	j+	0w1	-	-	110:12.12.49	-	-	Ob	4
	2	no code	C1948fw:np	--	C-	j+	+0w1	-	-	16:14.12.49	-	-	RH	4
	2	no code	C1948fw:np	--	C-	j+	8b1	-	-	122:5.1.51	-	-	Lpro	5
	2	proof 4	C1948fw:np	--	C-	j+	8w1	-	-	7.8.51	-	-	Lpro,CCS	6

Not published. The "overprint" colour is the black name plate. Unique numbers allocated: 3809,3810. See p.111. Lit: Hellyer (1992a), where a more extensive list of states, known and unknown, is provided

1. Sheet 1 with an inset outline map at 1:2,500,000 of Great Britain, north half: see the illustration on p.113. Copies existed in MTCP (listed in 1947 unpublished catalogue at T/G/7/5)
2. A special printing, with the title deleted, made for LM&S Railway Co, School of Transport, Derby
3. The job file has Crabb's signature on the above date, the map on 24.9.1949, but it was examined in June
4. These are probably copies of a special printing done for MOH. The layer boxes are divided into nine sections. The only copy of the non-contoured version seen was overdrawn in red by MHLG as **Hydrometric Areas** (a title glued over the original). This overprint is disregarded here
5. A special printing, with the title deleted, made for British Railways, Derby. The PRO copy is in OS 1/705
6. A proof version with specimen layer colours in Snowdonia and East Anglia only. The PRO copy is in OS 36/4

+Population Changes, 1921-31

"*The two sheets of this map are not so far advanced, and in view of the probable difficulties over Scotland, Dr Willatts asked for all work on these two sheets to be suspended for the time being.*" (PRO OS 1/162 117A, 3.4.1944). Eventually cancelled awaiting development of new specification. Sheet 1 was "*not printed*", Sheet 2 "*prepared only*", according to PRO OS 1/999 132A. Work on it ceased on 24.5.1944. Like the next, red would have been used for increases and blue for decreases in population

+■Population Changes, 1931-38 / ■Provisional Edition For Official Use Only

2	544 Ch	1942fx:np	1f	B-	g	0w3	1944:5/-	1931-38	500:19.9.44	-	-	CPT,Lpro	1
2	proof 2	1942fx:np	1f	B-	k	0b4	1944:5/-	1931-38	22.2.45	-	-	BSg	2
2	proof	1942fx:np	1f	B-	k	0b3	1944:5/-	1931-38	22.2.45	-	-	NTg,RH,PC	3

Prepared by MTCP. DHS disliked the map in this form, which led to the cancellation of Sheet 1 and the 1921-31 map (above) pending further experiments. Thus this was but a temporary edition, and a meeting of the Maps Advisory Committee on 17.7.1944 determined to add the superscription, to print only 500 copies, and to issue the whole edition to MTCP for distribution (PRO OS 1/556). Isle of Man is overprinted. With an inset map at 1:253,440 of London.

1. Overprinted in red, blue and black. The PRO copy is in OS 1/162. Copies have a typed note appended:

    This map (Sheet 2 only) was one of the earliest to be prepared in the series of national maps on the 1:625,000 scale. The statistics on which it was based contained certain inaccuracies, the majority of which were due to inadequate correction for the difference between resident and enumerated populations in 1931. The colouring of the map may therefore be a shade too deep or a shade too light in certain localities.

    It is intended to produce a revised map which will remedy these minor defects and at the same time cover the period 1931-39 instead of 1931-38 and be more fully comparable with maps now in preparation showing population changes during other periods. In the meantime, since the present map gives a useful indication of the broad changes which took place before the war, it is being given a limited and provisional circulation. October, 1944

2. Experimental proofs offering violet or buff administrative boundaries. The job also required versions with boundaries in brown (lacking) or black (see next). Green replaces the original's cold blue for population decreases. Lundy is no longer overprinted. Eight of each type were printed
3. Proof in red, green and black, the black plate including the administrative area boundaries

Population: Changes by Migration, and Population: Total Changes maps

Deficiencies in the specification of the previous two maps led to new editions with revised specifications prepared by MTCP and its successor, MHLG, with DHS, from information taken from various census returns and estimates (not Isle of Man). With an inset map at 1:253,440 of London. Explanatory Text 3 was written for eight Population maps

+■Population: Changes by Migration 1921-1931

1	p1:3834	1942fx:2/-	2i	B+	g*	0b6	1947:5/-	1921-31	24.5.49	-	-	Eg,Gg,CCS
1	3834	1942fx:2/-	2i	B+	g*	0b6	1949:5/-	1921-31	1135:16.11.49	-	12/49	CPT
2	proof	1942fx:2/-	2g	B+	g*	0b6	1947:5/-	1921-31	24.5.49	-	-	Eg,Gg,CCS
2	no code	1942fx:2/-	2i	B+	g*	0b6	1949:5/-	1921-31	1005:16.11.49	-	12/49	CPT

Unique numbers allocated: 3834,3835. Instructions for 3835 to appear were disregarded. The CCS are "Exam" copies

1	2	3	4	6	11	12	■13	■14	15	16	17	18	19

+■Population: Changes by Migration: Scotland 1931-38: England and Wales 1931-39 (Sheet 1) **1931-39** (Sheet 2)

1	p1:3834:■3838	1942fx:2/-	2i	B+	g*	Ob6	1947:5/-	1931-38	24.5.49,31.5.49	-	-	Eg,Eu,Gg,CCS	
1	p2:3834:■3838	1942fx:2/-	2i	B+	g*	Ob6	1949:5/-	1931-38	21.9.49	-	-	CCS	
1	3834	1942fx:2/-	2i	B+	g*	Ob6	1949:5/-	1931-38	555:18.5.50	-	6/50	CPT	
2	p1:■3839	1942fx:2/-	2g	B+	g*	Ob6	1947:5/-	1931-39	24.5.49,31.5.49	-	-	Eg,Eu,Gg,CCS	
2	no code	1942fx:2/-	2h	B+	g*	Ob6	1949:5/-	1931-39	1185:18.5.50	-	6/50	CPT	

The unique numbers allocated (3838,3839) appear on the proofs, but not the published maps. Evidently the wrong one was deleted from Sheet 1 in error. Since 3835 failed to appear on the 1921-31 map, it did not here either

+■Population: Changes by Migration: Scotland 1938-47: England and Wales 1939-47 (Sheet 1) **1939-47** (Sheet 2)

| 1 | 3897 | 1942fx:2/- | 2i | B+ | g* | Ob6 | 1954:5/- | 1938-47 | 630:19.7.54 | - | 9/54 | CPT | |
| 2 | ■3898 | 1942fx:2/- | 2h | B+ | g* | Ob6 | 1954:5/- | 1939-47 | 960:30.8.54 | - | 9/54 | CPT | 1 |

1. The white of the title panel is extended south to the 380km line

+Population Density, 1931

1	proof 2	1942fx:np	1h	B-	g	Ob3	>1944:5/-	1931	31.10.44	-	-	BSg	
1	2044 Ch	1944fx:np	1h	B-	g	Ob3	>1944:5/-	1931	2225:24.11.44	-	1944/4	CPT	
1	3805	1944fx:np	1i	B+	gh	Ob3	>1944:5/-	1931	1979:15.7.49	-	-	Bg,Lu,RH	
2	proof 2	1942fx:np	1f	B-	g	Ob3	>1944:5/-	1931	31.10.44	-	-	BSg	
2	2044 Ch	1944fx:np	1f	B-	g	Ob3	>1944:5/-	1931	2176:24.11.44	-	1944/4	CPT	
2	3806	1942fx:np	1f	B+	g	Ob3	>1944:5/-	1931	1995D:5.4.48	-	-	Bg,Lu,RH	

Compiled by MTCP. The need seen by the Maps Advisory Committee quickly to produce such a map overrode OS objections that to republish at 1:625,000 the pre-war 1:1,000,000 population maps based on the latest available census information (the war had prevented the 1941 census from taking place) could not be achieved without compromise. In the event the number of categories depicted on the 1:1,000,000 map was reduced. An extract of Sheet 2 is in DOSSSM 1947,1951,1957

+■Population of Urban Areas

1	proof 1	1944fz:2/-	2h	B+	k	Ob2	1945:5/-	[1938]	16.7.45	-	-	RH	
1	2045/Cr	1944fz:2/-	2h	B+	k	Ob2	1945:5/-	[1938]	2000:22.9.45	-	9/45	CPT	
2	proof 1	1945fx:np	1f	B+	k	Ob2	1945:5/-	1938	16.7.45	-	-	RH	
2	2045/Cr	1945fz:np	1f	B+	k	Ob2	1945:5/-	1938	2031:22.9.45	-	9/45	CPT	

Compiled by MTCP from Registrar-General's mid-1938 Estimate of the Residential Populations of Boroughs and Urban Districts. With inset map at 1:253,440 of London. A preparatory version of Sheet 1, on an outline base map, with the "spotted dog" dots inserted in black, but without area names, is in Gg. Unique numbers allocated: 3824,3825

+■Population of Urban Areas 1951

1	■4156	1944fz:2/-	2k	B+	g	Ob3	1957:5/-	1951	840:6.12.57	-	12/57	CPT	
1	■4156	1944fz:2/-	2k	B+	g	Ob3	1957:5/-	1951	803:28.6.60	-	-		
2	■3926	nd:fz:np	1i	B+	g	Ob2	1954:5/-	1951	1085:24.12.54	-	1/55	CPT	
2	■3926	nd:fz:np	1i	B+	g	Ob2	1954:5/-	1951	1130:4.11.59	-	-		

Compiled by MHLG and DHS from statistics of enumerated populations in the 1951 Census Reports. With inset map at 1:253,440 of London. Sheet 1 is the last known new publication using the 1942 base map

+■Population: Total Changes 1921-1931

1	proof 1	1942fx:2/-	2i	B+	g*	Ob6	1947:5/-	1921-31	21.11.47	-	-	Eu	
1	p2:3832	1942fx:2/-	2i	B+	g*	Ob6	1948:5/-	1921-31	12.10.48	-	-	Eg,Gg	
1	3832	1942fx:2/-	2i	B+	gh*	Ob6	1949:5/-	1921-31	2191:9.5.49	-	5/49	CPT	
2	proof 1	1942fx:2/-	2i	B+	g	Ob6	1947:5/-	1921-31	20.2.48	-	-	Eu	
2	3833	1942fx:2/-	2i	B+	gh*	Ob6	1949:5/-	1921-31	2051:9.5.49	-	5/49	CPT	

+■Population: Total Changes: Scotland 1931-38: England and Wales 1931-39 (Sheet 1) **1931-39** (Sheet 2)

1	proof 1	1942fx:2/-	2i	B+	g*	Ob6	1947:5/-	1931-38	5.4.48	-	-	Eu	
1	p2:3832:■3836	1942fx:2/-	2i	B+	g*	Ob6	1947:5/-	1931-38	12.10.48	-	-	Eg,Gg	
1	3832:■3836	1942fx:2/-	2i	B+	gh*	Ob6	1948:5/-	1931-38	2131:9.5.49	-	5/49	CPT	
1	■3836	1942fx:2/-	2i	B+	gh*	Ob6	1948:5/-	1931-38	do	-	-	Bg,NTg	1
2	■3837	1942fx:2/-	2i	B+	gh*	Ob6	1949:5/-	1931-39	2015:9.5.49	-	5/49	CPT	

IV.3 137

| 1 | 2 | 3 | 4 | 6 | 11 | 12 | ■13 | ■14 | 15 | 16 | 17 | 18 | 19 |

1. Some copies quite clearly have only one unique number, and not the 3836 superimposed on the 3832. There is no record of a second printing, so perhaps the base map number 3832 was noticed part way through the printing and removed

+■Population: Total Changes: Scotland 1938-47: England and Wales 1939-47 (Sheet 1) 1939-47 (Sheet 2)
1	3899	1942fx:2/-	2i	B+	g*	0b6	1954:5/-	1938-47	614:19.7.54	-	9/54	CPT	
2	proof 2	1942fx:2/-	2h	B+	g	0b6	1954:5/-	1939-47	17.6.54	-	-	CCS	
2	■3900	1942fx:2/-	2h	B+	g*	0b6	1954:5/-	1939-47	830:30.8.54	-	9/54	CPT	1

1. The white of the title panel is extended south to the 380km line (not so on proof)

+■Railways
1	proof 1	1942fv:2/-	2h	A-	g	0w4	1944:5/-	nd	11.5.44	-	-	BSg	
1	3046	1944fz:2/-	2h	B+	g	3b4	1946:5/-	2.46	2425:C22.8.46	-	10/46	CPT	
2	proof 2	1942fx:np	2f	B+	g	0w6	1946:5/-	1944	8.2.46	-	-	RH	
2	3046	1942fx:np	2f	B+	g	3b6	1946:5/-	2.46	2500:C21.8.46	-	10/46	CPT	
2	B	1942fx:np	2i	B+	g	3b6	1946:5/-	1946	500:10.8.53	-	-	Ng,Og	1

Compiled by MTCP and DHS from data supplied by the Railway Companies and MWT. With inset maps at 1:253,440 of the Glasgow and London areas. The map distinguishes between single, double and multiple track lines, as well as narrow gauge and electrified lines. Mineral only lines are shown only when owned by the main companies. The four principal companies established at grouping in 1923 are shown in their traditional colours of brown, red, blue and green. London Transport is in black, and other systems are in purple. Jointly owned lines are shown by pecks in alternate colours. No layer box. Unique numbers allocated: 3814,3815

1. Because of the number of plates required, the unit cost of this reprint was reportedly 10/8d! However, with the railways already nationalised, it was thought unethical to increase the cost to the public of an out of date map

+■Rainfall: Annual Average 1881-1915
1	3842	1942fx:2/-	2i	B+	g	0w2	1949:5/-	1881-1915	1250:30.10.49	-	11/49	CPT	
1	3852	1942fx:2/-	2i	B+	g	0w2	1949:5/-	1881-1915	4209:18.1.51	-	-	Bg,Lrgs,RH	
2	3843	nd:fx:np	2i	B+	g	0w2	1949:5/-	1881-1915	1244:31.10.49	-	11/49	CPT	
2	3843	nd:fx:np	2i	B+	g	0w2	1949:5/-	1881-1915	5113:18.1.51	-	-	Cg,Mg,RH	1

Compiled by MTCP and DHS from isohyets and statistics contributed by MO. With dispersion graphs of representative rainfall stations. Areas lacking overprint: Isles of Sule Skerry, Rona, St Kilda. With Explanatory Text 2. Though normally very cautious in their assessments of reprint runs, this was a case of OS gross over-estimation, and some 2500 copies of each sheet were surplus to requirements when superseded in 1967. MTCP ms maps dated from 1946

1. Instructions were given on 25.11.1950 to alter the unique number to 3853, but this was not done until 9.4.1951 - too late for this print run. 3843 was merely redrawn. The reprint is more easily distinguished by the repositioning of the legend "Key to position of representative rainfall stations..." from the top margin to the England & Wales diagram

■Retriangulation of Great Britain: Primary Observations: 1936-1937 & 1949-1951 (Sheet 1)
■Retriangulation of Great Britain: Primary Observations: 1936-1939 & Dec. 1950 (Sheet 2)
1	?no code	-	-	-	-	--1	fiuo	1936-51	107:28.5.51	-	-	not found	
1	3801	1942fx:2/-	2i	B+	s	0w1	fiuo	1936-51	112:8.6.51	-	-	CCS	
2	?no code	-	-	-	-	--1	fiuo	1936-50	107:28.5.51	-	-	not found	
2	3802	1942fx:2/-	2j	B+	s	0w1	fiuo	1936-50	112:8.6.51	-	-	CCS	

■Retriangulation of Great Britain: Primary Observations: 1936-1952 [title on Sheet 1 only]
| 1 | ■L525 | 1942fx:2/- | 2i | B+ | s | 0w1 | fiuo | 1936-52 | | - | - | PC | |
| 2 | ■L527 | 1942fx:2/- | 2j | B+ | s | 0w1 | fiuo | [1936-52] | | - | - | PC | |

■Retriangulation of Great Britain: Primary Observations: 1936-1957 [title on Sheet 1 only]
| 1 | ■L525 | 1942fx:2/- | 2i | B+ | s | 0w1 | fiuo | 1936-57 | | - | - | Bg | |
| 2 | ■L527 | 1942fx:2/- | 2j | B+ | s | 0w1 | fiuo | [1936-57] | | - | - | Bg | |

■Retriangulation of Great Britain: Secondary Blocks: December 1950
1	?■L182	-	-	-	-	--1	fiuo	12.50	107:28.5.51	-	-	not found	
1	3801:■L182	1942fx:2/-	2i	B+	s	0w1	fiuo	12.50	112:8.6.51	-	-	CCS	
2	?no code	-	-	-	-	--1	fiuo	12.50	107:28.5.51	-	-	not found	
2	3802	1942fx:2/-	2j	B+	s	0w1	fiuo	12.50	112:8.6.51	-	-	CCS	

The base map colour on all these retriangulation sheets is reduced Gribbletts brown. The control cards refer first to the subject 11.1945, with printing Retriangulation diagrams Sheets 1 and 2, probably at 1:633,600 (see IV.2)

1	2	3	4	6	11	12	■13	■14	15	16	17	18	19
+■Roads													
1	proof D	1944fz:2/-	2h	B+	g	Ob3	1945:5/-	10.44	16.7.45	-	-	BSg	
1	proof	1944fz:2/-	2h	B+	g	3b3	1945:5/-	10.44	undated	-	-	BSg	
1	proof	1944fz:2/-	2h	B+	g	3b3	1946:5/-	4.46	17.5.46	-	-	SOos	
1	25046	1944fz:2/-	2h	B+	g	3b3	1946:2/6	4.46	35100	96.3	9/46	CPT	
2	proof	1942fv:np	1f	A+	g	Ob3	1945:5/-	10.44	30.5.45	-	-	BSg	
2	proof	1942fv:np	1f	A+	g	3b3	1945:5/-	10.44	16.7.45	-	-	BSg	
2	25046	1942fx:np	1f	A+	g	3b3	1946:2/6	4.46	34000	96.3	8/46	CPT	
2	3817	1942fx:np	1i	A+	g	3b3	1946:2/6	4.46	30000:9.48	96.3	-	Lu,RH	1

Compiled by MTCP from data supplied by MWT. With inset maps at 1:253,440 of the Glasgow and London areas. Area lacking overprint: Isle of Man. The map distinguishes trunk roads (in purple) from "A" and "B" roads (in red and green). Since they were now lacking on the base map, "B" roads had to be specially drawn. AA and RAC telephone boxes are shown. No layer box. The size of the title panel was increased between proof and published map stage. It proved impossible after the war for the OS to resurrect their own ten-mile map, so, on 7.6.1945, "*It has been decided that the road edition of the 1/625,000 Town & Country Planning Series will be used in place of the old 10 mile map for the present.*" (PRO OS 1/789 26A - extract from DG's conference). Superseded by the 1956 **Roads** map. Used for GSGS 4813 (see Supplement 2). An extract of Sheet 2 is in DOSSSM 1947,1951. Unique numbers allocated: 3816,3817. In spite of the print codes and some inconsistent documentary evidence, it seems that the print runs given are accurate, in round figure terms. Also issued in the Ansell-fold back to back format (H.101.1, H.101.2 covers). The earlier yellow boards were replaced in 1948 after complaints from customers of the speed with which they became soiled.

1. 74 changes were made, mainly to AA and RAC telephone boxes. But some errors remained, and at least two new ones created: see SP 5519 and ST 1167 (PRO OS 1/432 63,106C)

+Sand and Gravel - see **Gravel including Associated Sands** on p.159

Topography - see **Great Britain: Topography**, the second base map in this list, on p.128

+■Types of Farming: Scotland....England and Wales (Sheet 1) England and Wales (Sheet 2)

n	544/Ch	nd-:np	1h	B-	j	Ow5	fouo	nd	743:7.9.44	-	-	Lpro	1
1	4044/Ch	1942fx:np	1h	B-	g	Ow6	1944:5/-	6.39	3938:22.8.44	-	1944/3	CPT	
1	3979	1942fx:np	1i	B-	g	Ow6	1944:5/-	6.39	801:20.10.55	-	-	Lrgs,RH	
1	p:3979	1942fx:np	1i	B-	j	Ow0	-	-	-	-	-	RH	3
1	3979	1942fx:np	1k	B-	g	Ow6	1944:5/-	6.39	754:3.11.60	-	-	not found	
2	4044/Ch	1942fx:np	2f	B-	g	Ow6	1944:5/-	6.39	4103:7.9.44	-	1944/3	CPT	2
2	3812	1942fx:np	2f	B+	g	Ow6	1944:5/-	6.39	3546:15.3.48	-	-	Bg,Lrgs,RH	2

The Scotland portion was compiled by LUS in conjunction with DAS, the England & Wales by MAF. Intended for pre-war publication (see p.125). Willatts and his MTCP Maps Office brought the map to publication. The Orkney and Shetland Islands inset map is reduced by 20km at its southern edge. Areas lacking overprint: Isles of Sule Skerry, Rona, St Kilda, Man, Lundy, Holy Island (Northumberland). Unique numbers allocated: 3811 (Sheet 1), 3812 (Sheet 2), 3813 (North Strip)

1. Not published, so carries no publication or copyright statement. MAF received 500 copies adjoined to Sheet 2: fifty of these were on cloth, while the remainder went to Messrs A.P.Taylor of Holborn Hall, Clerkenwell Road, London E.C. (PRO OS 1/556). The PRO copy is at ZOS 4/37
2. NG values are missing around the title panel. West coast water names relocated
3. A security proof used for the deletion of reservoir references during the making of the 3979 1960 printing

+■Vegetation: Grasslands of England & Wales [title on Sheet 2 only]

n	4046	1946-:2/6	1h	B-	g*	Ob8	-	nd	4000:26.8.46	-	1/47	CPT	1
2	proof 1	1945fx:2/-	2f	B+	g*	Ob9	1945:5/-	1940	28.12.45	-	-	BSg	
2	4046	1945fx:2/-	2f	B+	g*	Ob9	1946:5/-	1940	4064:16.9.46	-	9/46	CPT	

Compiled in 1940 by the Welsh Plant Breeding Station on behalf of MAF. Begun in 1939 as a reconnaissance survey on manuscript quarter-inch county maps classifying the grasslands in Wales by Sir R.George Stapledon and William Davies of the Grassland Research Institute, and extended in 1940 into England. It was made available to the LUS, who prepared for their own use a 1:625,000 coloured map and printed a monochrome map on the 20-mile scale. The former proved so interesting that it was brought to publication through Willatts and his MTCP Maps Office. Areas lacking overprint: Isles of Man, Lundy, Scilly, Holy Island (Northumberland). With Explanatory Text 5. Unique number allocated: 3831

1. Scotland is uncoloured. 3059 copies cancelled on 3.9.1953, the day after the printing of the replacement map

IV.3   139

| 1 | 2 | 3 | 4 | 6 | 11 | 12 | ■13 | ■14 | 15 | 16 | 17 | 18 | 19 |

+■Vegetation: Reconnaissance Survey of Scotland.....Survey of England
1	A	1942fx:2/-	2i	B+	k	Ow4	1953:5/-	1900-39	722:15.7.53	-	9/53	CPT	
1	A■/	1942fx:2/-	2j	B+	k	Ow4	1953:5/-	1900-39	631D:14.3.58	-	-	Og,CCS	1
1	A■/	1942fx:2/-	2j	B+	k	Ow4	1953:7/6	1900-39	750:5.7.63	-	-	Cg,Sg,CCS	1
1	A/	1942fx:2/-	2j	B+	k	Ow4	1953:np	1900-39	976:19.8.69	-	-	Eu,CCS	1,2

Scottish information was prepared in 1945 by Drs Arthur Geddes and L.Dudley Stamp on a 1:625,000 coloured manuscript map (reduced and reproduced in monochrome in Stamp (1962) 151). T.E.Williams prepared the English portion from the 1946 map. Willatts and his MTCP Maps Office brought the map to publication. With a "Relative Reliability" diagram. Areas lacking overprint: Isles of Sule Skerry, Rona, St Kilda, Man. With Explanatory Text 8. Unique number allocated: 3830
  1. Cliff tops separated from sand dunes and links.   2. Printed after the cancellation of Planning Series materials in 1967, using film positives from the Disaster Store. The print code letter is upright: the others are sloping

Supplement 1. Special sheets

+Maps of Proposed National Parks
| - | ■3847 | nd:-:np | f | B+ | j | Ob4 | 1947:np | nd | - | - | BPP |

Twelve extract maps (Lake District, North Wales, Peak District, Dartmoor, Yorkshire Dales, Pembrokeshire Coast, Exmoor, South Downs, Roman Wall, North York Moors, Brecon Beacons and Black Mountains, Broads), for use in MTCP's *Report of the National Parks Committee (England and Wales)* (BPP(HC) 1946-47 Cmd 7121, XIII, 443-454). Prepared by MTCP, and printed on one plate. Railway revision 2 (Sheet 1) and 1 (Sheet 2) areas (PRO OS 1/557, 30.7.1946 meeting)

■Appendix D: Jurisdiction of the Durham Palatine Court
| - | 3748 | nd:fx:np | 1i | B+ | g | Ow5 | 1952:np | nd | - | - | BPP |

Covering Northumberland, Durham and Yorkshire. In *Third Interim Report of the Committee on Supreme Court Practice and Procedure* (BPP(HC) 1951-52 Cmd 8617, XVI, 431)

Supplement 2. Military issues

| 1 | 2 | 3 | 4 | 6 | 9 | 10 | 11 | 12 | ■13 | ■14 | 16 | 17 | 18 | 19 |

GSGS 4676: probably not published until as M325 (see p.171)

Roads / Great Britain: Roads
| 1 | ■5000/11/540S | ■1954fz:np | 2i | B+ | NGm | 1954 | g | 3b3 | - | 4.46 | - | - | ABn,Lrgs,Ob |
| 2 | ■10000/11/540S | ■1954fx:np | 1i | A+ | NGm | 1954 | g | 3b3 | - | 4.46 | - | - | ABn,Lrgs,Ob |

GSGS 4813 Edition 1-GSGS, 1946. Published by War Office, 1954. On the OS **Roads** map, Sheet 1 in a revision without civilian use, Sheet 2 on the 3817 printing. The base map title panel is new. The series continues as M322 (see p.171)

IV.4. 1:1,250,000 Great Britain, to "D" edition, 1947. *NB*: An alphabetical listing follows the base map

| 2 | 3 | 6 | 8 | 11 | 12 | ■13 | ■14 | 15 | 17 | 18 | 19 |

Great Britain: Scale: 1/1,250,000 or about Twenty Miles to One Inch [outline base map]
1046	1946sa:2/-	A1	CCR	j	Ow0	-	-	1060D:30.12.46	1/47	CPT	
1046	1946sa:2/-	A1	CCR	k	Ow0	-	-	200D:30.12.46	-	[MTCP]	1
3363	1946sa:2/-	A1	CCR	j	Ow0	-	-	1008:1.11.48	-	Lpro MFQ 596	
3363	1946sa:2/-		CCR	kh	Ow0	-	-	206:26.7.50	-	[MTCP]	1
3363	1946sa:2/-	B1	CCR	j	Ow0	-	-	2548:24.4.51	-	SOos	2
A	1946sa:2/-		CCR	kh	Ow0	-	-	218:29.10.53	-	[MTCP]	1
A/	1946sa:2/-		CCR	g	Ow0	-	-	208:7.8.56	-	[MTCP]	1
A/	1946sa:2/-	B2	CCR	j	Ow0	-	-	2004:21.1.57	-	DRg,NTg	
A//	1946sa:2/3	A4	CCR	j	Ow0	-	-	3000:CANCELLED	-	-	3
B	1962sa:2/3			1962	j	Ow0	-	-	3010:21.11.61	-	Bg,Lse,CCS,RH	4

2	3	6	8	11	12	■13	■14	15	17	18	19
C	1964sb:np		1964	j	Ow0	-	-		-	En,Lrgs	5
C	1964sb:np		1964	kh	Ow0	-	-		-	SOrm	5
C	1964sb:np		1964	j	Ow0	-	-	4036:6.5.65	-	CCS	5
proof:D	1969td:np		1969	j	Ow0	-	-	25.3.69	-	SOos,CCS	6
proof:D	1969td:np		1969	j	Ow0	-	-	24.6.69	-	CCS	7
D	1969td:np		1969	j	Ow0	-	-	6115:10.7.69	8/69	CPT	8
D	nd:td:np		1969	j	Ow0	-	-	2190:23.4.74	-	CCS	8

NB: For "E" edition, see section V.4 on p.173

Column 6 revision codes: **A1**: Tunbridge Wells (original state, 1947). **A2**: Grassholm Island, The Smalls, Haskeir I. named (by 11.1952). **A3**: Flintshire (det) set within Welsh Border (23.7.1954), and St Aldhelm's Head name redrawn, the apostrophe cutting the grid (23.8.1954). **A4**: Price 2/3. The longitude value in bottom right corner of the Orkney and Shetland Isles inset panel corrected to 0°30' (15.8.1960). **B1**: Royal Tunbridge Wells (by 25.4.1951). **B2**: Flintshire (det) set within Welsh Border (23.7.1954), and St Aldhelm's Head name redrawn, the apostrophe below the grid (23.8.1954)

There was a serious railway error in Scotland, uncorrected until the "E" edition (V.4), in that the East Coast Main Line between Arbroath and Kinnaber Jcn was never shown, whereas the original ex-Caledonian route from Arbroath, using the Friockheim to Glasterlaw link (closed in 1909!) was - though not the still open route between Friockheim and Guthrie. The fault was compounded in the "D" edition when Arbroath to Friockheim was closed, leaving no East Coast route to the north at all. For notes on the "tc" state, see **1:10,000-1:10,560 National Grid Series** below. The map was priced 2/3d in DOSSSM 1951 and 1957, which makes 1957 printing at 2/- the more remarkable

    1. Printings for MTCP (not seen). Pre-publication printings for MTCP have not been listed here

    2. A second negative would appear to have been used, with Royal Tunbridge Wells (see "B" revision codes *supra*)

    3. A print run of 3000 was ordered in April 1961, but then cancelled. It is noted here because this revision state was used in overprinted issues (see "A4" revisions elsewhere)

    4. Beccles-Yarmouth railway closed. Much bigger towns. Leamington to Royal Leamington Spa, St Anne's on the Sea to Lytham St Anne's. Some placenames relettered, suggesting population reclassification (eg Cardiff, Bristol, Grantham). Royal Tunbridge Wells redrawn. Holy Isd to Holy I (both)

    5. Lundy Island to Lundy

    6. "td" title but with symbols for towns to the left of the symbols column

    7. "td" title with symbols for towns to the right of the symbols column, with "Crown Copyright"

    8. "Crown Copyright" altered to "Crown copyright". A revised map (derived from RPM "E" edition) with a newly designed title panel with legend and revised National Grid notes. Full stops after "St" deleted. Some new placenames, eg Bude, Skegness. Some revision of county names: eg Huntingdonshire to Huntingdon & Peterborough, Yorkshire Ridings deleted. Further extension to built-up areas. A more fully developed network of roads, including motorways. Offshore lighthouse locations added - Wolf Rock, Eddystone, Inchcape or Bell Rock. Dubh Artach named, Skerryvore added and named

Many railways closed: Swindon-Cheltenham, Pulborough-Petersfield, Poole-Salisbury, Lampeter-Aberystwyth, Hereford-Merthyr, Bridgnorth-Shrewsbury, Denbigh-Barmouth, Caernarvon-Afonwen, Whitchurch-Chester, Stafford-Wellington, Aylesbury-Leicester (& GW links), Rugby-Luffenham, Workington-Keswick, Darlington-Tebay, Rillington-Grosmont, Hexham-Riccarton Jcn, North Berwick line, Arbroath-Glasterlaw, Dunblane-Crianlarich, Balquhidder-Gleneagles, Boat of Garten-Forres. Liverpool-Southport altered to the coast route

■**Index showing Sheets of the New Popular[,] 5th Edition and Popular Edition One Inch Maps of Great Britain for compiling sheets to be drawn, reconstituted and provisional edition**
■3393   1946sa:2/- A1 CCR g Ow3 fiuo nd   -   SOrm
    Probably not published, but printed in 1949

■**Index showing Sheets of the Popular, New Popular and 7th Series One Inch Maps of Great Britain**
■3794   1946sa:2/- A2 CCR g Ow2 fiuo 20.11.52   -   Lpro,Sg,SOrm
    Marked "For Internal Use Only", and "As approved by D.G. 20.11.52." The PRO copy is in OS 1/477

■**Index to Sheets of the Ordnance Survey 1:25,000 Second Series (20 Km x 15 Km Sheet Lines)**
■773   1946sa:np A3 CCR k Ow1 nd   nd   -   SOos

National Grid letters appear in grey on the face of the map. The SB square is filled with 10km figure references. The date of issue is unknown, but was presumably in advance of the one sheet (856) to be published in 1960, since it appears here with sheet lines 10km too far west. The total number of sheets would have been 977, this total including five (440-444) covering the Isle of Man. The sheet lines of the Isles of Scilly sheet as published in 1964 were 10km west and south of those given here for Sheet 977

IV.4   141

| 2 | 3 | 6 | 8 | 11 | 12 | ■13 | ■14 | 15 | 17 | 18 | 19 |

■Index to Sheets of the Ordnance Survey 1:25,000 Second Series
■Index to the 1:25,000 Provisional Edition & 1:25,000 Second Series [from "D" edition]
■Index to the 1:25 000 First & Second Series [from "M" edition]
■Index to the 1:25 000 First & Second Series: Index to the 1:25 000 Outdoor Leisure Maps [from "T" edition]

C:■A	1964sb:np		1964	k	Ow1	nd	nd		-	Eg,En,Sg	
C:■B	1964sb:np		1964	k	Ow1	nd	?1.1.65		-	not found	
C:■C	1964sb:np		1964	k	Ow1	nd	1.1.66		-	BFq,En,NTg,Ob	
C:■D	1964sb:np		1964	k	Ow1	nd	1.1.67	50600:23.1.67	-	Lse,Ob,Sg,CCS	1
C:■E	1964sb:np		1964	k	Ow1	nd	1.4.68		-	SOrm	
:■F				k	Ow1	nd			-	not found	
D:■G	1969td:np		1969	k	Ow1	nd	1.7.69		-	BFq,Cg,En,SOrm	
D:■H	1969td:np		1969	k	Ow1	nd	1.1.70		-	Yu,RH,PC	
D:■J				k	Ow1		?1.7.70		-	not found	
D:■K	nd:td:np		1970	k	Ow1	1971	1.1.71		-	SOos	
D:■L	nd:td:np		1970	k	Ow1	1971	1.7.71		-	Ng	
D:■M	nd:td:np		1970	k	Ow1	1972	1.1.72		-	NTg	
D:■N	nd:td:np				Ow1		?1.7.72		-	not found	?2
D:■P	nd:td:np		1969	s	Ow1	1973	1.1.73		-	Cu,NTg	2
D:■Q	nd:td:np		1969	s	Ow1	1973	1.7.73		-	Gg,Lse,SOrm,RH	
D:■R	nd:td:np		1969	s	Ow1	1974	1.1.74		-	SOos,PC	
D:■S	nd:td:np		1969	s	Ow1	1974	1.7.74		-	BFq,SOos,PC	
D:■T	nd:td:np		1969	s	Ow1	1975	1.1.75		-	Cu,Lgh,RH	
D:■U	nd:td:np		1969	s	Ow1	1975	1.7.75		-	BFq,NTg	
D:■V	nd:td:np		1969	s	Ow1	1976	1.1.76		-	BFq,SOrm	3
D:■W	nd:td:np		1969	s	Ow1	1976	1.7.76		-	RH	3

1. This print run must have sufficed for many later editions as well
2. The overprint colour changed at this point from red to black
3. Some if not all issues are double sided with the six-inch index

■1:50,000 Sheet Lines: Number of sheets 204
■Index to the 1:50,000 Scale Map [1974]

C	1964sb:np		1964	k	Ow1	nd	12.70	150:2.4.71	-	SOos,CCS	
D	nd:td:np		1969	k	Ow1	nd	4.72	960:26.4.72	-	SOrm	
D:■C	nd:td:np		1969	k	Ow1	1974	nd	2021:2.4.74	-	Lgh,NTg,CCS	
D:■C	nd:td:np		1969	k	Ow1	1974	nd	3790:30.12.74	-		

Showing National Parks, sheet lines and numbers of the 1:50,000 series, together with a list of sheet names

■1:50,000 Sheet Lines: Number of sheets 204 [red]: One Inch Seventh Series Sheet Lines [blue]
■Index to the 1:50,000 Scale Map [red] and to the One-Inch Seventh Series Map [blue] [1974]

proof:D	nd:td:np		1969	k	Ow2	nd	4.72	27.3.72	-	CCS	
D	nd:td:np		1969	k	Ow2	nd	4.72	530:26.4.72	-	SOrm,CCS	
p:D:■B	nd:td:np		1969	k	Ow2	1973	nd	18.5.73	-	CCS	
D:■B	nd:td:np		1969	k	Ow2	1973	nd	10390:13.6.73	-	Lgh,SOrm	
p:D:■C	nd:td:np		1969	k	Ow2	1974	nd	22.2.74	-	CCS	
D:■C	nd:td:np		1969	k	Ow2	1974	nd	3250:28.2.74	-	SOrm,CCS	
D:■C	nd:td:np		1969	k	Ow2	1974	nd	10300:2.4.74	-		

Showing National Parks, sheet lines and numbers of the two series, and a list of sheet names of the 1:50,000 series

Land required by Service Departments for Training
p1:3252 1946sa:2/- A1 CCR g Ob4 nd 6.47 14.6.47 - SOos

Compiled and drawn by MTCP for the use of civilian planners rather than the military. Specification was discussed on 11.6.1947 (PRO OS 1/558). The base and sea tint are as on **National Parks**. The additional colours used are pink (for War Department land), violet (for Air Ministry), electric deep blue (for Admiralty) and green (for Ministry of Supply), the intention being to select colours of adequate clarity so that the larger areas would not overpower the smaller ones. The land requirements of Territorial and Auxiliary Force Associations for local training and small arms ranges are not shown. Publication was blocked for security reasons

2	3	6	8	11	12	■13	■14	17	18	19

■National Parks and Conservation Areas: Proposed by The National Parks Committee (England and Wales)

■3846	1946sa:2/-	A1	CCR	g	Ob3	nd	1938	C1.5.47	-	BPP
■3846	1946sa:2/-	A1	CCR	g	Ob3	nd	1938		-	BPP	1
■3846	1946sa:np	A3	-	g	Ob3	nd	1938	500C:19.11.58	-	SOrm	2

Used in MTCP's *Report of the National Parks Committee (England and Wales)* (BPP(HC) 1946-47 Cmd 7121, XIII, 455). Since this was to be a map relevant to England and Wales only, proposals were made to cut the map at 660km N, or reduce Scotland to an outline, like Ireland. Layered hills could be included. The OS tested the latter by reducing the hill layering of the 1:625,000 map, but in the end resisted both suggestions (PRO OS 1/557, minutes of 30.7.1946 meeting), and in the event the areas recommended in the *National Parks: A Scottish Survey* (BPP(HC) 1944-45 Cmd 6631, V, 341-367) as suitable for National Parks in Scotland were also shown in order of preference. The population symbols are reduced from the 1:625,000 **Population of Urban Areas** map. There is some evidence that the unique number allocated was 3851

1. A version with superscript "Map No. 1" in MTCP's *Conservation of Nature in England and Wales* (BPP(HC) 1946-47 Cmd 7122, XIV, 681). 2. The title and unique number are redrawn. This issue was for HMSO

■Scale 1:50,000 40 Km x 40 Km: Layout No 6

D	nd:td:np		1969	b	Ow1	fiuo	nd			SOos (photo)

■The Second Land Utilisation Survey of Britain: Index to 1:25,000 Series

C:■C	1964sb:np		1964	k	Ow2	nd	nd	-	CPT

■Six-Inch National Grid Series
■1:10,000-1:10,560 National Grid Series [from "113" edition]

■69	1946sa:2/-	A3	CCR	k	Ow1	nd	1.7.56	-	RH
■70	1946sa:2/-	A3	CCR	k	Ow1	nd	?1.10.56	-	not found
■71	1946sa:2/-	A3	CCR	k	Ow1	nd	1.1.57	-	Lgh
■72	1946sa:2/-	A3	CCR	k	Ow1	nd	1.4.57	-	Lpro OS 1/1136
■73	1946sa:2/-	A3	CCR	k	Ow1	nd	?1.7.57	-	not found
■74	1946sa:2/-	A3	CCR	k	Ow1	nd	?1.10.57	-	not found
■75	1946sa:2/-	A3	CCR	k	Ow1	nd	?1.1.58	-	not found
■76	1946sa:2/-	A3	CCR	k	Ow1	nd	1.4.58	-	RH
■77	1946sa:2/-	A3	CCR	k	Ow1	nd	?1.7.58	-	not found
■78	1946sa:2/-	A3	CCR	k	Ow1	nd	?1.10.58	-	not found
■79	1946sa:2/-	A3	CCR	k	Ow1	nd	?1.1.59	-	not found
■80	1946sa:2/-	A3	CCR	k	Ow1	nd	?1.4.59	-	not found
■81	1946sa:2/-	A3	CCR	k	Ow1	nd	?1.7.59	-	not found
■82	1946sa:2/-	A3	CCR	k	Ow1	nd	?1.10.59	-	not found
■83	1946sa:2/-	A3	CCR	k	Ow1	nd	?1.1.60	-	not found
■84	1946sa:2/-	A3	CCR	k	Ow1	nd	1.4.60	-	Lpro OS 5/51
■86	1946sa:2/-	A3	CCR	k	Ow1	nd	1.7.60	-	Lpro OS 5/52
■87	1946sa:2/-	A3	CCR	k	Ow1	nd	1.10.60	-	Lpro OS 5/53
■88	1946sa:2/-	A3	CCR	k	Ow1	nd	1.1.61	-	Lpro OS 5/54
	1946sa:		CCR	k	Ow1	nd	?1.4.61	-	not found
	1946sa:		CCR	k	Ow1	nd	?1.7.61	-	not found
■93	1946sa:2/3	A4	CCR	k	Ow1	nd	1.10.61	-	Lu
■94	1946sa:2/3	A4	CCR	k	Ow1	nd	1.1.62	-	PC
■96	1946sa:2/3	A4	CCR	k	Ow1	nd	1.4.62	-	Eg
■97	1946sa:2/3	A4	CCR	k	Ow1	nd	1.7.62	-	BFq
B:■98	1963sa:2/3		1963	k	Ow1	nd	1.1.63	-	LDg
B:■99	1963sa:2/3		1963	k	Ow1	nd	1.7.63	-	En,Mg
:■100				k	Ow1	nd	?1.1.64	-	not found
:■101				k	Ow1	nd	?1.7.64	-	not found
:■102				k	Ow1	nd	?1.1.65	-	not found
:■103				k	Ow1	nd	?1.7.65	-	not found
C:■104	1964sb:np		1964	k	Ow1	nd	1.1.66	-	En
C:■105	1964sb:np		1964	k	Ow1	nd	?1.7.66	-	not found
C:■106	1964sb:np		1964	k	Ow1	nd	1.1.67	-	BFq

IV.4   143

2	3	6	8	11	12	■13	■14	17	18	19
C:■107	1964sb:np		1964	k	Ow1	nd	1.7.67	-	En	
C:■108	1964sb:np		1964	k	Ow1	nd	1.1.68	-	BFq	
C:■109	1964sb:np		1964	k	Ow1	nd	1.7.68	-	BFq	
C:■110	1964sb:np		1964	k	Ow1	nd	1.1.69	-	BFq	
C:■111	1964sb:np		1964	k	Ow1	nd	1.7.69	-	BFq	
D:■112	1969td:np		1969	k	Ow1	nd	1.1.70	-	BFq,Lkg,PC	
D:■113	1969td:np		1969	k	Ow1	nd	1.7.70	-	BFq	
D:■114	1969td:np		1969	k	Ow1	nd	1.1.71	-	SOrm	1
D:■115	1969 :np		1969	k	Ow1	nd	1.7.71	-	Gg	
D:■116	1969tc:np		1969	k	Ow1	nd	1.1.72	-	BFq,Lkg	
D:■117	tc:np			k	Ow1	nd	?1.7.72	-	not found	
D:■118	nd:tc:np		1969	k	Ow1	nd	1.1.73	-	NTg	2
D:■119	1969tc:np		1969	k	Ow1	nd	1.7.73	-	BFq,Lse,RH	3
D:■120	1969tc:np		1969	k	Ow1	nd	1.1.74	-	SOos	
D:■121	1969tc:np		1969	k	Ow1	nd	1.7.74	-	BFq,RH,PC	
D:■122	1969tc:np		1969	k	Ow1	nd	1.1.75	-	Lgh	
D:■123	1969tc:np		1969	k	Ow1	nd	1.7.75	-	Bg	
D:■124	1969tc:np		1969	k	Ow1	nd	1.1.76	-	Ng,SOos	4
D:■125	1969tc:np		1969	k	Ow1	nd	1.7.76	-	NTg,RH	4

An issue with unique number 68 is also possible, overprinted 1.4.1956: to date the number has no known use
1. An overprint date of 1.7.1970 is blocked out and replaced by 1.1.1971.   2. Lacks the "c" in the CC circle
3. The change to the "tc" state suggests a reversion to a preparatory state of the "D" edition, after all revision had been completed except placenames, roads and title panel. Most of the roads and placenames added to the "D" edition are absent here, though, eg, the Lincoln-Skegness road is present.   4. Known issues double sided with 1:25,000 index

Specification of One Inch Seventh Series: Appendix No 3: Index to Sheets
no code   1946sa:2/-   A3   CCR   g   Ow1   fiuo   nd   -   SOrm,CCS
Two copies in One Inch Seventh Series Master File: Miscellaneous R1000. Showing sheet lines, numbers and titles

Specification of One Inch Seventh Series: Appendix No 3: Index to Sheets [red]: Administrative Diagrams [blue]
C   1964sb:np   1964   k   Ow3   fiuo   nd   -   SOos
The third overprint colour (black) is mainly for National Grid references

Specification of One Inch Seventh Series: Appendix No 3: Index to Sheets [red]: Scale 1:50,000 40 Km x 40 Km: Index to Sheets [blue]
C   1964sb:np   1964   k   Ow2   fiuo   nd   -   SOos

Supplement 1. Military issue

■Adjustment Curves: UK 1936 Datum to UK 1955 Datum
■250/5/56 nd:sa:np   A3   -   g   Ow2   nd   1936-55   -   Ob
GSGS (Misc.) 1698, Published D Survey, WO and AM, 1956. Printed by No 1 SPC RE 1956. The base map National Grid is deleted, and replaced by a graticule. The National Grid appears as a blue overprint, and the adjustment curves in red

---

Supplement 2. Editions not published by the Ordnance Survey

■Administrative Areas: Great Britain
D   nd:td:np   1974   k   Ow1   1974   1.4.74; 16.5.75   -   Lse
D   nd:td:np   1979   k   Ow1   1979   1.4.74; 16.5.75   -   Lse
Published by DOE

■United Kingdom Atmospheric Corrosivity Values 1986
no code   nd:-   1986   k   Ow2   -   LEg
Published by MAFF

Supplement 3. Skeleton maps published by the Ordnance Survey and others

2	3	6	8	11	12	■13	■14	17	18	19

National Nature Reserves Geological Monuments and Areas of Outstanding Scientific Value: Proposed by The Wild Life Conservation Special Committee (England and Wales)
3851   -   -   j   Ow1   1947:np   nd   -   BPP
A photolithographed skeleton map of England and Wales used as **Map No. 2** in MTCP's *Conservation of Nature in England and Wales* (BPP(HC) 1946-47 Cmd 7122, XIV, 683)

Administrative Areas: England & Wales
L1581   1966   1966   j   Ow0   -   -   SOrm
A diagram, also overprinted as follows:-

■**Cases for Trial at Courts of Quarter Sessions in 1965** / Administrative Areas: England & Wales
L1581   1966   1966   j   Ow1   fouo   1965   -   SOrm
■**Civil Actions tried and Judgment Summonses heard in County Courts 1965** / Administrative Areas: England & Wales
L1581   1966   1966   j   Ow1   fouo   1965   -   SOrm
■**Criminal Cases for Trial and Civil Actions tried at Assizes in 1965** / Administrative Areas: England & Wales
L1581   1966   1966   j   Ow1   fouo   1965   -   SOrm
Printed by OS for MHLG. The overprints (red) from the Royal Commission on Assizes and Quarter Sessions

Ministry of Technology Assisted Areas for Industrial Development at 26.2.1970
no code -   1970   j   Ow1   fouo   26.2.70; 30.6.69 -   NTg,Ob
A diagram, published by MHLG, printed by OS

Assisted Areas as defined by the Department of Trade and Industry at 22.3.72 (or 14.8.74) / Department of Employment Local Office Areas
no code -   1972   j   Ow2   fouo   22.3.72   -   NTg
no code -   1975   j   Ow2   fouo   14.8.74   -   Og
Assisted Areas as defined by the Department of Industry at 18.7.79 (or 1.8.80): Subsequent amendments from 1.8.80 (or 1.8.82) / Manpower Services Commission (Employment Service Division) Local Office Areas
B   -   1979   j   Ow1   fouo   18.7.79   -   NTg
B   -   1979   j   Ow1   fouo   1.8.80   -   NTg
Assisted Areas as defined by the Department of Industry: To take effect from 1.8.82 subject to review / Manpower Services Commission (Employment Service Division) Local Office Areas
B   -   1979   j   Ow1   fouo   1.8.82   -   NTg
Diagrams, compiled by DOE. The last three with provisional boundaries, and with DOE Cartographic Services code P13. Several other issues, even annual ones, are probable

Local Authorities: ■**Administrative Areas as at 1.4.74**>: Administrative Areas as at 31.3.74
no code -   1974   k   Ow1   fouo   1.4.74   -   Lbl
Printed by DOE Cartographic Services, June 1974. Boundaries are as at 1.4.71

IV.5.   1:1,900,800 Great Britain, 1954

Great Britain: Scale: 1:1,900,800 or about Thirty Miles to One Inch
no code 1946sa:2/- A3   CCR   jh   Ow0   -   -   -   CCS
Experimental Kodaline positive supplied on 29.5.1954, incorporating the St Aldhelm's Head change not added to the 1:1,250,000 until August. In 1:1,250,000 Job File R1798. The many other maps at this scale have been disregarded here

---

IV.6.   1:2,500,000 **Great Britain**

Unpublished. The north half only is known from its appearance as an inset map on the 1946 1:625,000 **Physical Map** proof, where regional names were added for experimental reasons (Copy OS). Lit: Hellyer (1992a)

CHAPTER V

TEN-MILE MAPS SINCE 1955

1. "Ten Mile" Map of Great Britain[1]

The Ordnance Survey produced the 1942 1:625,000 map as an emergency war-time measure, to satisfy Central Planning Authority requirements for a base map for the MTCP Series. There is no evidence that it was ever conceived as the replacement to the OS statutory ten-mile map - quite the reverse: this aspect seems never even to have been considered. But as the crisis of the war faded, the OS was very conscious of its duty to replace its pre-war ten-mile map, and as early as 8 March 1945, it was determined at the DG's conference that:

Preparatory work on all editions of the 10-mile map should be put in hand with a view to work being started as soon as draughtsmen can be made available. It is desirable that this map should be published soon after the publication of the first blocks of 1-inch and quarter-inch.

But it quickly became apparent that nothing could be done in the short term, and so, by 7 June 1945, in order that the OS might fulfil its statutory requirements: "*It has been decided that the road edition of the 1/625,000 Town and Country Planning series will be used in place of the old 10 Mile map for the present.*" Other priorities increasingly delayed the creation of a new map: on 11 January 1946: "*No action to be taken about the production of this map until 1-inch and ¼-inch programmes are nearing completion*", and on 19 December 1947: "*No action at present to be taken to redraw base.*"

It only proved possible, in fact, to give positive consideration to the ten-mile map following top level discussions in August 1948. Coloured and outline maps, employing new sheet lines conforming naturally with National Grid co-ordinates and offering up-to-date topographical detail, were by now urgently required. In October a preliminary specification was written for the drawing of planning and coloured editions of a new 1:625,000 map. This is a resumé of it:

Transverse Mercator Projection; Origin 2°W, 49°N
Sheet lines: 60km E-670km E and 000km N-980km N in two sheets dividing at 490km N
Drawing: at 1:422,400
Drawings for Planning Series:
  a. Outline (Names, Railways, Towns, Villages, Boundaries, Borders and Notes)
  b. Road Casings, Class I and II
  c. Water (Rivers etc. and Names)
  d. Water tint
  e. Town filling and grid
Drawings for Coloured edition:
  f. Roads Class I and II and a selection of good motoring roads
  g. Contours. Interval as 1/625,000 Physical Map
  h. Layers. Green, 2 buffs (8 in all)
Drawings at a,c,d,e can be used for both Editions, and d,e,f,h will be drawn at final scale
Drawings at enlarged scale will be a,b,c,g
Drawing: Intermediate remarks of 10 mile Edition will be added when drawing, and outline of large and medium towns will be taken from the ¼" series.

This specification was sent to MTCP and DHS for comment. Many details would change before publication, not least the sheet lines, which on 24 May 1949 were altered to 50km E to 670km E, and 10km N to 990km N, with the joining line at 500km N. This last was fractionally further north than on the 1942 edition, and the sheet lines 10km further north and 10km wider than the original specification. With either joining line, the Isle of Man was split between the sheets, but the decision to find a way of depicting the entire island on Sheet 2 may well have triggered this adjustment, since at 500km N this became possible by means of an extrusion into the top border.

---

1. This section is largely based on PRO OS 1/789.

At the same time, the Isles of Scilly extruded into the bottom border.

On 19 August 1949, Director General R.L. Brown had a policy paper prepared to consider the OS mandate for producing the map, and the requirements it was intended to meet, as a base for the Planning Series, for defence, and general use.[2] The mandate question was deemed vital because the decision had already been taken within the OS to supersede its statutory 1:633,600 scale (in fact out of print since the war) with one for which it had no authority - the 1:625,000. In this assessment, the 1942 map was, of course, irrelevant. This contradicted the recommendation of the Davidson Committee that maps at scales smaller than the one-inch should remain mathematically related to that scale, not a metric one.

The policy document was ready by 9 September 1949.[3] It recounted the history of OS ten-mile maps, and with particular emphasis described how the metric 1942 Planning Series map came to be made. It was accepted by the OS until the end of 1945 that its commitment to produce an imperial ten-mile map still existed. Two courses of action then became possible: either to retain the 1:625,000 scale purely for the Planning Series and to produce the old pre-war range of maps at 1:633,600, or to make either an imperial or a metric map to cover both. It was quite clear that the mandate for the production of a ten-mile map had not been revoked. But having now been out of print for ten years, a map at that scale was no longer missed. On the other hand, the 1:625,000 scale of the Planning Series was now well known. To undertake both scales would be wasteful. Now that redrawing was essential, a map at the 1:625,000 scale should replace the 1:633,600. With this policy determined, Treasury approval was sought - and received on 22 October 1949. In the event OS officials need not have caused themselves so much anguish, because it was the Treasury's opinion that its authorisation was not even required for this project.

It was now possible to proceed towards a formal specification. On 4 November 1949 it was decided that:

The specification should aim at producing the best possible general topographical map, but the need to produce thereafter a satisfactory base map for the planning series from the various printing plates must be kept in mind.

It was completed by 3 December 1949. Anxious to clarify the objectives of the new map before map making commenced, Brown initiated a detailed discussion. Recommendations reached him on 3 March 1950. It should serve three main purposes: as a general reference map, for motorists' long distance planning, and for military use. For its first function, the map would need to show the main road and railway systems, the large centres of population, main river systems and relief shown by layers. Motorists would require the main road system, large towns, and vehicle ferries. The military would require everything that could be shown at this scale without cluttering up the map. The military requirement was thus the yardstick of what could be included. Map making should therefore begin with the quarter-inch map, and omit most minor roads, small villages, streams, woods and antiquities, county boundaries and names, and the names of smaller hills. This would leave main roads and most class II roads, railways, towns and villages, rivers, lakes and reservoirs, contours and relief, names of main physical features including principal mountains, main headlands and bays, and vehicle ferries. Nine colours would probably be required: black, two blues, red, brown, three layers and possibly grey for grid and town filling. These recommendations satisfied Brown, with some reservations, especially in connection with county boundaries, and he permitted matters to proceed.

A resumé of the objectives and specification of the new map were drawn up by May 1951, and were sent out to all manner of interested parties for consideration. A very large number of responses were received, and the opinions sifted and collated. By 3 November 1952, what became the final specification was in hand, and was approved at the DG's conference on 16 January 1953. The differences to the pre-war map, beyond the obvious ones of sheet lines, border, grid and marginalia are interesting. There would be one-sixth fewer villages, the shading of one side of the road would be abolished, the railway gauge would be reduced, no stations were to be marked, foreshore stipple would be changed from blue to brown (its third colour, having been black in 1903), town infill was to be grey, not hatched, there were to be no lighthouses or lightships. The grid would be grey. There would be an increase from four to six categories of town classification, listing for the first time the population numbers that governed them. Roman alphabets would be used for medium size towns, with capital letters only for those with over 20,000 inhabitants. Letterpress made its first appearance on the ten-mile map, and Gill Sans, Times Roman and Italic faces were used. Having been metric on published 1:625,000 maps since 1943, contours were to revert to imperial, using the contour plates prepared from the quarter-inch map for the Physical Features map still in preparation on the old 1942 sheet lines.

By 20 February 1953 the decision was made to draw the map, not at 1:422,400 as originally suggested, but at final scale in one piece for each sheet, using as source material the outline and water plates of the old imperial ten-mile map, still available in GSGS 3993 form, coupled with the National Grid used on its current Second Military Edition.[4] With the original borders masked out, a combined negative of these elements was

---

2. PRO OS 1/432 177.
3. PRO OS 11/42.
4. The 1942 map, with so much information such as minor roads and spot heights deleted, was useless for this purpose.

photographically enlarged to 1:625,000. Production now followed much the same pattern as that of the One-inch Seventh Series as described in Oliver (1991). Prints were made from the negative in ferro-prussiate blue ink, which will not photograph, on enamelled-surface Whatman's drawing-paper. These "FP blues" served as drawing keys for three sets of drawings in black ink. The first included border, graticule, grid figures, title panel, and outline detail, without coastline. The second included all water features, coastline to high water mark, and submarine contours. The third included contours and values, to be read uphill, and sand stipple around the coast. These were then photographed at scale, and from them, a further set of five FP blues were printed, for roads (all MOT trunk and First Class Roads), for town filling (the outline of large and medium towns was taken from the quarter-inch map), for names (both detail and water, in monotype), for sea tint (both sea up to the high water mark and inland water, including lakes and reservoirs), and for the National Grid. These were all photographed, the names drawing twice in order to separate the detail names from the water names by duffing.

With the addition of four for layers, there were now thirteen negatives in all. This number was reduced to eleven by combining outline with detail names, and water with water names. From these, printing plates were made. Drawing was at first limited to the portion of Sheet 2 above the 400km line. With the title **Great Britain**, this was printed in eleven colours[5] on 22 October 1953, and inspected at the DG's conference on 11 November. The few adverse comments referred chiefly to the title panel and title. Others concerned colour contrasts, and the recommendation that streams should stop at the high water mark. Second proofs were prepared in December, and also mock-ups to illustrate the positions of the panels and insets proposed for the Planning Series Maps which were to use the base.[6] Copies were sent to Willatts at MHLG, and he replied with his comments, generally favourable, on 9 April 1954.[7] He sent a supplementary letter on 13 April giving his views on the thorny question of the status of county towns: neither MHLG nor the OS seemed certain of the legal definition of these, and it was in the end decided to omit them from the specification.

It was now possible to begin the drawings of Sheet 1 and to complete those of Sheet 2 below the 400km line. First proofs of both sheets were printed on 29 November 1954, with the legend in its published form and the **"Ten Mile" Map** title in place. This title would be common to all maps in the family, including the overprinted members. Up to this point the bottom layer had been uncoloured, and contours had been in place. Many experimental proofs followed, with and without contours, with different strength contours, and using various shades of emerald green in place of the uncoloured bottom layer. Even yellow was tried instead. A price was added: 5/-, which on publication was reduced to 4/-.

A further important question to be decided was which titles were to constitute the new OS 1:625,000 series titles, as distinct from titles in the Planning Series which would remain entirely the responsibility of MHLG. The possibility that the OS should take over the **Administrative Areas** map had been addressed in 1950, but, while it was seen as the obvious course, with the OS the principal Government authority on boundaries, it was decided not to revise the old map but keep the *status quo* until the arrival of the new base map. The preliminary recommendation made on 18 December 1954 was that the OS series should include six titles - the coloured topographical map, the outline base map, a layered base map (the equivalent of the current **Topography** title), a **Roads** map, and an **Administrative Areas** map. These five would have a common base. Only at this stage was it proposed that the physical map in production since 1942 be cancelled and the title enter this series as a sixth member. It had previously not been thought necessary to include it since other than in its frame and map outline it would have little in common with the other members anyhow, requiring as it would separate drawings.[8]

It was put to Willis on 22 January 1955 that five of the six be published. The layered base map, which did not sell well then and would fare even worse against the competition of a physical map, should be discontinued. The question of whether the OS needed to consult with MHLG over the future of the titles on which they had collaborated since the war seems conclusively answered by these recommendations: they did not. Against the Roads map is the comment: "*This is at present a MHLG map. It sells well and since we have a road map in the past I think we should take this over as an OS map.*" Confirmation of the new policy appears in the official circular of 7 March 1955 which, while giving notice of the forthcoming OS 1:625,000 series and listing the five titles, continues: "*The responsibility for the information and design of these maps will rest entirely with the OS Department. They will not be regarded as sponsored by any other department.*" Final proofs of the coloured topographical map were printed on 22 April 1955, and Willis was presented with a choice of four layer colour variants, all in versions with and without contours. He made his choice on 25 April. Some concern was expressed that road map and topographical map sales would overlap. It was agreed that they probably would, but this could not delay the topographical map which was

---

5. See p.152 for details, except that the emerald green ground layer mentioned there was not yet in place, but contours in contour brown were.
6. PRO OS 1/433 125.
7. PRO OS 1/433 132.
8. PRO OS 36/4.

urgently needed. It was in print by June 1955. The outline version followed in October, and **Administrative Areas** and **Roads** the following year. Yet more problems with names held up the **Physical Map** until 1957.

I will do no more than make the briefest mention of topographical revision. Added roads and closed railways became obvious signs of change, and placenames also have come and gone. Some spot heights were revised. The most significant change was the deletion of reservoir references following a 1957 security directive (remarkably, their survival on the **Geological Map** was overlooked). Their gradual restitution began no later than 1963. An important adjustment was made on 1 February 1961 to Sheet 1 when three of the values given in the "Difference to Grid North" panel were corrected.

2. Physical Map of Great Britain[9]

We have noted above that attempts to publish the 1942 **Physical Map** continued until 1954. Only on 18 December 1954 was it finally buried with the decision to make a new one on sheet lines conforming with those of the new topographical map. Its layering and general design would also require overhaul and its specification that of a new map. Production was recommended to Willis on 22 January 1955. He was persuaded that it had potential additional merits for educational reasons and as an alternative base map. It was announced on 7 March 1955 that the old Planning Series title would become the **Ordnance Survey Physical Map**, and dealt with in the Small Scale Drawing Department under separate instructions.

Although Willatts at MHLG was apprised of the new OS plans on 23 May 1955,[10] for some reason no mention was made of the new Physical Map. He therefore remained in ignorance until March 1956 when he was surprised to receive proof copies of the new map - presumably from the same printing as those now in the RGS dated 18 January 1956. Acknowledgement of DHS and MHLG assistance was not displayed. Displeased, Willatts wrote to Willis on 5 March 1956 complaining of the lack of consultation on a map on which MHLG and DHS co-operation had in the past been considerable. This was a unilateral change of policy. Willis investigated the whole sorry saga of the post-war physical maps. He found nothing in writing in the files confirming Willatts's official involvement in or responsibility for it, but plenty of corroborative references to his assistance. He replied on 22 March, noting that when it had been agreed in 1952 that the Physical Map was indeed an OS map, it had also been agreed that MHLG and DHS would be consulted. He could only apologise that this had not occurred, and now invited their comments, though at this stage he was not seeking major changes.

Nonetheless, as regards names, he seems to have got them. Willatts and his MHLG team submitted a long paper on 25 June 1956 recommending changes. Tautological labels should be expunged, names applied to waterways longer than 20km (so Peover Eye, not named on the proof, was added to the published map), summit heights on more elevated peaks were preferable, (for instance Blackstone Tor [1765ft] for Dovestone Tor [1656ft]). The OS acted upon most of these recommendations, and added MHLG and DHS acknowledgement to the map, so delaying publication into 1957. The Physical Map differs in many ways from its 1955 parent coloured topographical map. National Grid 100km squares only were drawn, with 10km ticks at the neat line. Both have nine layers, though different browns were used on the Physical Map. The more extensive river systems, physical names and grid derive from the earlier physical map, and the submarine contours are based on Admiralty charts. And to this day the physical names shared with the **Routeplanner** vary, with the latter still preferring the tautological "Cotswold Hills". The Physical Map has remained in print ever since, only once going to a readily identifiable reprint, when the publication address was altered to Southampton, and the price removed. There have been several others since. But the expectation that it might be used as a base map seems only twice to have been fulfilled, and these are noted in the cartobibliography.

3. Route Planning Map

The 1956 **Roads** map was not successful, and incurred enough public criticism of its lack of clarity for the OS to permit it to fall out of print in 1961. For the next three years the topographical map had to double as a road map, and for this reason it was vital to the OS that it too should not at any time be out of print.

The creation of an entirely new road map began on 8 March 1961 with a specification for a prototype Sheet 2. The starting point was not the 1956 Roads map, but the "Ten Mile" base map, from which a six-colour road map was to be produced. This would entail a large injection of new information, much of it derived from the Quarter-inch Fifth Series map to be completed in 1962, including inset maps at 1:250,000 of London (in preparation since 3 February 1961), Birmingham and Manchester, telephone call boxes and gradient symbols. Also on the black plate would be an incidence diagram of the Quarter-inch Fifth Series, trunk roads and motorways in solid black, stippled built-up areas and redesigned title panels, with road information removed from the traditional title panel and assembled in a second one at the foot of the sheet. On the blue plate would be stippled sea, frying pan symbols for distance measuring with associated mileages, ferry symbols and details in the inset boxes. All road numbers were to be on the red plate, those of

---

9. This section mostly derives from PRO OS 1/705. For a fuller account, see also Hellyer (1992a).
10. PRO OS 1/433 177A.

motorways and "A" roads in upright numerals, and of "B" roads in italic. The grid and a layer over 1000ft would be brown, woods would be green and "B" roads yellow.

A significant adjustment to this specification was proposed on 24 October: the base map should be in grey leaving overprinted information in black. Already the solid black motorways and trunk roads was questioned, and by 8 November 1961 it had been decided to leave them as on the topographical map. The yellow plate could be eliminated by leaving "B" roads uncoloured. There would be no "B" road numbers. Two days later the inclusion of dual carriageways was proposed: following discussion, they would be depicted by means of a red line outside the road casing, a method already adopted on Automobile Association (AA) maps.

First proofs were printed on 21 May 1963, entitled **Roads**, in six colour permutations, half with motorways, motels and point to point mileage apparatus in blue and half in a stipple of red and blue, combined with three black plate alternatives: in best black, in best black except krystal black outline and town fillings, or with base names in krystal black as well. In each case the grid was in krystal black. Woods were in block green, main roads in red, and there was a buff layer above 1000ft. New road information included motels, gradients, tolls, level crossings, air and ferry services, telephone call boxes and road numbers for motorways and "A" class roads. In addition to the panels noted above, there were a mileage chart and information on Car Freightage Facilities.

The proof specification was discussed in detail on 13 June 1963, and many recommendations were made.[11] The second of the three base map options noted above was preferred, though with black stipple town infill. Woods should be excluded, and block green used instead for areas of outstanding natural beauty, national parks and national forest parks. Relief should be enhanced with a second layer from 200ft to 1000ft. Mileages and the associated frying pan markers, level crossings and motels should be deleted. The mileage chart and many symbols required revision. The inset maps should have a scale-bar. Blue was preferred for motorways, with a pecked line for motorways under construction.

A policy statement based on these recommendations was sent to Director General A.H.Dowson on 5 July 1963 for his approval. The map should be maintained in a state of revision and sold in dated annual issues beginning on 1 January 1964. The intention should be to provide an up-to-date map showing main roads, all dual carriageways, motorway access points, a selection of minor roads, sea and air ferries, customs airports, AA and RAC telephone boxes, relief, areas of outstanding natural beauty, national forest parks and national parks, and enlargements of conurbations. The title "Main Route Map" was proposed. Dowson replied the same day, approving the greater part of the schema, though he confirmed that minor roads should be "B" roads, unnumbered. He suggested adding motorway service areas and ordinary telephone boxes in isolated areas. He was strongly in favour of annual editions. He proposed as a title "(Main) Route Planning Map", implying a map for prior study than for a motorist to use in following a route.

The title **Ordnance Survey Route Planning Map** was confirmed on 9 July 1963. MHLG and DHS were sent proofs on 11 July, and the policy of the new map explained:

....to provide private motorists and transport operators with an up-to-date map showing main roads....at a smaller scale than the quarter-inch series, and to present users with such additional information as will enable them to plan routes to suit best their individual requirements.

Willatts's comments from MHLG arrived in time for the next specification meeting on 7 August. The War Office had already expressed its interest in the map. Second proofs carrying the new title were printed on 6 September. There were marked changes to the first ones. Altitude layers were increased from one to three, at 200ft, 600ft and 2000ft, with block green for national parks, national forest parks and areas of outstanding natural beauty. Toll, ferry and aerodrome symbols were altered. Blue was adopted for motorways (except on the town plans where the change from red did not occur until after the third proof stage). Motels, level crossings, point to point mileages and the Car Freightage panel were dropped. Trunk road numbers were altered from red to black, followed by "(T)": this would be dropped on third proofs, and the size of the numerals increased. The mileage chart was improved. On the original title panel, map title and sheet number were in black, and this was again altered in the published version, where only the sheet number remained in black. "B" road numbers and colour infill and layer colour were removed from the inset town maps, and scale references added. Third proofs were printed on 5 November 1963, and at the same time first proofs of Sheet 1. These were examined by the Advisory Committee on Surveys and Mapping on 7 November. Few further changes were required afterwards, but road information was updated as late as possible, and the mileage chart improved yet again.[12]

There had been an intensive discussion on print runs, the earlier estimates of 5000 and 7000 (Sheets 1 and 2) finally being reduced to 3000 and 4000, with a further 5000 of each sheet for the War Office. Printing was completed by 12 December 1963. What followed staggered the Ordnance Survey. As early as 7 January 1964, demand for the map had proved so heavy that a reprint was necessary. Until the position was clearer, it was decided to use half the War Office allocation still lying at Crabwood without its military overprint. The balance of this order could be made up later. Emergency

---

11. PRO OS 1/1136 is the principal source for the following information, together with the job files.
12. See the copy in PRO OS 1/1136 91A.

reprints in February and March resulted in some 20,000 copies of Sheet 1 and 25,000 of Sheet 2 by the end of the first quarter, with the remaining 2500s for the War Office in addition. In June still more copies of Sheet 2 were printed. Public demand of this magnitude came as a complete surprise both to the OS and its dealers Stanford and Nelson, who had earlier made soundings as to potential sales, and it was this gross underestimate that resulted in so many expensive emergency returns to the printing presses. The surprise was the greater because the public had already been advised of the intention to publish the map in annual updated editions. This demand would rise to a peak with the "Z" edition which saw a first print run of over 100,000.

Whether or not planned as such, the RPM became the base map from which all other 1:625,000 maps subsequent to it, except those requiring a physical base, were to be derived. Brigadier Gardiner met Willatts on 1 October 1964 to discuss the Planning Series,[13] and explained that in future overprinted maps would be based on the RPM. To this Willatts expressed satisfaction. As early as October the future of the 1955 topographical map was already in doubt.[14] With its function now largely overlapped by the RPM, it was considered too expensive to maintain and it was soon allowed to lapse. By 1 April 1965 the victory of the RPM, now into its "B" edition, was complete, when its new derivatives were launched. These were the **1/625,000 Outline Map of Great Britain**, to replace the 1955 uncoloured topographical map, and a new **Administrative Areas** map. To those of us fascinated by OS policy on print codes, it may be of interest to recount that the new Outline Map caused a minor skirmish: it was derived from the "B" edition of the RPM, but current outline editions were "B" (North Sheet) and "A/*" (South Sheet), but (on the third hand), it was felt that a new map should have an "A" print code. This last opinion prevailed, the more so since it was realised that even if placed in step with the RPM at this stage, it quickly would not remain so.

I do not propose to weary readers with a lengthy discussion of the revisions made each year to the RPM, and most of the significant alterations will be found in the notes to the map itself. I have entirely ignored revision to base map topographical detail, such as road network development and changes to ferry routes, which could form the subject of a future study. Here I would merely point out some of the most significant design developments, and refer briefly to some of the experimental versions, surviving examples of which are noted on p.164. Until 1986, the RPM was revised and published annually, the revision also being used for quarter-inch special revision. New editions of the outline or administrative maps were periodically taken from the most recent RPM edition, but it was not felt necessary to revise these maps with anything like the same frequency, even through the 1960s and 1970s when annual editions of administrative maps were normal. Until 1971, all RPM revision was done on glass negatives, but experimental work led to its transfer to film positives on the "J" (1971) edition, and to a computer database with the "Y" (1984) edition. Since then it has been marketed as the 1:625,000 Digital Database, and is offered to the public in both paper and digital form. OS Catalogues state:

This is a fully structured national topographic database of maps at 1:625,000 scale. The structure of the data allows information to be extracted for a named area, feature selection by location, type or name, and analysis of road or river networks. Data is available in two (later five) *datasets (Coastline data, Hydrological data, Administrative boundaries, Communication data, Settlement data) which can be purchased separately in national or regional (100 x 100km square) coverage on a "Licence to use" basis.*

On the "B" edition (published 1965), inset maps and panels were repositioned in order to avoid their falling on the folds of covered maps; this was carried a stage further on the "G" (1969) edition when the National Parks panels were relocated to form one map on sheets mounted together. Following experiments on a Midlands area extract, green primary routes, consistent with the colour of the road sign posts in use by MOT, were introduced with the "D" (1967) edition.[15] Layers were changed from brown to yellow. New title panels were introduced with the "D" and "M" (1974) editions which affected also the derived maps. Advancing publication from January to November in order to attract Christmas trade came with the "E" edition: though published in 1967, it was called the 1968 RPM. Experiments with distance markers and road colours were made with "G" edition South Sheets: the former were added on the "J" (1971) edition. At the same time road casings were deleted, which led to a lack of clarity which, it was anticipated, would be improved on the "K" (1972) edition by replacing layers with hillshading. To this end experimental "J" edition South Sheets were printed with hillshading offering ten different colour options. Separate sets of "J" edition South Sheet extracts were printed for further hillshading and lettering experiments[16]. The latter were necessary because the Times Roman, Gill Sans and Italic names were too broken down to be capable of another printing. On the "K" edition they were replaced by Univers sanserif style. This was the only RPM to be published with grid and border (as well as hillshading) in a true grey. After publication, further experimental pulls of the South Sheet were printed, with this plate in a variety of grey tints. The krystal black option was preferred.

---

13. PRO OS 1/999 190B.
14. The remainder of this paragraph is drawn from PRO OS 1/706.
15. A copy of the approved format of the extract sheet (dated 28 September 1966) is in the job file.
16. A copy of the hillshading experimental extract sheet (dated 4 May 1972) is in the job file.

Main road categories were rescribed on the "L" (1973) edition. This was essential for the new outline ("C") edition to be derived from it if the imperfections in road widths inherent in the existing system (ie the pre-1971 road fillings before the casings had been removed) were not to appear exaggerated. Experiments for the "M" edition involved altering the type size of town names in congested areas, and showing inset areas by boxes on the main map: the one was done, but the latter was not achieved until the "S" edition. Some "J", "K" and "L" edition sheets had been printed back to back, but the first mass printing for the covered map market occurred with the "M" (1974) edition. Though a continuing requirement of single sided sheets for wall mounting was recognised, back to back printing of folded maps cost little more, a cost offset by a marked reduction in paper and distribution costs. Tumble turning came in a year later. Experiments to the "M" edition involved road gauges, and colour contrasts. Four colour process printing was introduced with the "R" (1978) edition. But the hill plates were now very worn, and following experiments, new hillshading was incorporated in the "S" (1979) edition.[17] At the same time the title was altered to **Routeplanner**, a yellow base tint introduced, and the green primary routes displaced by magenta, though still with green road numbers and destination names. Railways were deleted. "B" road numbers were at last introduced on the "T" (1980) edition. With the "V" (1982) edition, the hillshading was changed to orange.

An outline specification for the Routeplanner database was formulated at a meeting held on 24 September 1981, when decisions were taken on how the 28 separate components that at the time constituted the map were to be incorporated. To be excluded (and accommodated manually where necessary) were the accumulation of supplementary boxes surrounding the main map, and hillshading, which may explain its absence since the "Y" edition. On the other hand, to be included in the database, though not shown on the map, were railways and metric contours. Road mileages, a very complex database item, were later excluded from it, and inserted on the map manually. Another later development was the reduction of placename categories from five to three: major conurbations, primary towns, and small towns and villages.

As part of the development process, Routeplanner extracts were printed in 1980-82 for market research purposes. Some were simplified versions lacking standard characteristics, including hillshading and grid lines but for "0" and "5". One included an experimental additional symbol (green highlighting of one side of designated routes) indicating scenic routes, which also appears on experimental "X" edition sheets. It may have been the results of this survey that led to the omission from the first digitised edition ("Y" in 1985) of single line rivers and canals, unclassified roads, and their associated names. All were reinstated on the "Z" (1986) edition, but hillshading has never reappeared. The railway passenger network was revived on the "AA" (1988) edition. Digitalisation also permitted experiment with an enlarged sheet size, with the whole of England and Wales on one side with Scotland and a placename index on the other. A placename index was indeed introduced with digitalisation, but limited to primary towns and major conurbations (those with names in capital letters) only, which a conventional sheet size could accommodate. Thus an enlarged sheet size, and other early 1980s ambitions, such as cyclist's and railway Routeplanners, have now been dropped: its future appears to be as Sheet 1 of a new Travelmaster series combined with the 1:250,000 map.

4. 1:1,250,000 Great Britain, "E" edition

It will be recalled that in 1951 current policy for the 1:1,250,000 map was seen to be to do nothing[18]. No uses appear to have been discovered for it in the subsequent 24 years, beyond its function as an index, and even that would virtually peter out in 1976. Yet against this background, the map was not just revised in 1975, but completely renewed, with a new specification entailing sheet lines, layout, outline, lettering and characteristics. This action may have been prompted by the revision of county boundaries throughout Great Britain. In spite of all this, it was given only an "E" print code, in sequence with its predecessors, and was omitted from OSPR, with the result that most of the copyright libraries do not even own copies of it.

The map falls on the true National Grid co-ordinates of the 1:625,000 **Route Planning Map**, and derives from the RPM "M" edition of 1974. This would, no doubt, have been photographed, then reduced, and unwanted elements cleared. What survives is a much more detailed map than preceding 1:1,250,000 editions, with more rivers, lakes, railways (including, at last, Arbroath to Kinnaber Jcn), roads and islands, though some actually disappeared.[19] Rockall was added on an inset. Built-up areas and roads were screened for easy updating from RPM material. The map was lettered in Univers sanserif, with more placenames which are often in alphabets different to earlier editions. It is not clear whether this was because the populations or the specifications altered. There are more geographical names.[20] The border was redesigned, with no diced graduation, and all National Grid values written upright. The map was published in two forms, with the conventional 10km grid, and another version carrying the 100km National Grid squares only.

---

17. Several hillshading proof sheets are in the job file.
18. PRO OS 11/5 2. See p.112.
19. Dubh Artach, for instance, (NM 01) was even named on the "D" edition, but it is lacking altogether here.
20. The change in specification of symbols and lettering can be examined in more detail in the table on p.9.

Standard attributes

V.1. "Ten Mile" Map of Great Britain.   V.2. Physical Map of Great Britain

NB: Features irrelevant to the Physical Map (V.2) are noted
NB: Titles marked "+" are members of the Planning Series. Details of Planning Series titles may be exceptional

National Grid co-ordinates: 50km E - 670km E; 10km N - 500km N - 990km N
Mapped area per sheet: 620km by 490km = 303,800 square km

1. Map border showing:
   a. Latitude and longitude values at 30' intervals, related to a diced graduation outside them at 5' intervals. There are usually associated graticule intersections shown on the face of the map (not V.2)
   b. The values at 10km (V.2 100km) intervals of the National Grid
   c. The start or end of names situated near the edge of the map (not V.2)
2. The title panel, usually top centre (Sheet 1) and top right (Sheet 2)
   a. Title: ORDNANCE SURVEY "TEN MILE" (V.2 PHYSICAL) MAP OF GREAT BRITAIN, with SHEET 1 (or 2): from 1965 NORTH (or SOUTH)
   b. Scale: 1/625,000 or about Ten Miles to One Inch, with imperial (10+50 miles) and metric (10+70km) scale-bar
   c. Transverse Mercator Projection, Origin 49°N 2°W (moved to the panel bottom after "ga" states)
   d. Legend: "ga","gb","gc","gd","pa","ra" types in four boxes
   e. THE INCIDENCE OF GRID LETTERS ON THIS SHEET
   f. *Index to Sheets* (INDEX TO SHEETS on "gc" and "gd" Sheet 1 states only)
   g An explanatory note on THE NATIONAL GRID (not V.2)
   h. Magnetic Variation arrow (not V.2), with Difference from Grid North calculations, with date (not V.2). Three of the values on Sheet 1 found to be in error were corrected on 1.2.1961. This affected states derived from the outline "B" edition priced at 4/-
   j. Nine-section imperial layer box, with HEIGHTS IN FEET AND METRES. An associated altitudes statement is on V.2 only: *The altitudes are given in feet above Ordnance Survey Datum (Mean Sea Level)*
   k. REVISION DIAGRAM, with dates (on "gc" and "gd" Sheet 1 states only)
   l. Six-section marine layer box, with DEPTH IN FATHOMS, with submarine contour statement (V.2 only): *The Submarine Contours are given in fathoms and are based on Admiralty Charts*
3. In the bottom margin:
   a. Bottom left:
   1. Copyright statement: Crown Copyright Reserved; from 1958: (c) Crown Copyright 19xx
   2. Print code, often a mixture of base map and overprint elements
   b. Bottom centre: PRICE statement (various wordings, the coloured and outline map prices increased from 4/- and 2/6d to 4/6d and 3/- on 1 April 1958, and to 5/- and 4/- respectively, probably on 1 January 1961)
   c. Bottom right: publication details. Instructions were given in May 1958 to alter *Printed and published* for all but facsimile reprints to *Made and published by the* Director General *of the* ORDNANCE SURVEY, CHESSINGTON, SURREY (until 1967), 19xx, perhaps followed by details of *minor corrections*
4. The National Grid across the map at 10km intervals, emphasised at 100km intervals (V.2 at 100km intervals only)
5. Symbols: see col."M" on p.9 for a detailed analysis of the 1955 situation: changes thereafter may be found listed in the notes below
6. Framed inset maps of Sule Skerry, Rona, St Kilda (with a location diagram) and the Orkney and Shetland Islands with Fair Isle are on Sheet 1. The Isles of Scilly and Man extrude into the border. The Irish and French coasts are shown in outline, with some coastal placenames and appropriate symbols

Colour plates (V.1): *NB*: Colour strengths 1:0 unless noted otherwise
1. Black combined outline and name plates (50% Lorilleux & Bolton, 50% Parson Fletcher) for everything except:
2. Fishburn Monastral Blue: coastline, inland water, water feature names
3. Fishburn Monastral Blue (1:96 spray): sea and inland open water
4. Flemings Concentrated Scarlet: trunk and first class roads
5. Light Emerald (1:50): ground layer to 200ft
6.7.8. Buff: 3 strengths (1:24, 1:10, 1:6) for 1st layer, 2nd layer, 4th layer (ie layers above 200ft), sand and mud
9. Contour Brown (1:6): 3rd layer. *NB*: Not used for contours
10. Winstone Blue Grey (1:0): grid
11. Winstone Blue Grey (1:8): town fillings

Outline edition: plates 1,2,10, all black (with hatched town infill)

Administrative Areas: plates 1,2,10, all krystal black (with hatched town infill)

Roads: plates 1,2,10 (base in krystal black), with 3,4 (trunk roads separated out in black), with second class roads in emerald green, and warm buff for layers

Physical Map (V.2): lacking plates 4,10,11. Plate 1 has border, names and grid only. There are three blue "sub" layers (1:72, 1:72, 1:50). The lower two layers are in buff (1:10, 1:4), the upper two in contour brown (1:4, 1:1)

NB: For RPM "A" edition, see V.3 below

Lettering: Monotype: Times Roman, Gill Sans and Italic faces: see col."M" on p.11 for a detailed analysis

Cover types recorded: H.96.1a2, H.96.1a3, H.96.2b1, H.108, H.181, H.194.2, H.194.3

V.3. Route Planning Map from "B" edition

National Grid co-ordinates as V.1

NB: For detail changes after 1965, check also title panel notes and RPM notes below

1. Map border showing:
    a. Latitude and longitude values at 30' intervals, related to a diced graduation outside them at 5' intervals. There are associated graticule intersections shown on the face of the map
    b. The values at 10km intervals of the National Grid; from the "J" edition all National Grid values shown were multiplied by ten
    c. The start or end of names situated near the edge of the map
2. The title panel, top right of centre (North Sheet), and top right (South Sheet) (top left from "G" edition)
    a. Title: ORDNANCE SURVEY ROUTE PLANNING MAP, with NORTH (or SOUTH). See the Outline Map *infra* for its title variants. As a base map, these titles were used: ORDNANCE SURVEY 1/625,000 MAP OF GREAT BRITAIN [gap for overprint title, which would include "North" or "South"] ("hd" type, based on Outline "A" edition (RPM "B" edition), noted as "1/625,000" in the lists below), ORDNANCE SURVEY [gap for overprint] NORTH (or SOUTH) ("jf" type, unused as an outline map). The outline "B" edition (from RPM "F" edition) was little used. Maps with "km" codes derive from the Outline "C" edition, constituted from both RPM "L" and "M" editions, with titles as quoted in the list of maps
    b. Scale: 1/625,000 or about Ten Miles to One Inch, with a combined imperial (10+50 miles) and metric (10+70km) scale-bar, from "F" edition: Scale 1:625,000 etc, from "G" edition: Scale 1:625,000 or about one inch to ten miles, with two scale-bars - imperial (10+50 miles) and metric (10+80km), from "K" edition: Scale 1:625 000 etc
    c. Legend: "rb"-"rz","raa","rbb","hd","jf","km","ks" types
    d. THE INCIDENCE OF GRID LETTERS ON THIS SHEET. *NB*: Outline "B" edition on special panel; not on RPM
    e. INDEX TO SHEETS (to RPM "L" edition)
    f. Four section layer box, with HEIGHTS IN FEET (to RPM "J" edition)
    g. Publication statement: *Made and Published by the* Director General *of the* ORDNANCE SURVEY, CHESSINGTON, SURREY (until 1967, SOUTHAMPTON from 1968), 19xx. The change was physically made to the outline negatives in December 1967, and to the Administrative Areas base maps by duplication. Some maps remained 1966 with updated information, and were changed to Southampton without a revised Made and Published date. Exceptional cases are marked C-hessington or S-outhampton. RPM publication statement altered, from "R" edition, to: Revised and published annually by the Director General of the Ordnance Survey, Southampton, from "V" edition to: Revised and published annually by the Ordnance Survey, Southampton, and from "AA" edition to: Revised and published by the Ordnance Survey, Southampton
3. Below the map:
    a. Bottom left:
    1. Copyright statement: usually (c) Crown Copyright 19xx
    2. Print code, often a mixture of base map and overprint elements
    b. Bottom right: publication details of non-RPM issues. For wording see section 2 above. New titles from 1969 have the publication statement on the title panel
4. The National Grid across the map, at 10km intervals, emphasised at 100km intervals (from 1974 usually to grid boundaries in the Irish Sea, North Sea and English Channel)
5. Symbols: see col."M" on p.9 for a detailed analysis (V.1), and for changes thereafter in the lists below
6. Layer box: see title panel
7. Framed inset maps of Sule Skerry, Rona, St Kilda (with a location diagram) and the Orkney and Shetland Islands with Fair Isle are on the North Sheet. The Isles of Scilly and Man extrude into the border. The Irish and French coasts are shown in outline, with some coastal placenames and appropriate symbols

Colour plates (RPM "A" to "J" editions): *NB*: Colour strengths 1:0 unless noted otherwise
1. Krystal black: grid, railways, road casings, map frame, county and national boundaries, town and village symbols and area limits of large towns, inset maps, title panel on "A" edition
2. Best black: most numbers and letters except water names, graticule intersections, diagrams, road numbers, telephone and ferry symbols, tolls, gradient signs, large town infill (stipple)
3. Monastral blue: inland water, coastline and coastal water features, and water names, sea and inland open water (by blue stipple), motorways, motorway numbers, ferry routes, air ferry symbols
4. Concentrated scarlet: main roads, information boxes
5. Light emerald green (1:5): national parks, etc (until "C" edition)
   From "D" Edition, primary routes used a mixture of trichromatic blue and yellow, from "F" edition emerald green
6. Buff (2nd layer buff, 1:3): layers above 200ft (lemon yellow (1:6) from "D" edition), sand or mud

Colour plates (RPM "K" to "Q" editions): *NB*: Colour strengths 1:0 unless noted otherwise
1. Krystal black (1:1): grid and border combine, and hill shading (from "L" to "Q" editions: Shuck McLean's blue-grey was used on the "K" edition only)
2. Best black: town names combine
3. Monastral blue: motorways, water and water names, including stipple for open water
4. Concentrated scarlet: main roads combine
5. Malachite green: primary routes, national parks
6. Buff (1st layer buff): town fillings

Colour plates (RPM from "R" edition): the trichromatic process, plus black (now known as four-colour process printing)
1. Black. 2. Magenta. 3. Yellow. 4. Cyan. *NB*: These are variously used: for detail changes, see notes

Lettering: Monotype: Times Roman, Gill Sans and Italic faces, replaced on RPM "K" edition by Univers sanserif

Cover types recorded: H.84, H.108 (first Sheet 1 issues have the royal, not Scottish, coat of arms), H.127, H.128, H.129, H.130, H.131, H.178, H.187, H.203.1, H.203.2, H.304.1c, H.401.2, H.403.1, H.404.4, H.404.6a, H.404.6b, H.406.1, H.407.6, H.407.7a, H.407.7b, H.407.7c, H.407.8

V.4. 1:1,250,000 Great Britain, "E" edition

National Grid co-ordinates: 50km E - 670km E; 10km N - 990km N
Area mapped: 620km by 980km = 607,600 square km

1. Map border showing:
   a. Latitude and longitude values at 30' intervals
   b. The values at 10km intervals of the National Grid, drawn upright in all borders including corners. A second version has 100km values only
2. The title panel, top right
   a. Title: Ordnance Survey GREAT BRITAIN
   b. Scale: 1:1 250 000, with and imperial (10+50 miles) and metric (10+80km) scale-bar
   c. Legend (headed SYMBOLS): "ue" type
   d. NATIONAL GRID REFERENCE SYSTEM (example Dover), with gridded sheet index
3. In the bottom margin:
   a. Bottom left: print code "E", and Crown Copyright 1975, with copyright symbol
   b. Bottom right: Made and Published by the Director General of the Ordnance Survey, Southampton.
4. The National Grid across the map at 10km intervals, to the neat line or the official limits between Great Britain, Ireland, France and in the North Sea, emphasised at 100km intervals (second version 100km lines only)
5. Symbols: see col."P" on p.9 for a detailed analysis
6. Framed inset maps of the Shetland, Orkney (separated), Skerry, Rona, Rockall, and St Kilda groups (with a location diagram). The Isles of Scilly extrude into the border. The Irish and French coasts are shown in outline, with some coastal placenames and appropriate symbols

Colour: only black, and two grades of black stipple are used

Lettering: see col."P" on p.11 for a detailed analysis

V   155

Variable attributes (V.1,2,3,4)

1. Sheet number or name. *NB*: North of England strip maps appear as **n**. Sheet number changed to sheet name in 1965
2. Print code
3. Publication date of base map. Legend. Price.   Published at Chessington to 1967, Southampton from 1968: exceptions are preceded **C** or **S**, "Printed and published" was used until 1958, "Made and published" from 1958, without necessarily altering the publication date printed on the map: 1958 and exceptions are preceded **p** or **m**
Legend types: "g*","h*","j*","k*" sequence for outline and base maps, "r*" sequence for RPM map, the second letter being identical to the print code, "pa" for the Physical Map, and "ue" (in sequence from IV.4) for the 1:1,250,000 map (V.4). *NB*: Sample alterations are given below. Local differences between north and south sheet legends are disregarded. No reference indicates a non-standard legend, such as may appear on proof copies. Other RPM differences may be deduced from the notes to the map.

    ga   MOT Roads: "Trunk & Class 1", "Class 2", "Other Tarred Roads", "Ferries (Vehicular)", "Railways" (showing "Tunnel"), "National" and "County" "Boundaries", "Canals", "Lakes, Reservoirs & Rivers", "Coastline with sand or mud", three grades of community, "Heights (in feet)", six population classes, Italic type

    gb   MOT Roads: with "Motorways", (and "Junction"), "Motorways Under construction", "Reservoirs" deleted

    gc   MOT Roads: with "Narrow Class A roads with passing places" (Sheet 1), or reference to "Isle of Man" roads (Sheet 2), Sheet 1 with revision diagram (1947-62)

    gd   Road references deleted, Sheet 1 with revision diagram (1947-62)

      ra   Secondary legend for RPM 1964 with reformatted road information, with, in addition, "Service areas" on motorways, "Gradients (1 in 7 or steeper) & Tolls", "Telephone Call Boxes" (AA, RAC, PO), "Transport for Vehicles": "By Air", "By Water" (drive on and lift on ferries distinguished), "Aerodromes" (with and without Customs facilities), "National Parks, Forest Parks & Designated Areas of Outstanding Natural Beauty", Class 1 and Class 2 roads also classified "A" and "B" respectively, Roman type

      rb   New layout: boundaries, railways, population samples, town symbols, water plate symbols incorporated from "gd" legend, singular and not plural descriptions preferred, road information again reformatted, and reference to "A" and "B" roads dropped, "National Nature Reserve (selected)" added to National Parks

      rb/   Reference to "National Nature Reserve" deleted

    hd   Four "gd" boxes in one, North Sheet revision diagram deleted

      rc   "Tarred" deleted from "Other Roads", National Park wording layout altered

      rd   New layout, italic type, trilingual, "Tarred" reinserted, "Primary Routes" introduced, no "sand or mud"

      re   Motorway junction numbers, primary route numbers added

  jf **rf**   Class 1 and 2 to Main and Secondary roads, distinct N and S samples, thus Glasgow to Birmingham (S), sample road numbers altered, MOT legend simplified (**jf**: no road information, Made and published as "rh")

      rg   Gills sans title and reworded scale statement, separate imperial and metric scale-bars, roads reordered, narrow roads within list (N), some brackets and upper case deleted, layers and index panels reversed

      rh   Sheet index diagram includes Ireland, Gills sans publication notice, MOT information to foot of panel

      rj   1 mile = 1.6093 kilometres added, road casings deleted, so all roads now coloured lines and numbers in boxes, motorway limited interchange junction sample introduced, motorway service area and junction under construction added, trunk roads deleted, road omitted in error from toll and gradient example (N)

      rk   Univers panel (1:625 000), layer box deleted, large town sample coloured, primary route legend rewritten

      rl   Primary route legend rewritten, dual carriageway gauge reduced, aerodrome name samples altered, main road colour lacking in error on large town symbol (S)

  km **rm**   New layout, road information (to 7.1974) reinstated on non-RPM versions, population samples deleted, sheet index deleted, town symbols brought into list, motorway service area with limited access introduced, airport, ferry symbols revised, primary route legend rewritten

      rn   Wider gauge roads on panel, height references given in metres

      rp   Glasgow to Dundee, German primary route legend rewritten

      rq   Title dated

      rr   Publication statement altered (see p.153), new title alphabet introduced

  ks **rs**   Motorways shown as double line, primary routes coloured red with green number, railways deleted, airport symbol black, and air transport information deleted

      rt   New symbols designed for car ferries, "B" roads numbers added

      ru   Primary route legend rewritten

      rv   Publication statement altered (see p.153), service areas added to primary routes, National Traveline Service details added (no "rw" state)

      rx   Main road under construction symbol added, coastline, river and lake absorbed into one symbol, HR (S) symbols added, National Traveline Service information removed from panel

ry Intaglio service and junction symbols abolished, motorway limited interchange numbers in red, town and village absorbed in one, height example altered

rz "Canal" added to canal symbol

raa Publication statement altered (see p.153), motorway symbol altered to thick single line, gradient percentage added, passenger railway added

rbb Telephone call box reference deleted, "ss" to "β" in German legend

pa Legend of Physical Map. Diagrams: scale-bar, sheet index, layer box, and, uniquely, marine layer box

ue Legend of 1:1,250,000 map, "E" edition, with "Motorways", "Trunk roads", "Other selected roads", "Railways", "Large towns", "Other towns or villages", "National Boundaries", "County and Region Boundaries"

6. Minor Correction date (perhaps with road revision date included)
8. Copyright statement or date (some ■), u-nivers type, d-igital, - graticule intersections, + reservoirs (V.1)
10. Magnetic variation date, + corrected figures in "Difference from Grid North" panel (V.1 Sheet 1 only)
11. Colour of base map, usually on 92gsm paper. Weight of paper or colour of grid if uncharacteristic, + contours
12. Number of coloured layers. Colour of sea. ■Number of basic colours in overprint. By default inland water and water names and town hatched infill are on the base map plate, except on coloured topographical maps where water is blue, towns solid grey or stipple black, and roads red
13■ Overprint publication date. Price. ■14. Date of overprinted information. 15. Print run. Date of printing
16. Cover type. 17. OSPR or GSPR. 18. Location of copies. 19. Notes

---

V.1. "Ten Mile" Map of Great Britain, 1955. *NB*: An alphabetical listing follows the base map

1	2	3	6	8	10	11	12	■13	■14	15	16	17	18	19

NB: Unless noted otherwise, all V.1 maps have a "Ten Mile" title panel

Ordnance Survey "Ten Mile" Map of Great Britain: Sheet 1 (or Sheet 2)

1	proof 1	1954ga:5/-	-	CCR+	1954	jg+	8b0	-	-	29.11.54	-	-	Lpro	2
1	proof 3	1954ga:5/-	-	CCR+	1954	jg+	8b0	-	-	8,11.2.55	-	-	Lpro	5
1	proof	1955ga:5/-	-	CCR+	1955	jg±	9b0	-	-	22.4.55	-	-	Lpro	6
1	3970	1955ga:4/-	-	CCR+	1955	jg	9b0	-	-	2015:13.6.55	96.1a3	6/55	CPT	
1	B	p1958ga:4/-	-	CCR	1958	jg	9b0	-	-	3490:12.3.58	96.1a3	-	Bg,Lse,Mg	7
1	C	1963gc:5/-	-		1963	6.63+	jg	9b0	-	4964:11.10.63	96.1a2	11/63	CPT	7,8
2	proof 1	1954 :np	-	CCR+	1954	jg+	8b0	-	-	22.10.53	-	-	Lpro	1
2	proof 2	1954 :np	-	CCR+	1954	jg+	8b0	-	-	15.12.53	-	-	Lpro	1
2	proof 1	1954ga:5/-	-	CCR+	1954	jg+	8b0	-	-	29.11.54	-	-	Lpro	2
2	proof	1954ga:5/-	-	CCR+	1954	jg+	9b0	-	-		-	-	Lpro	3
2	proof	1954ga:5/-	-	CCR+	1954	jg+	8b0	-	-		-	-	Lpro	4
2	proof 3	1954ga:5/-	-	CCR+	1954	jg+	9b0	-	-	8,11.2.55	-	-	Lpro	5
2	proof	1955ga:5/-	-	CCR+	1955	jg±	9b0	-	-	22.4.55	-	-	Lpro	6
2	3971	1955ga:4/-	-	CCR+	1955	jg	9b0	-	-	3187:31.5.55	96.1a3	6/55	CPT	7
2	A/*	1955gb:4/6	1960	CCR	1960	jg	9b0	-	-	1838:15.8.60	96.1a3	9/60	CPT,RH	
2	A/*	1955gb:5/-	1960	CCR	1960	jg	9b0	-	-	1322:23.10.61	96.1a	-	Bg,RH	
2	proof:B	1961gc:5/-	-	CC-	1959	jg	9b0	-	-	13.11.61	-	-	CCS	
2	B	1962gc:5/-	-	CCR	1962	jg	9b0	-	-	4900:7.8.62	96.1a2	12/62	CPT	7

This topographical map finally replaced the 1:633,600 two-sheet coloured map (IV.2) which went out of print during the war. Issued in covers from 9.1956 (OSPR 9/56). An extract of Sheet 2 appears in DOSSSM 1957. Lapsed soon after the publication of the RPM which thereafter fulfilled its function

1. Entitled **Great Britain**, with an early form of legend. Mapping down to 400km, plus the coastline east to 300km. In PRO OS 36/4

2. Now with "Ten Mile" title, and "ga" legend. In PRO OS 36/4

3. Coloured copies, some with bottom layer in emerald or light emerald green down to 400km, are in PRO OS 36/5

4. Five copies with different strength brown layers, and no green layer. In PRO OS 36/5

5. Proofs in PRO OS 36/5 with different strength layers. Most copies of Sheet 2 have an emerald green bottom layer

6. Four pairs of proofs, with and without contours, with different colour strengths (one with yellow lowest layer), are in PRO OS 36/6. On 25.4.1955 that coded L2 was approved for publication

7. These print runs were extended for GSGS (Misc.) 1820 and M325 issues (see Supplement 1)

8. Values for both True North and Magnetic North are given in the "Difference from Grid North" panel

1	2	3	6	8	10	11	12	■13	■14	15	16	17	18	19
	Ordnance Survey "Ten Mile" Map of Great Britain: Sheet 1 (or Sheet 2) [outline edition]													
1	3970	1955ga:2/6	-	CCR+	1955	jh	Ow0	-	-	3021:3.10.55	-	10/55	CPT	
1	3970	1955ga:2/6	-	CCR+	1955	gh	Ow0	-	-	200:2.2.56	-	-	[MHLG]	
1	B	p1958ga:3/-	-	CCR	1958	jh	Ow0	-	-	1942:18.9.58	-	-	RH	
1	B	p1958ga:4/-	-	CCR	1958+	jh	Ow0	-	-	2970:17.5.62	-	-	Lrgs,Ng,Og	
1	no code	nd:gc:np	-	-	6.71+	gh	Ow0	-	-		-	-	RH,PC	3
2	proof	1954 :np	-	CCR+	1954	g	Ow0	-	-		-	-	Lpro	1
2	3971	1955ga:2/6	-	CCR+	1955	jh	Ow0	-	-	3985:3.10.55	-	10/55	CPT	
2	3971	1955ga:2/6	-	CCR+	1955	gh	Ow0	-	-	200:2.2.56	-	-	[MHLG]	
2	A/*	1955gb:3/-	1960	CCR	1960	j	Ow0	-	-	5045:24.11.60	-	12/60	CPT	2
2	A/*	1955gb:4/-	1960	CCR	1960	jh	Ow0	-	-	5085:9.1.61	-	-	Lrgs,Ng,CCS	
2	no code	nd:gc:np	-	-	6.71	gh	Ow0	-	-		-	-	RH,PC	3

With the publication of this outline (or base) map, the 1942 1:625,000 base map was withdrawn. This map was superseded by **1/625,000 Map of Great Britain (Outline Style)**. Tracing paper printings in grey were made in 1961 and 1963

1. Entitled **Great Britain**, with an early form of legend. Mapping down to 400km, plus the coastline east to 300km. In PRO OS 36/4

2. This edition had to be redone because it was printed on the wrong weight paper (92gsm instead of 152gsm)

3. The greater part of Ireland is blanked off on Sheet 1 (?for an overprint title panel). It is replaced with the same handwritten inset map of Northern Ireland (with War Office Irish Grid) as GSGS (Misc.) 502 and 505 (IV.2). With National Grid Military System. The hatched town infill is omitted. Values for both True North and Magnetic North are given in the "Difference from Grid North" panel

NB: The 1955 "Ten Mile" outline map was also used in an atlas *The National Trust Maps* (London, National Trust, 1957, code: 31.12.56 CP T1889). It was reprinted in forty sections, with green overprint, at 1:506,880. The 1:625,000 scale might also be reduced, eg to 1:750,000 for the DOE map **Developed Areas 1969 England and Wales**, printed by Cartographic Services DOE in 1978 (copy DRg).

	■Administrative Areas													
1	no code:■A	1956ga	-	CCR+	1956	k	Ow7	1956:4/-	1.4.56	1400C:26.5.56	-	7/56	CPT,Lrgs	
1	no code:■A	1956ga	-	CCR+	1956	k	Ow7	1956:4/-	1.4.56	1000C:31.7.56	-	-		
1	no code:■A	1956ga	-	CCR+	1956	k?h	Ow7	1956:4/-	1.4.56	200C:31.7.56	-	-	[MHLG]	
1	B:■B	m1958ga	-	1958	1958	k	Ow7	1958:4/-	1.4.58	1500C:19.5.58	-	12/58	CPT,Lrgs	
1	B:■B	1958ga	-	1958	1958	k	Ow7	1958:4/-	1.4.58	1000C:8.9.59	-	-		
1	B:■B/	1958ga	-	1958	1958+	k	Ow7	1958:5/-	27.2.62	1600C:2.3.62	-	5/62	CPT,Lrgs	
1	B:■B//	1958ga	-	1958	1958+	g	Ow1	1958:5/-	27.2.62	2:10.62	-	-	Lkg	2
1	B:■B//	1958ga	-	1958	1958+	k	Ow7	1958:5/-	9.7.63	2800C:11.7.63	-	9/63	CPT	3
2	no code:■A	1956ga	-	CCR+	1956	k	Ow7	1956:4/-	1.4.56	3800C:5.6.56	-	7/56	CPT,Lrgs	1
2	no code:■A	1956ga	-	CCR+	1956	k	Ow7	1956:4/-	1.4.56	2500C:31.7.56	-	-		
2	no code:■A	1956ga	-	CCR+	1956	k?h	Ow7	1956:4/-	1.4.56	200C:31.7.56	-	-	[MHLG]	
2	A/:■B	m1956ga	1958	CCR	1958	k	Ow7	1958:4/-	1.4.58	1800C:19.5.58	-	12/58	CPT,Lrgs	
2	A/:■B	1956ga	1958	CCR	1958	k	Ow7	1958:4/-	1.4.58	2500C:8.9.59	-	-		
2	A/:■B/	1956ga	1958	CCR	1958	k	Ow7	1958:5/-	27.2.62	3000C:2.3.62	-	5/62	CPT,Lrgs	
2	A/:■B//	1956ga	1958	CCR	1958	k	Ow7	1958:5/-	9.7.63	3500:19.9.63	-	9/63	CPT	3

The same three foreground and three infill overprint colours are used as the IV.3 map, the additional colour being the black Administrative Areas title. With inset map at 1:253,440 of London. An extract of Sheet 2 is in DOSSSM 1957

1. Copies were also supplied with *The Municipal Year Book and Public Utilities Directory 1957*

2. The overprint boundaries and names plates are combined in black. The control cards refer to similar combined pulls of base and boundaries of both sheets in 1959, 1963 and 1964. Both these copies are in Lkg

3. This reprint was planned to serve until mid-1965

	■Agreed Areas 10/5 Metre Contouring / Index to the Ordnance Survey Maps at Scale of 1:10,560 (Six Inches to the Mile) on National Grid Sheet Lines													
1	proof:B	m1958ga	-	1958	1958+	gh	Ow2	nd	nd	8.11.68	-	-	CCS	1
1	B	m1958ga	-	1958	1958+	gh	Ow2	fiuo	nd	32:4.2.69	-	-	SOrm,CCS	
2	proof:■B	m1956ga	1958	CCR	1958	gh	Ow2	nd	nd	8.11.64	-	-	CCS	1
2	no code:■B	m1956ga	1958	CCR	1958	gh	Ow2	fiuo	nd	32:31.1.69	-	-	CCS	

1. The proof title is **Agreed Areas 10/5 Metre Contouring Elements**

1	2	3	6	8	10	11	12	■13	■14	15	16	17	18	19

Large Scale Field Programme

■Basic Scales Planning and Progress Map: Edition of [date] [from "B" edition]

1	■4194					k	Ow5	fiuo	3.58	281:24.7.58	-	-	not found	
1	B:■A	p1958ga:np	-	CCR	1958+	k	Ow8	fiuo	3.62	392:27.12.62	-	-	RH	
1	B:■A	p1958ga:4/-	-	CCR	1958+	k	Ow8	fiuo	3.62	53:27.7.65	-	-	CCS	
1	B	p1958ga:np	-	CCR	1958+	gh	Ow4	fiuo	31.3.73		-	-	SOrm	
2	■4195					k	Ow5	fiuo	3.58	247:30.7.58	-	-	not found	
2	B:■A	1962gc:4/-	-	1962	1962	k	Ow8	fiuo	3.62	416:27.12.62	-	-	CCS	
2	B:■A	1962gc:4/-	-	1962	1962	k	Ow8	fiuo	3.62	50:27.7.65	-	-	RH	1
2	B	1962gc:np	-	1962	1962	gh	Ow4	fiuo	31.3.73		-	-	SOrm	

1. This state is identifiable from the previous in the use of an upright rather than sloping "B" letter code

■Continuous Revision: Group and Section Areas of Responsibility: Edition of [date]

1	B	p1958ga:3/-	-	CCR	1958	k	Ow4	fiuo	5.59	200C:17.9.59	-	-	RH	
1	B	p1958ga:np	-	CCR	1958+	k	Ow5	fiuo	1.63	C27.2.63	-	-	RH	
1	B	1958ga:np	-	CCR	1958+	k	Ow5	fiuo	1.63	20C:8.6.65	-	-	not found	
2						k	Ow4	fiuo	5.59	200C:17.9.59	-	-	not found	
2	B	1962gc:np	-	1962	1962	k	Ow5	fiuo	1.63	C27.2.63	-	-	RH	
2	B	1962gc:np	-	1962	1962	k	Ow5	fiuo	1.63	20C:8.6.65	-	-	not found	

■Division of 1:10,000/Six-inch Mapping / Index to the Ordnance Survey Maps at Scale of 1:10,560 (Six Inches to the Mile) on National Grid Sheet Lines

1	proof:B:■B	m1958ga	-	1958	1958+	gh	Ow3	fiuo	nd	8.11.68	-	-	CCS	
1	B:■B	m1958ga	-	1958	1958+	gh	Ow3	fiuo	nd	32:4.2.69	-	-	CCS	
2	proof:■B	m1956ga	1958	CCR	1958	gh	Ow3	fiuo	nd	8.11.64	-	-	CCS	
2	no code:■B	m1956ga	1958	CCR	1958	gh	Ow3	fiuo	nd	32:31.1.69	-	-	SOrm,CCS	

The map distinguishes areas covered by six-inch maps with 25-foot vertical contours from those mapped at 1:10,000 with metric contours, and those still with imperial contours

■Geological Map of Great Britain: Sheet 1: Scotland & England (North of National Grid Line 500 Km N)
■Geological Map of Great Britain: Sheet 2: England & Wales (South of National Grid Line 500 Km N)

1	■5000/57	1955ga	-	CCR+	1955	kh	Ow12	1957:12/6	nd	194.2	note	CPT	
1	■5000/59	p1955ga	-	CCR+	1955	kh	Ow12	1957:12/6	nd	194.2	-	Bg	
1	■6000/64	1957ga	-	1957+	1955	k	Ow12	1964:15/-	nd	194.2	-	CPT	
1	■12500/66	1957ga	-	1957+	1955	k	Ow12	1966:15/-	nd	194.2	-	Eg,Gg,Ng	
1	■10000/71	1957ga	-	1957+	1955	k	Ow12	1971:np	nd	194.2	-	Lse	
1	■18680/74	1957ga	-	1957+	1955	k	Ow12	1971:np	nd	194.2	-	LDg,RH	
1	■3000/78	1957ga	-	1957+	1955	k	Ow12	1971:np	nd	?194.2	-	Bg	
2	■5000/57	1955ga	-	CCR+	1955	kh	Ow12	1957:12/6	nd	194.2	note	CPT	
2	■7500/59	p1955ga	-	CCR+	1955	kh	Ow12	1957:12/6	nd	194.2	-	Bg	
2	■8000/64	1957ga	-	1957+	1955	k	Ow12	1964:15/-	nd	194.2	-	CPT	
2	■15000/66	1957ga	-	1957+	1955	k	Ow12	1966:15/-	nd	194.2	-	Eg,Ng,Sg	
2	■12000/71	1957ga	-	1957+	1955	k	Ow12	1971:np	nd	194.2	-	Sg	
2	■18680/74	1957ga	-	1957+	1955	k	Ow12	1971:np	nd	194.2	-	DRg,RH	
2	■5000/77	1957ga	-	1957+	1955	k	Ow12	1964:np	nd	?194.2	-	SOos	

Prepared by GS. Second Edition, 1957 (GSPR 1957/4), reprints 1964,1966,1971. Number of overprint colours uncertain. "Ten Mile" 3970,3971 editions were used, and the reservoir references were never removed

■Geological Map of Great Britain (as above) [with index overprint]

1	■6000/64	1957ga	-	1957+	1955	k	Ow13	1964:16/6	nd	194.3	-	Bg,Lse	1
1	■18680/74	1957ga	-	1957+	1955	k	Ow13	1971:np	nd	?194.3	-	LDg	
2	■8000/64	1957ga	-	1957+	1955	k	Ow13	1964:16/6	nd	194.3	-	Bg,Lse	1
2	■18680/74	1957ga	-	1957+	1955	k	Ow13	1971:np	nd	?194.3	-	DRg	

Prepared by GS. Second Edition, 1957, reprinted 1964. Issued with a violet overprint to show the sheet lines of the one-inch Geological Series and Special District Maps. Number of overprint colours uncertain. Base map as above

1. The first price state is 15/- deleted and replaced by 16/6d

V.1   159

1	2	3	6	8	10	11	12	■13	■14	15	16	17	18	19

■Geological Map of Great Britain (as above) [outline edition]
| 1 | ■1250/57 | 1955ga | - | CCR+ | 1955 | gh | Ow1 | 1957:5/- | nd | | - | note | Lrgs,on sale | |
| 2 | ■1250/57 | 1955ga | - | CCR+ | 1955 | gh | Ow1 | 1957:5/- | nd | | - | note | Lrgs,on sale | |

Prepared by GS. Second Edition, 1957 (GSPR 1957/4). Base map as above. Still on sale (1991)

+■Gravel: including Associated Sands / [RPM, "B" edition] (n), "Ten Mile" [actually RPM, "A" edition] (S)
n	proof 1	-	-	■1969	-	g	Ob4	1969	1959	13.12.68	-	-	CCS	
n	no code	-	-	■1969	-	g	Ob4	1969	1959	1000:18.2.69	-	4/69	BSg,Lse,NTg	1
2	proof 1	nd:gd	-	-	1959	g	Ob4	1964	1959	13.1.65	-	-	CCS	
2	A:■A	nd:gd	-	■1965	1959	g	Ob4	1965	1959	2994:24.8.65	-	9/65	CPT	

Compiled by MHLG. An index of Geological Survey Drift Maps is on Sheet 2. The French coast is excluded. Earlier known as **Sand and Gravel**, MTCP ms versions date back to 1944. It was listed as in preparation in the 1946 press release and on 8.7.1949 (PRO OS 1/432 3). Print run and other information from PRO OS 1/1224. Treasury approval granted 1961
    1. The England & Wales northern strip (to 670km N) was printed by OS for MHLG

■Index to the Ordnance Survey Maps at Scale of 1:10,560 (Six Inches to the Mile) on National Grid Sheet Lines
1	no code	1956ga	-	CCR+	1956	g	Ow1	1957:2/-	nd		-	-	Lu,Og	
1	B:■no code	m1958ga	-	1958	1958	g	Ow1	nd:np	nd		-	-	DRg,SOrm	
1	B:■B	1958ga	-	1958	1958+	g	Ow1	nd:np	nd		-	-	Ob	
1	B:■B/	1958ga	-	1958	1958+	g	Ow1	nd:np	nd		-	-	SOrm	
1	B:■B//	1958ga	-	1958	1958+	g	Ow1	nd:np	nd		-	-	Yu	
2	no code	1956ga	-	CCR+	1956	g	Ow1	1957:2/-	nd		-	-	Lkg,Lu	
2	no code	m1956ga	1958	CCR	1958	g	Ow1	1957:np	nd		-	-	Mg	
2	nc:■B	1956ga	1958	CCR	1958	g	Ow1	nd:np	nd		-	-	Ob	
2	nc:■B	1956ga	1958	CCR	1958	g	Ow2	nd:np	nd		-	-	SOrm	1
2	nc:■B/	1956ga	1958	CCR	1958	g	Ow1	nd:np	nd		-	-	SOrm,Yu	
2	nc:■B/	1956ga	1958	CCR	1958	g	Ow2	nd:np	nd		-	-	SOos	2
2	nc:■B//	1956ga	1958	CCR	1958	g	Ow1	nd:np	nd		-	-	RH	

With an inset map at 1:250,000 of London. Permatrace copies were made in 1967, 1968 and 1970
    1. The second overprint colour for the one-inch Geological series. A Sheet 1 state is also likely
    2. Version overprinted "Plate 2B" in black in top right hand corner, which implies a Sheet 1 "Plate 2A"

Large Scale Field Programme - see Basic Scales Planning and Progress Map

+■North (or South) Sheet: Population Change 1951-1961: by Wards and Civil Parishes
■N	p1	p-	-	■1966	-	k	Ok5	1966:np	1951-61	2.3.66	-	-	CCS	
■N	B:■A	p-	-	■1966	-	k	Ok5	1966:np	1951-61	1954:22.6.66	-	7/66	CPT	
■S	p1:A/*:■A	p-	-	■1965	-	k	Ok5	1965:np	1951-61	21.1.66	-	-	CCS	
■S	p2:A/*	p-	-	■1966	-	k	Ok5	1966:np	1951-61	20.4.66	-	-	CCS	
■S	p3:A/*	p-	-	■1966	-	k	Ok5	1966:np	1951-61	1.6.66	-	-	CCS	
■S	A/*:■A	p-	-	■1966	-	k	Ok5	1966:np	1951-61	2966:6.66	-	7/66	CPT	

Compiled by SDD and MHLG from the 1961 Census. Willatts recommended print runs of 3000 for each sheet on 11.3.1966. Treasury approval granted on 6.1.1965 (PRO OS 1/999). Lacking "Ten Mile" title panel, perhaps because by publication the base map had already been superseded. An earlier England & Wales version (with later base map) is in Supplement 3

+Population Density, 1951
1	B4288	m1958ga	-	■1961	1958+	g	Ob2	1961:7/6	1951	3200:11.10.61	-	12/61	CPT	1
1	B:B4288	1958ga	-	■1961	1958+	g	Ob2	1961:np	1951	2300:6.5.69	-	-	DRg,Sg	
2	proof	p1955ga:4/-	-	CCR	1955	g	Ob2	p1959:np	1951	28.7.59	-	-	CCS	
2	p2:4266	p1955ga	-	■1960	1955	g	Ob2	m1960:5/-	1951	23.8.60	-	-	CCS	
2	4266	p1955ga	-	■1960	1955	g	Ob2	1960:5/-	1951	1000:8.9.60	-	10/60	CPT	
2	4266	p1955ga	-	■1960	1955	g	Ob2	1960:5/-	1951	1734:24.2.61	-	-		
2	4266	p1955ga	-	■1960	1955	g	Ob2	1960:7/6	1951	3200:11.10.61	-	-	Bg,Cg,Mg	
2	4266	1955ga	-	■1960	1955	g	Ob2	1960:?np	1951	3600:2.5.69	-	-	not found	

Compiled by DHS and MHLG from the County Reports of the 1951 Census. The job file also contains five grey pulls of the outline map (3971) overprinted with experimental colour wedges - two (proof 2) dated 8.8.1958, two (proof 3) dated

1	2	3	6	8	10	11	12	■13	■14	15	16	17	18	19

16.10.1958, and the last (proof 4 - without unique number) dated 16.11.1959
1. On some copies (eg CCS file copy) the top of the base map "B" print codes survives

+■Rainfall: Annual Average 1916-1950 / ["Ten Mile" "B" editions]

1	p1:B:■A	p1958ga:4/-	-	■1963	1958+	kh	Ow1	1963:np	1916-50	14.11.63	-	-	CCS
N	B	nd:ga	-	■1967	1958+	k	Ow4	1967:np	1916-50	3000:7.2.67	-	12/67	CPT
2	proof	nd:gc	-	■1963	1962	kh	Ow1	1963:np	1916-50	14.11.63	-	-	CCS
2	proof 1:B	nd:gc	-	■1966	1962	k	Ow4	1966:np	1916-50	19.7.66	-	-	CCS
S	B	nd:gc	-	■1967	1962	k	Ow4	1967:np	1916-50	4090D:6.2.67	-	12/67	CPT

Prepared by MHLG in consultation with SDD from the detailed maps and statistics of rainfall compiled by the Meteorological Office. Dispersion graphs and other diagrams are included. Coloured by the new trichromatic or three colour masking system, plus black (PRO OS 1/721 71A). Treasury approval granted 4.1962. With Explanatory Text 2A

■The Ray Society Publication No 146: Watsonian Vice-counties of Great Britain

| N | B | 1958ga | - | ■1969 | 1958+ | j | Ow1 | 1969 | nd | | x | - | SOrm,PC |
| S | B | 1962gc | - | ■1969 | 1962 | j | Ow1 | 1969 | nd | | x | - | SOrm,PC |

Preparation of this map seems to have begun with pulls of the 1942 base map taken in November 1948

■Roads

1	proof	1955ga	-	CCR+	1955	g	Ob3	nd:np	nd	21.12.55	-	-	Lpro	1
1	proof	1955ga	-	CCR+	1955	g	7b3	nd:np	nd	29.12.55	-	-	Lpro	2
1	proof	1955ga	-	CCR+	1955	k	7b3	nd:np	nd	20.2.56	-	-	RH	3
1	nc:■A	1956ga	-	CCR+	1956	k	7b3	1956:2/6	5.56	4378D:4.9.56	96.2b1	9/56	CPT	
1	B:■A	p1958ga	-	CCR	1958	k	7b3	1956:2/6	5.56	220:3.1.63	-	-	Lrgs,Ob,RH	4
2	proof	1955ga	-	CCR+	1955	g	Ob3	nd:np	nd	21.12.55	-	-	Lpro	1
2	proof	1955ga	-	CCR+	1955	g	7b3	nd:np	nd	29.12.55	-	-	Lpro	2
2	proof	1955ga	-	CCR+	1955	k	7b3	nd:np	nd	20.2.56	-	-	RH	3
2	nc:■A	1956ga	-	CCR+	1956	k	7b3	1956:2/6	5.56	6804:15.8.56	96.2b1	9/56	CPT	
2	nc:■A	p1956ga	-	CCR	1958	k	7b3	1956:2/6	5.56	220:3.1.63	-	-	Lrgs,Ob,RH	4

With inset map at 1:250,000 of London. Sheet 2 "Northern Test Strips", with title area (Great Britain, and MV 1954) are in OS 1/709 3A,7A. They are lettered A-O (E is missing and I is omitted) and AA and BB. A,B,C,D,F,G are stamped "First Proof" and dated 5.1.1955; H,J 12.5.1955; K 24.5.1955. The others are undated. They show different shades for base map, sea and layers, and different permutations of colours for trunk, A and B class roads. N and O have the names of main centres on the black plate. Full Sheet 1 proofs based on O (with yellow minor roads) were produced, but missing. Yellow was then changed to emerald green for subsequent proofs. Security corrections were made to the plates in 1958. The map lapsed in 1961, and was not replaced until 1964. Uncoloured to 400ft. An extract of Sheet 1 appears in DOSSSM 1957. Print run information and proofs in PRO OS 1/709. Used for M322 (ex GSGS 4813) Edition 2-GSGS, 1958 (see Supplement 1)

1. Proof copies in PRO OS 1/709 15A (N), 15B (S). Two copies in each envelope, lettered A and B, with different shades of grey base. The file also refers to C proofs (not present), and to other alternatives on each base: yellow land, yellow layers, buff layers. These also are lacking
2. Proof copies in PRO OS 1/709 16A. These were approved for publication by the Director General
3. Road information was corrected at a very late stage: eg the Newbury-Basingstoke road is still shown on this proof as an "A", not a trunk road. The water and layer colours were also subsequently altered
4. The lack of an in-print Road Map embarrassed MOT who used it as an index for road classification (PRO OS 1/1136 53). Late in 1962 they requested a special printing of 200 copies. The OS printing of 220 included 20 copies of each sheet for their own use (PRO OS 1/1136 30,40,41). Sheet 2 has a Roman code letter "A". The evidence of the military printing (qv) is that a base map code of "A/" was applied, apparently deleted here

■Route Planning Map / ["Ten Mile" "C" edition (Sheet 1), "B" edition (Sheet 2)]

1	proof:C:■A	1963gd:5/-	-	■1964	6.63+	k	3b0	1964	nd	5.11.63	-	-	SOos,CCS,RH	3,5
1	nc:■A	nd:gd	-	■1964	1964+	k	3b0	1964ra	11.63	3156D:10.12.63	108	1/64	CPT	5
1	nc:■A	nd:gd	-	■1964	1964+	k	3b0	1964ra	11.63	2500D:8.1.64	108	-		5,6
1	nc:■A	nd:gd	-	■1964	1964+	k	3b0	1964ra	11.63	5000:10.2.64	108	-		5,7
1	nc:■A	nd:gd	-	■1964	1964+	k	3b0	1964ra	11.63	4000:1.3.64	108	-		5,7
1	nc:■A	nd:gd	-	■1964	1964+	k	3b0	1964ra	11.63	5004:19.3.64	108	-		5,7

1	2	3	6	8	10	11	12	■13	■14	15	16	17	18	19
2	proof	nd:gd	-	■1963	1959	k	1b0	1963	9.62	21.5.63	-	-	RH	1
2	proof 2	nd:gd	-	■1964	1959	k	3b0	1964	nd	6.9.63	-	-	RH	2
2	proof 3	nd:gd	-	■1964	1959	k	3b0	1964	nd	5.11.63	-	-	CCS	3
2	proof	nd:gd	-	■1964	1959	k	3b0	1964ra	nd	undated	-	-	Lpro	4
2	nc:■A	nd:gd	-	■1964	1959	k	3b0	1964ra	11.63	3800:12.12.63	108	1/64	CPT	
2	nc:■A	nd:gd	-	■1964	1959	k	3b0	1964ra	11.63	2500D:8.1.64	108	-		6
2	nc:■A	nd:gd	-	■1964	1959	k	3b0	1964ra	11.63	5000:10.2.64	108	-		
2	nc:■A/	nd:gd	-	■1964	1959	k	3b0	1964ra	11.63	5000:28.2.64	108	-	NTg,Og,RH	8
2	nc:■A/	nd:gd	-	■1964	1959	k	3b0	1964ra	11.63	7500:17.3.64	108	-		8
2	nc:■A/	nd:gd	-	■1964	1959	k	3b0	1964ra	11.63	4134:19.6.64	108	-		8

For later editions, without "Ten Mile" title panel, see section V.3 below. For further details on specifications see p.148. The published map has inset maps at 1:250,000 of Edinburgh, Glasgow, Manchester, Birmingham and London. Used for M322 Edition 3-GSGS (see Supplement 1). The Newbury-Lambourn railway (closed 1960) appears - it survived until the general deletion of the railway system on the "S" edition RPM.

1. 6 proofs in various colour combinations, entitled **Roads**. No layer box. With an early form of additional legend
2. Second proof, entitled **Route Planning Map**. Legend lacks motorway service area. Trunk and main roads reversed
3. Two proofs with different strength layers
4. In PRO OS 1/1136 91A. With a revised mileage chart in proof. Roads in the large towns symbol are coloured
5. Values for both True North and Magnetic North are given in the "Difference from Grid North" panel
6. This printing represents emergency use of half the M322 Edition 3-GSGS supply. It was made up later in the year
7. The unique letter was moved from 4mm bnl to 1mm bnl on these reprints
8. The motorway service station shown on the A1(M) west of Doncaster was deleted

V.2. Physical Map of Great Britain, 1957. *NB*: An alphabetical listing follows the base map

Ordnance Survey Physical Map of Great Britain Sheet 1 (or Sheet 2)

1	proof	1955pa:4/-	-	CCR-	-	J	9b0	-	-	18.1.56	-	-	Lrgs	1
1	proof	1955pa:4/-	-	CCR-	-	J	9b0	-	-	20.2.56	-	-	CCS	1
1	A	1957pa:4/-	-	CCR-	-	J	9b0	-	-	4026:15.1.57	-	2/57	CPT	3
1	A	pS1957pa:np	-	CCR-	-	J	9b0	-	-	1800:18.5.72	-	-		
1	A	pS1957pa:np	-	CCR-	-	J	9b0	-	-	1400:1.6.76	-	-	CCS	
1	A	pS1957pa:np	-	CCR-	-	J	9b0	-	-	2479:5.4.77	-	-	CCS	
1	A	pS1957pa:np	-	CCR-	-	J	9b0	-	-	1957:24.7.87	-	-	on sale	
2	proof	1955pa:4/-	-	CCR-	-	J	9b0	-	-	18.1.56	-	-	Lrgs	1
2	proof	1955pa:4/-	-	CCR-	-	J	9b0	-	-	20.2.56	-	-	CCS	1
2	proof:A	1956pa:4/-	-	CCR-	-	J	9b0	-	-	15.10.56	-	-	CCS	2
2	A	1957pa:4/-	-	CCR-	-	J	9b0	-	-	5188:16.1.57	-	2/57	CPT	
2	A	pS1957pa:np	-	CCR-	-	J	9b0	-	-	2300:22.5.72	-	-		
2	A	pS1957pa:np	-	CCR-	-	J	9b0	-	-	2000:5.4.77	-	-	CCS	
2	A	pS1957pa:np	-	CCR-	-	J	9b0	-	-	2100:24.9.82	-	-	on sale	

Prepared by OS in consultation with DHS and MHLG. With six submarine layers and contours based on Admiralty charts. See also p.148. Lit: Hellyer (1992a)

1. These proof copies lack DHS and MHLG acknowledgement. Second proofs were printed on 28.1.1956 and 18.2.1956
2. Proofs incorporating MHLG's suggested alterations, with DHS and MHLG acknowledgement, and unique letter "A" (Sheet 1, not found, printed on 25.9.1956)
3. Only the top of the sloping "A" unique letter is visible; the unique letter on Sheet 2 is upright

■The Distribution in Scottish Rivers of the Atlantic Salmon, <u>Salmo salar</u> L.

| [N] | no code | nd:pa:np | - | 1985- | - | J | 9b1 | - | nd | | x | - | on sale | |

Made and printed by the OS. On the 1957 Physical Map (qv for specification), with the original sheet title, number and index deleted, and most of the original black name plate replaced. The watershed south of the Tweed and the Sark (off the Solway Firth) river systems provides the southernmost limit of the overprinted area. With accompanying text, prepared by Ross Gardiner and Harry Egglishaw for the Department of Agriculture and Fisheries for Scotland, Freshwater Fisheries Laboratory, Pitlochry, 1986. A Scottish Fisheries Publication, obtainable from the Scottish Office, Edinburgh (ISBN 0 903386 10 0).

1	2	3	6	8	10	11	12	■13	■14	15	16	17	18	19

■Ordnance Survey Map of Southern Britain in the Iron Age

■S	proof:A	1960-:np	-	1960-	-	g	9b4	-	nd	1.7.60	-	-	SOrc	
■S	no code	-	-	■1962-	-	g	9b4	1962:7/6	nd		181	7/62	CPT	
■S	A/	-	1967	■1962-	-	g	9b4	1962:np	nd		181	-	CPT	
■S	A//	-	1975	■1962-	-	g	3b4	nd:np	nd		181	-	RH	

Compiled by A.L.F.Rivet. With letterpress. With the unique letter "A" attached to the proof copy, it seems the OS avoided coding the first published issue at all! The proof follows the **Physical Map** in the position of its marginalia and title, and the grid, being numbered at only 100km intervals, though the 10km squares were restored. The published map has ornamental borders, graticule values placed inside the neat line, no layer box, and changed overprint colours. The grid is rouletted, the title repositioned. Lacking marine contours. Lit: Hellyer (1987),(1988),(1989d),(1992a)

V.3. Route Planning Map, from "B" edition", 1965. *NB*: An alphabetical listing follows the base map

1	2	3	8	10	11	12	13	14	15	16	17	18	19

Titles. Both overprint and (abbreviated) base map titles are given. "/" is used to separate title areas, ":" the various elements on the same title area. "■" denotes titles on the overprint plate, and persists until cancelled by "/" or ">". Base map titles in "[]" are deduced, not given. "+" denotes maps in the Planning Series. The RPM "B" edition was the first base map. For outline base map titles, see p.153. See p.161 for earlier RPM editions (in section V.1)

NB: On the North Sheet dates of overprinted information often apply for Scotland that differ from that for England & Wales. Dates given here are for Scotland: England & Wales dates should be checked from the partner South Sheet

Ordnance Survey **Route Planning Map** North (or South) ["B" to "R" editions: "Q" and "R" titles dated **1978** and **1979**]
Ordnance Survey **Routeplanner** North (or South) [from "S" edition: "S" title dated **1980**]

N	B	1965rb	1965	-	k	3b0	[1965]	11.64	16558:3.12.64	108	1/65	CPT	1
N	C	1966rc	1966	-	k	3b0	[1966]	11.65	16300:14.12.65	108	1/66	CPT	1
N	C	1966rc	1966	-	k	3b0	[1966]	11.65	4227:4.7.66	108	-	RH	
N	D	1967rd	1967	-	k	3b0	[1967]	11.66	20600:16.12.66	108	1/67	CPT	1
N	E	1967re	1967	-	k	3b0	[1968]	7.67	17871:18.9.67	108	11/67	CPT	
N	F	1968rf	1968	-	k	3b0	[1969]	[1.6.68]	18800:8.10.68	127	11/68	CPT	
N	G	1969rg	1969	-	k	3b0	[1970]	9.69	17307:30.10.69	127	11/69	CPT	
N	H	nd:rh	1970	-	k	3b0	[1971]	7.70	25000D:17.8.70	127	10/70	CPT	2
N	H	nd:rh	1970	-	k	3b0	[1971]	7.70	4440:5.7.71	127	-	RH	
N	J	nd:rj	1971	-	k	3b0	[1972]	7.71	32500:14.9.71	127	10/71	CPT	5
N	J	nd:rj	1971	-	k	3b0	[1972]	7.71	12300:26.4.72	127	-	RH	
N	K	nd:rk	1972u	-	g	Rb0	[1973]	7.72	48400:8.9.72	127	9/72	CPT	6
N	K	nd:rk	1972u	-	g	Rb0	[1973]	7.72	5770:13.3.73	127	-	RH	
N	L	nd:rl	1973u	-	k	Rb0	[1974]	7.73	30250:14.9.73	127	10/73	CPT	
N	L	nd:rl	1973u	-	k	Rb0	[1974]	7.73	7428:26.3.74	127	-	RH	
N	M	nd:rm	1974u	-	k	Rb0	-	7.74	5546:14.9.74	-	10/74	CPT	
N	N	nd:rn	1975u	-	k	Rb0	-	7.75	3829:12.9.75	-	10/75	CPT	
N	P	nd:rp	1976u	-	k	Rb0	-	6.76	3619:15.9.76	-	10/76	CPT	
N	Q	nd:rq	1977u	-	k	Rb0	1978	6.77	3815:2.9.77	-	10/77	CPT	
N	R	nd:rr	1978u	-	k	Rb0	1979	6.78	3945:12.10.78	-	10/78	CPT	
N	S	nd:rs	1979u	-	k	Rb0	1980	6.79	3800:6.11.79	-	9/79	CPT	
N	T	nd:rt	1980u	-	k	Rb0	-	6.80	271:3.9.80	-	10/80	CPT	
N	U	nd:ru	1981u	-	k	Rb0	-	6.81	210:17.9.81	-	10/81	CPT	
N	V	nd:rv	1982u	-	k	Rb0	-	8.82		-	10/82	Ob	
N	W	nd:rv	1983u	-	k	Rb0	-	8.83	127:28.9.83	-	10/83		
N	X	nd:rx	1984u	-	k	Rb0	-	11.84	110:13.10.84	-	10/84	Ob	
N	Y	nd:ry	1985d	-	k	1b0	-	11.85	248	-	11/85		
N	Z	nd:rz	1986d	-	k	1b0	-	11.86	119:28.10.86	-	10/86		
N	AA	nd:raa	1988d	-	k	1b0	-	11.88	300	-	3/89		
N	BB	nd:rbb	1991d	-	k	1b0	-	11.90	300	-			

1	2	3	8	10	11	12	13	14	15	16	17	18	19
S	B	1965rb	1965	-	k	3b0	[1965]	11.64	22750:4.12.64	108	1/65	CPT	1
S	B/*	1965rb/	1965	-	k	3b0	[1965]	11.64	8200:21.4.65	108	4/65	CPT	
S	B/*	1965rb/	1965	-	k	3b0	[1965]	11.64	3100:13.7.65	108	-	RH	
S	C	1966rc	1966	-	k	3b0	[1966]	11.65	34000:25.11.65	108	1/66	CPT	1
S	D	1967rd	1967	-	k	3b0	[1967]	11.66	39076:11.1.67	108	1/67	CPT	1,3
S	E	1967re	1967	-	k	3b0	[1968]	7.67	35893:19.9.67	108	11/67	CPT	
S	F	1968rf	1968	-	k	3b0	[1969]	[1.6.68]	36100:14.10.68	127	11/68	CPT	
S	G	1969rg	1969	-	k	3b0	[1970]	9.69	35003:28.10.69	127	11/69	CPT	
S	H	nd:rh	1970	-	k	3b0	[1971]	7.70	45700:11.8.70	127	10/70	CPT	2
S	H	nd:rh	1970	-	k	3b0	[1971]	7.70	5000:5.7.71	127	-	RH	4
S	J	nd:rj	1971	-	k	3b0	[1972]	7.71	55100:21.9.71	127	10/71	CPT	5
S	J	nd:rj	1971	-	k	3b0	[1972]	7.71	10627:26.4.72	127	-	RH	
S	K	nd:rk	1972u	-	g	Rb0	[1973]	7.72	40200:20.9.72	127	9/72	CPT	
S	K	nd:rk	1972u	-	g	Rb0	[1973]	7.72	6143:21.11.72	127	-	RH	7
S	K/*	nd:rk	1972u	-	g	Rb0	[1973]	7.72	35757:15.3.73	127	4/73	CPT	8
S	L	nd:rl	1973u	-	k	Rb0	[1974]	7.73	45271:8.10.73	127	10/73	CPT	
S	L	nd:rl	1973u	-	k	Rb0	[1974]	7.73	27300:29.3.74	127	-	RH	
S	M	nd:rm	1974u	-	k	Rb0	-	7.74	5443:23.9.74	-	10/74	CPT	
S	N	nd:rn	1975u	-	k	Rb0	-	7.75	3700:15.9.75	-	10/75	CPT	
S	P	nd:rp	1976u	-	k	Rb0	-	6.76	4323:15.9.76	-	10/76	CPT	
S	Q	nd:rq	1977u	-	k	Rb0	1978	6.77	4300:23.8.77	-	10/77	CPT	
S	R	nd:rr	1978u	-	k	Rb0	1979	6.78	4466:5.10.78	-	10/78	CPT	
S	S	nd:rs	1979u	-	k	Rb0	1980	6.79	4320:29.10.79	-	9/79	CPT	
S	T	nd:rt	1980u	-	k	Rb0	-	6.80	470:11.9.80	-	10/80	CPT	
S	U	nd:ru	1981u	-	k	Rb0	-	6.81	284:21.9.81	-	10/81	CPT	
S	V	nd:rv	1982u	-	k	Rb0	-	8.82	275:5.10.82	-	10/82	RH	
S	W	nd:rv	1983u	-	k	Rb0	-	8.83	172:6.10.83	-	10/83		
S	X	nd:rx	1984u	-	k	Rb0	-	11.84	160:18.10.84	-	10/84		
S	Y	nd:ry	1985d	-	k	1b0	-	11.85	106:11.10.85	-	11/85		
S	Z	nd:rz	1986d	-	k	1b0	-	11.86	100R:13.10.86	-	10/86		
S	AA	nd:raa	1988d	-	k	1b0	-	11.88	300	-	3/89		
S	BB	nd:rbb	1991d	-	k	1b0	-	11.90	300	-			

Back to back printings (see also notes 2,5,6,7,8)

1	2	3	8	10	11	12	13	14	15	16	17	18	19
NS	L	nd:rl	1973u	-	k	Rb0	-	7.73	5000:20.9.73	-	-	RH	
NS	M	nd:rm	1974u	-	k	Rb0	[1975]	7.74	60000:25.9.74	128	10/74	RH	
NS	M	nd:rm	1974u	-	k	Rb0	[1975]	7.74	12982:19.6.75	128	-	RH	
NS	N	nd:rn	1975u	-	k	Rb0	[1976]	7.75	70685:15.9.75	128	10/75	RH	9
NS	N	nd:rn	1975u	-	k	Rb0	[1976]	7.75	8583:13.8.76	128	-		
NS	P	nd:rp	1976u	-	k	Rb0	[1977]	6.76	70400:14.9.76	128	10/76	RH	
NS	Q	nd:rq	1977u	-	k	Rb0	1978	6.77	71120:9.9.77	128	10/77	RH	
NS	R	nd:rr	1978u	-	k	Rb0	1979	6.78	72342:12.10.78	128	10/78	RH	
NS	S	nd:rs	1979u	-	k	Rb0	1980	6.79	65560:7.11.79	129	9/79	RH	
NS	T	nd:rt	1980u	-	k	Rb0	[1981]	6.80	82640:18.9.80	130	10/80	RH	
NS	U	nd:ru	1981u	-	k	Rb0	[1982]	6.81	72661:23.9.81	131	10/81	RH	
NS	V	nd:rv	1982u	-	k	Rb0	[1983]	8.82	62296:12.10.82	403.1	10/82	CPT	
NS	W	nd:rv	1983u	-	k	Rb0	[1984]	8.83	61623:29.9.83	404.6a	10/83	CPT	
NS	W	nd:rv	1983u	-	k	Rb0	[1984]	8.83	25000:3.2.84	404.6a	-		
NS	X	nd:rx	1984u	-	k	Rb0	[1985]	11.84	90000:15.8.84	404.6b	10/84	CPT	
NS	Y	nd:ry	1985d	-	k	1b0	[1986]	11.85	75301:23.10.85	406.1	11/85	CPT	
NS	Z	nd:rz	1986d	-	k	1b0	-	11.86	108401:27.10.86	407.6	10/86	CPT	
NS	AA	nd:raa	1988d	-	k	1b0	-	11.88	70000	407.7a	3/89	CPT	
NS	AA	nd:raa	1988d	-	k	1b0	-	11.88	30000	407.7a	-		
NS	BB	nd:rbb	1991d	-	k	1b0	-	11.90	75000	407.7c		CPT, on sale	

NB: From the "E" edition onwards, the copyright date is the year preceding the year given on the cover. Annual reprints and dated covers ceased with the "Z" edition. Used for M322 Edition-4 GSGS onwards (see Supplement 1)

Also surviving are proof and experimental sheets: "G" editions (S) showing roads in 20%, 50%, 80% concentrated scarlet; 12.1969 (S) Road Distances; 3.1971 (N); (S) "Motoring Map" title: 5.1971 interchanges solid blue, 6.1971 interchanges screened blue; 9.1971 "J" edition (S) hillshading in moss green, limpid green, violet layer No 2, 3rd layer buff, contour brown, 2nd layer buff, warm brown 1-0, warm brown 1-1, krystal black; 12.1972 (S) names of towns in green and black; "K" edition proofs (N&S), still with "J" code: 7.1972, 2nd proofs (7.1972), 3rd proofs (8.1972); 2.1973 (N&S) "K" edition with hillshading in krystal black, grey offset, grey, coal grey; 2.1973 (S) "L" edition proof; 1.1975 (S) "M" edition experiments; 1.1978 "Q" edition (S) proofs with green, blue, yellow, black relief; 9.1978 (N&S) "R" edition proofs; 9.1979 (N) "S" edition proof; 7.1983 (S) "W" edition proof; 7.1984 (N,S) "X" edition proofs; "X" edition experimental sheets showing scenic routes; "AA" edition proof 4.11.1988 (S) (copies SOos,RH). Several RPM extracts were printed for market research purposes in 1980-82.

1. These issues were also supplied in yellow OS-AA covers
2. Fifty copies were experimentally printed (fiuo) back to back as part of this print-run (copy SOos)
3. Preceded by many experimental proofs for green primary routes and lemon yellow layers
4. Unique letters from this year changed to Gills sans. However this reprint has an "H" with serifs
5. There were in addition back to back "run-on" printings for Reise- und Verkehrsverlag (copy RH)
6. There was in addition a "run-on" printing for Reise- und Verkehrsverlag, for pairing with 2550 copies of the first South Sheet "K" edition: the remainder were intended for the K/*, but in the event returned to stock. See note 8
7. There was in addition a "run-on" printing of 5000 for Falkplan, Holland, backed on to copies of the first North Sheet "K" edition printing
8. There were in addition "run-on" printings of 4039 for back to back mounting for Reise- und Verkehrsverlag, and 757 for Falkplan, Holland, to replace damaged copies. Both backed on to North Sheet "K" edition second printings
9. The south sheet now printed upside down, for tumble turning

Notes by print code. *NB*: For alterations in the legend, see variable attributes on p.155

B. "North" and "South" replace sheet numbers. There was an additional printing of 5000 of each sheet for MOD. The AA also placed an order for 3000 (N) and 5000 (S), of which by October they still had 5000 copies remaining. Nonetheless they renewed their order, albeit reduced, for the next two issues as well, each of which was placed inside their own OS-AA covers. With an inset map at 1:250,000 of Tyneside, and the Edinburgh map enlarged to encompass the Forth Bridge; the inset maps of St Kilda, Rona, and Sule Skerry are excluded. Areas shown in block green now include a selection of National Nature Reserves. With diagram of British Railways Car-carrying services. Mileage chart redesigned. Major National Parks named. Car ferry and gradient symbols reduced in size. Inset panels repositioned to avoid folds on mounted and folded copies

B/*. Problems caused to Nature Conservancy by including some of their National Nature Reserves on the "B" edition led to their immediate deletion here: the Forest of Bowland and Hayling Island remained as additional overprinted areas

C. AA issue (1000 N, 2000 S) included, but not MOD (3000 N, 4000 S). London inset called Greater London. County boundaries and names revised. North Sheet mileage chart includes more Scottish and fewer English names. Eddystone added. Gyfyylliog to Cyffylliog

D. The introduction of super primary and primary routes with associated town names in green caused the omission of National Parks etc, and the change to lemon yellow layers. Motorway junction numbers added (not on legend). Sand and mud deleted (not from outline derivatives), though the names survive. London inset again called London. County names to krystal black plate. With a redrawn trilingual mileage chart. Eddystone coastline added. Llanelly to Llanelli

E. All primary routes now recognised as trunk roads. National Parks etc now inset on a diagram. Incidence of NG letters box added: the other diagrams are relocated. Swansea added to mileage chart. Glanamman to Glanaman

F. The NG incidence box made trilingual; Published....Southampton. National Park box colours altered to single rulings. Road classification note simplified. MOT date dropped. Wolf Rock added. Towyn to Tywyn

G. South Sheet title panel moved to top left. Quarter-inch "Fifth Series" index diagram and NG incidence box deleted. North Sheet mileage panel reduced. Other panels relocated: the National Parks to form one map when sheets are mounted together, and "designated" added to its key. MOT date restored. Aberayron to Aberaeron

H. Grey house infill removed from town diagrams. Crown Copyright date transferred to the bottom margin. Unique letter to Gills sans. Panels further revised and relocated. Motorway numbers of motorways under construction added

J. Road casings deleted and road numbers coloured and boxed. Minor road categories combined as "other roads (selected)" and the symbol altered to a single red line. Motorways with limited interchange added, with diagrams. Shetland and Orkney Islands divided on to two inset maps and relocated. Newly designed mileage chart with associated point to point mileages added. Newly designed panel for change to BR Motorail Services. Panels relocated, and National Parks panel extended to place Shetland and Orkney Islands geographically. NG letters added to it. Revision of NG reference system. Latitude and longitude measurements and NG values to Gills sans. NG values shown multiplied by ten

1	2	3	8	10	11	12	■13	■14	15	16	17	18	19

    K. Layers replaced by grey hillshading, grey town fillings by buff. Times Roman, Gill Sans, Italic types changed to Univers. Density of names and river systems reduced. Super primary route classification abolished, destinations deleted and colour strengthened on primary routes. Inset maps redrawn at c.1:153,846 in town-plan style and entitled

    K/*. "Reprinted with the addition of new major roads"

    L. NG to official limits between Great Britain, Ireland and France. The alternate coloured segments of the diced graduation in the border cleared. Motorways, primary and main roads rescribed. Primary route destinations reinstated. E&W county names revised and removed to National Parks diagram (though they are retained on the map on outline derivatives). With inset map at c.1:444,444 of S.W.London in RPM style. Portmadoc to Porthmadog, Aberdovey to Aberdyfi

    M. Motorway service areas with limited access added. Type size of some towns in congested areas reduced. New airport symbols. NG limit description rewritten. Road numbers in inset borders enlarged. Mile-kilometre conversion table extended. Motorail diagram moved. M5 completed by extrusion on Birmingham inset. Route to airport shown on Birmingham and Manchester insets. "North Yorkshire" realigned, Scotland county boundaries revised and renamed. Conway to Conwy

    N. Heights in metres. Manchester inset enlarged. Primary routes added to Edinburgh inset. Wicklow Head deleted

    P. Caernarvon to Caernarfon

    Q. Long distance paths added to National Parks diagram. New numeral font on National Grid

    R. Black plus three trichromatic (magenta, cyan and yellow) colours used, giving the appearance of five. Built up area background colour changed from orange to yellow. Yellow background given to dual carriageways. More solid tones of others colours used. Inset maps redrawn in RPM style: with an inset map at c.1:153,846 of Leeds and Bradford

    S. New hillshading, and yellow background colour used. Motorways drawn with double blue line. Primary routes are magenta with green road number. Other roads are orange. Point to point mileages in three colours dependent on category of road. Railways deleted. Inset areas boxed on map. There is now a single black airport symbol

    T. Route destinations added in the border. With internal and external car ferry symbols. "B" road numbers added. Birmingham inset enlarged. Some panels relocated

    U. Leeds and Bradford inset renamed Bradford and Leeds

    V. Grey hill tints changed to orange. Service areas on primary routes added

    W. National scenic areas (Scotland) added to National Parks etc panel

    X. Main road under construction added. National Traveline information relocated. Farmers to Ffarmers

    Y. Hillshading, Distance and BR Motorail diagrams (N), unclassified roads and associated placenames (eg Holbeach St Matthew), canals, single line rivers and associated names, coastal mud and sand names, many minor hydrographic features and names, English Channel name, geographical names except those in capitals and some island names all deleted (though Strumble Head, Dinas Head survived). Town and village reduced to one symbol. Placenames reduced from five to three categories. New condensed typeface for placenames. Motorway limited interchange numbers coloured red. Main road service area symbol open red. NG eastings in red, northings in blue: a placename index of primary towns and major conurbations dependent on these introduced on North Sheet. London insets combined. Route destinations added to inset maps

    Z. Rivers and river names reinstated. Minor roads and associated placenames reinstated

    AA. Motorways depicted by thick single line on map and insets. Passenger railways reinstated in black. Bont Newydd replaced by Brithdir

    BB. Telephone call boxes deleted. Selected motorway junction illustrations replaced by list of "Restricted Motorway Junctions". English Channel and North Sea names reinstated on South Sheet. Channel Tunnel added to limit of National Grid. National Traveline Service information deleted. National Parks inset title altered, the symbol for long distance path transferred to its own panel, the paths being divided into "National trails", "Long distance path" and "Long distance routes". Motorway proposed opening date omitted from South Sheet symbols panel

Ordnance Survey 1/625,000 Map of Great Britain (Outline Style North [or South] Sheet) / [RPM, "B" edition]
Ordnance Survey Outline Map of Great Britain: North (or South) [from "B" edition]

N	proof:B	nd:hd	-	-	j	Ow0	-	-		-	-	Lpro	1
N	A	1965hd	1965	-	jh	Ow0	-	-	5000R:10.2.65	-	6/65	CPT,Lrgs	
N	B	1969rf	1969	-	jh	Ow0	-	-	6500R:17.2.69	-	8/69	CPT	
N	C	nd:km	1974u	-	jh	Ow0	-	7.74; 16.5.75		-	4/75	CPT	2
N	D	nd:ks	1980u	-	jh	Ow0	-	6.79			-	7/80	CPT, on sale	2
S	proof:B	nd:hd	-	-	j	Ow0	-	-		-	-	Lpro	1
S	A	1965hd	1965	-	jh	Ow0	-	-	6000R:10.2.65	-	6/65	CPT,Lrgs	
S	B	1969rf	1969	-	jh	Ow0	-	-	7500R:17.2.69	-	8/69	CPT	
S	proof:C	nd:km	1975u	-	jh	Ow0	-	7.74	9.1.75	-	-	RH	
S	C	nd:km	1975u	-	jh	Ow0	-	7.74		-	4/75	CPT	2
S	D	nd:ks	1980u	-	jh	Ow0	-	6.79		-	7/80	CPT, on sale	2

1	2	3	8	10	11	12	■13	■14	15	16	17	18	19

The "A" and "B" editions include town maps at 1:250,000 of Glasgow, Edinburgh, Newcastle upon Tyne, Manchester, Birmingham and London. See p.153 for base map titles. The "A" edition derives from the RPM "B" edition, the "B" edition from the RPM "F" edition, and the "C" edition has elements of both RPM "L" and "M" editions. The "D" edition derives from the RPM "S" edition, and has not been used as a base map. Print runs from PRO OS 1/706 and OS 1/646

1. In PRO OS 1/646 1A. The printed title is **Ordnance Survey 1/625,000 Map of Great Britain**, below which there are manuscript emendations. The "B" print codes are a product of their RPM origins which were later deleted

2. Roads are now stippled black. Overprinted derivatives are usually krystal black. The coastline sand and mud is absent, though it remains on A/ and A// Administrative Areas etc. derivatives where it is essential to the defining of official boundaries

■**Administrative Areas** North (or South)>: 1/625,000
■**Administrative Areas**>: Ordnance Survey North (or South) [from "B L1619","B L1620" editions]

■N	B:■aaa	1965hd	1965	-	kh	0w3	-	1.1.65	1500R		4/65	CPT,Lrgs
■N	A/:■bbb	1966hd	1966	-	kh	0w3	-	1.4.66		-	6/66	CPT,Lrgs,RH
■N	A/:■bbb	1966hd	1966	-	k	0w3	-	1.4.67		-	8/67	CPT,Lrgs
■N	A:■ccc	1968hd	1966	-	k	0w3	-	1.4.68		-	6/68	CPT,Lrgs,RH
■N	A//:■ddd	1969hd	1966	-	k	0w3	-	1.4.69		-	7/69	CPT,Lrgs
N	B L1619:■eee	nd:jf	1970	-	k	0w3	-	1.4.70		-	7/70	CPT,Lrgs
N	J11/003c:■fff	nd:jf	1970	-	k	0w3	1971	1.4.71	2685:10.8.71	-	9/71	CPT,Lrgs
■S	B:■--a	1965hd	1965	-	kh	0w3	-	1.1.65	2500R		4/65	CPT,Lrgs,RH
■S	A/:■no code	1966hd	1966	-	kh	0w3	-	1.4.66		-	6/66	CPT,Lrgs
■S	A/:■bbb	1966hd	1966	-	k	0w3	-	1.4.67		-	8/67	CPT,Lrgs 1
■S	A/:■ccc	1968hd	1966	-	k	0w3	-	1.4.68		-	6/68	CPT,Lrgs,RH
■S	A//:■ddd	1969hd	1966	-	k	0w3	-	1.4.69		-	7/69	CPT,Lrgs
S	B L1620:■eee	nd:jf	1970	-	k	0w3	-	1.4.70		-	7/70	CPT,Lrgs
S	J11/003c:■fff	nd:jf	1971	-	k	0w3	1971	1.4.71	3750:10.8.71	-	9/71	CPT,Lrgs

The same three foreground and three infill (now stipple) overprint colours are used, their print codes listed here in the sequence red, purple, green. With inset maps at 1:250,000 of Glasgow, Edinburgh, Tyneside, Manchester, Birmingham and London areas. Print run figures, now anticipating the expectation of annual printings, from PRO OS 1/1135

1. The "/" beneath the "A" is sometimes indistinct to the point of invisibility

■**Administrative Areas** [red]>: Ordnance Survey North (or South)

N	81900c:■gg	nd:jf	1970	-	k	0w2	1973	1.4.72	-	-	EXg
S	81900c:■gg	nd:jf	1971	-	k	0w2	1973	1.4.72	-	-	EXg

No green plate. Municipal and London Borough information now transferred to the purple and red plates

■**Administrative Areas** [red] **and 1974 Changes** [green]>: Ordnance Survey North (or South)

N	81900c:■ggg	nd:jf	1970	-	k	0w3	1973	1.4.72; 1.12.72	2520:20.2.73	-	2/73	CPT,Lrgs 1
N	81900c:■hhb	nd:jf	1970	-	k	0w3	1973	1.4.73; 1.8.73		-	11/73	CPT,Lrgs
N	81900c:■hhb	nd:jf	1970	-	k	0w3	1973	1.4.73; 1.8.73		-	- book	2
N	81900c:■iic	nd:j-	1970	-	k	0w3	1973	1.11.73; 1.11.73		-	2/74	CPT,RH
N	81900c:■jjd	nd:j-	1970	-	k	0w3	1975	1.1.75; 16.5.75		-	4/75	CPT
S	81900c:■gga	nd:jf	1971	-	k	0w3	1973	1.4.72; 1.12.72	3770:20.2.73	-	2/73	CPT,Lrgs
S	81900:■hhb	nd:jf	1971	-	k	0w3	1973	1.4.73; 1.8.73		-	11/73	CPT,Lrgs,RH
S	81900:■hhb	nd:jf	1971	-	k	0w3	1973	1.4.73; 1.8.73		-	- book	2
S	81900:■hhc	nd:jf	1971	-	k	0w3	1974	1.4.73; 1.3.74		-	6/74	CPT,Lrgs
S	81900:■hhd	nd:jf	1971	-	k	0w3	1975	1.4.73; 1.1.75		-	4/75	CPT

The green plate now used for the newly proposed boundaries. Each sheet includes a note composed of segments relevant to it of the legend: "*Local Government Act 1972 & Local Government (Scotland) Act 1973. Interim Map of Administrative Areas to be operative, England [& Wales] 1974, Scotland 1975 with full county and district information. Changes on this map affect England only*". With inset maps at 1:250,000 of Edinburgh, Glasgow, Tyneside, Manchester, Birmingham and London. The red and purple print codes continue in sequence from the previous map, the green starts anew

1. With the function of the green plate changed in England, "ggg" in col.2 would appear to be a misprint for "gga"

2. In *The Municipal Year Book and Public Utilities Directory 1974*, and overprinted with the footnote "*Supplied with the compliments of the Municipal Year Book 1974*"

1	2	3	8	10	11	12	■13	■14	15	16	17	18	19

Ordnance Survey 1:625 000 Map of Great Britain (Local Government Areas) North (or South): ■Administrative Areas

N	A/:■a	nd:km	1975u	-	k	Ow1	1976	16.5.76		-	9/76	CPT	
N	A/:■b	nd:km	1975u	-	k	Ow1	1977	1.4.77		-	10/77	CPT	
S	A/:■a	nd:km	1975u	-	k	Ow1	1976	1.4.76		-	9/76	CPT,Lrgs	
S	A/:■b	nd:km	1975u	-	k	Ow1	1978	1.4.78		-	8/78	CPT	
S	A/:■C	nd:km	1975u	-	k	Ow1	1979	1.4.79		-	8/79	CPT,Lrgs	
S	A/:■D	nd:km	1975u	-	k	Ow1	1980	1.4.80		-	7/80	CPT,Lrgs	

With an inset map at 1:250,000 of London

Ordnance Survey 1:625 000 Map of Great Britain (Local Government Areas) North (or (Local Government Areas & European Constituencies) South): ■Administrative Areas [red]: European Constituencies [green]

N	A/:■ca	nd:km	1975u	-	k	Ow2	1985/85	1.10.84; 1.10.84		-	4/85	CPT	
N	A/:■da	nd:km	1975u	-	k	Ow2	1986/85	1.10.86; 1.10.84		-	4/87	CPT, on sale	
S	A//:■EA	nd:km	1975u	-	k	Ow2	1983/83	1.4.83; 1.4.83		-	7/83	CPT,Lrgs	
S	A//:■FB	nd:km	1975u	-	k	Ow2	1985/85	1.11.85; 1.4.85		-	2/86	CPT,Lrgs	
S	A//:■GC	nd:km	1975u	-	k	Ow2	1987/87	1.10.87; 1.10.87		-	2/88	CPT,Lrgs	
S	A//:■hd	nd:km	1975u	-	k	Ow2	1990/90	1.4.90; 1.4.90		-	8/90	CPT, on sale	

With an inset map at 1:250,000 of London. The red plate print codes continue in sequence from the previous map, the green ones start a new sequence

■Aeromagnetic Map of Great Britain: Sheet 1 (or Sheet 2).....(North (or South) of National Grid Line 500 Kms,N)

■1	1500/72	1967ga	■1972	1967	k	0g3	1972	1959-65	-	-	on sale	
■1	1500/72	1967ga	■1972	1967	jf	0-1	1972	1959-65	-	-	Sg	
■2	A:■1500/65	1965hd	■1965	1955	k	0g3	1965	1955-61	-	-	on sale	
■2	A:■1500/65	1965hd	■1965	1955	jf	0-1	1965	1955-61	-	-	Sg	

Compiled by GS, (IGS by 1972). On the outline map, "A" editions. Sheet 1 is unconventional, with the base labelled "Outline Style Extended North Sheet" in parentheses. It is extended northwards to include the Shetland Isles (National Grid squares HL-HP-NZ-NV), though with large marginal areas blanked out. Its sheet index and National Grid diagram are also extended northwards. Note also its unconventional symbols panel, and the presence on both sheets of magnetic variation diagrams. Sheet 2 has an inset southern extension to cover the Channel Islands

■Ordnance Survey Ancient Britain: North (or South) Sheet: Third (or Fourth) Edition / [Outline, "C", RPM, "AA" edns]

NS	C	-	■1982u	-	k	3b3	nd	-1066		404.4	6/82	CPT	1
NS	A	-	1990d-	-	kj	6b4	nd	-1066		407.8	9/90	CPT, on sale	2

A map of the major visible antiquities of Great Britain older than 1066. Lit: Hellyer (1987),(1988),(1989a),(1989d)
  1. The sheet sizes are reduced to allow for tabular and illustrative material. Thus the Scilly Isles are inset
  2. Published in conjunction with RCHM England, RCAHM Wales and RCAHM Scotland. Railways are omitted. All RPM elements are krystal black except inland open water, sea and sea names. Thus "Bristol Channel" is blue, "Mouth of the Severn" krystal black. The National Grid (in black dots) again extends across Irish and French mapping. County boundaries and names are added in green (Wales is divided into North and South, Isle of Man is not named)

■Ordnance Survey Britain before the Norman Conquest (871 AD to 1066 AD) North (or South) Sheet / [RPM, "L" edition]

■N	proof:■A		■1973u-	-	k	3b3	nd	871-1066	25.10.72	-	-	SOos
■N	■A		■1973u-	-	b	3b3	nd	871-1066		187	7/74	CPT, on sale
■S	proof		■1973u-	-	k	3b3	nd	871-1066	25.10.72	-	-	SOos
■S	■A		■1973u-	-	b	3b3	nd	871-1066		187	7/74	CPT, on sale

Compiled by C.W.Phillips. With an inset map of Welsh Cantrefs. With letterpress. Base map blue plate only used. The proof copies entitled Northern (or Southern) Britain before the Norman Conquest. Lit: Hellyer (1987),(1988),(1989d)

■The Camping Club's Camping and Caravan Site Map / [RPM, "Q","R","T" editions]
■The Camping and Caravanning Club's Camp Site Map / [RPM, "V", "X" editions]

NS	Q:■A	nd:rq	■1978u	-	k	Rb3	1978	6.77		-	-	SOos,SOrm,RH	1
NS	R:■B	nd:rr	■1979u	-	k	Rb3	1979	6.78		-	-	RH,PC	1
NS	T:■C	nd:rt	■1981u	-	k	Rb3	nd	6.80; 15.10.80		-	-	RH	2,3
NS	V:■D	nd:rv	■1983u	-	k	Rb3	nd	8.82; 15.10.82		-	-	RH,PC	2,3
NS	X:■E	nd:rx	■1984u	-	k	Rb3	nd	11.84; 15.10.84		-	-	RH	2,3,4

1	2	3	8	10	11	12	■13	■14	15	16	17	18	19

Route Planning Maps overprinted with campsites and 1:50,000 series sheet lines
1. With brown 1:50,000 sheet lines. 2. With green 1:50,000 sheet lines
3. With inset maps of Northern Ireland at 1:800,000, with Irish Grid, and of all Ireland at 1:2,000,000
4. Made and printed for the Camping and Caravanning Club Ltd, by the OS; with the clubs's logo in black and magenta

■Ministry of Labour Local Office Areas / ■Administrative Areas North (or South): 1/625,000
■Department of Employment and Productivity Local Office Areas / ■Administrative Areas North (or South)>: 1/625,000 [from 30.6.1968]
■Department of Employment Local Office Areas / ■Administrative Areas>: Ordnance Survey North (or South) [from 1971]

■N	A/:■bb	S1966hd	1966	-	k	Ow3	nd	30.6.67; 1.4.67	-	4/68	CPT	
■N	A/:■ddd	1969hd	1966	-	k	Ow3	nd	30.6.68; 1.4.69	-	10/69	SOrm	1
■N	A/:■ddd	1969hd	1966	-	k	Ow3	nd	30.6.69; 1.4.69	-	10/69	Lkg,SOrm	1
N	J11/003	nd:jf	1971	-	k	Ow3	fouo	30.6.71; 1.4.71	-	-	Lse,NTg,Ob	2
■S	A/:■bb	S1966hd	1966	-	k	Ow3	nd	30.6.67; 1.4.67	-	4/68	CPT	
■S	A/:■dd	1969hd	1966	-	k	Ow3	nd	30.6.68; 1.4.69	-	10/69	SOrm	
■S	A/:■dd	1969hd	1966	-	k	Ow3	nd	30.6.69; 1.4.69	-	10/69	Lkg,SOrm	
S	J11/003	nd:jf	1971	-	k	Ow3	fouo	30.6.71; 1.4.71	-	-	Lse,NTg,Ob	2

The green plate information of the Administrative Areas map is transferred to the red plate. A new black plate is added with Ministry of Labour information. With inset maps at 1:250,000 of Manchester, Birmingham and London areas. Prepared by MHLG in consultation with SDD from data supplied by DEP
1. "ddd" in col.2 is red, red, purple. 2. Made and printed for DOE by OS

■Ministry of Labour Local Office Areas [outline edition] / 1/625,000
■Department of Employment and Productivity Local Office Areas [outline edition] / 1:625,000 [from 30.6.1968]
■Department of Employment Local Office Areas [outline edition] / Ordnance Survey North (or South) [from 1971]

[N]	B:■CBH 7615	1965hd	1965	-	k	Ow1	fouo	30.6.66	-	-	Bg	1
[N]	A/	S1966hd	1966	-	k	Ow1	nd	30.6.67	-	4/68	CPT	
[N]	A/	1969hd	1966	-	k	Ow1	nd	30.6.68	-	10/69	SOrm	
[N]	A/	1969hd	1966	-	k	Ow1	nd	30.6.69	-	10/69	Lkg,SOrm	
N	J11/003	nd:jf	1971	-	k	Ow1	fouo	30.6.71	-	-	Lse,LDg,NTg	2
[S]	B:■CBH 7615	1965hd	1965	-	k	Ow1	fouo	30.6.66	-	-	Bg	1
[S]	A/	S1966hd	1966	-	k	Ow1	nd	30.6.67	-	4/68	CPT	
[S]	A/	1969hd	1966	-	k	Ow1	nd	30.6.68	-	10/69	SOrm	
[S]	A/	1969hd	1966	-	k	Ow1	nd	30.6.69	-	10/69	Lkg,SOrm	
S	J11/003	nd:jf	1971	-	k	Ow1	fouo	30.6.71	-	-	Lse,LDg,NTg	2

Prepared by MHLG in consultation with SDD from data supplied by DEP. With inset maps at 1:250,000 of Manchester, Birmingham and London areas. Overprints on film dated 30.6.1965 are in Lse, though whether by MHLG or OS is uncertain
1. Overprint title panel not boxed. Printed in August 1967   2. Made and printed for DOE by OS

■Department of Employment and Productivity Local Office Areas: **Local Employment Bill 1969** / ■Administrative Areas North (or South)>: 1/625,000

| ■N | A/:■ddd | 1969hd | 1966 | - | k | Ow4 | fouo | 30.6.69; 1.4.69 | - | - | SOrm |
| ■S | A/:■dd | 1969hd | 1966 | - | k | Ow4 | fouo | 30.6.69; 1.4.69 | - | - | SOrm |

■Department of Employment and Productivity Local Office Areas: **Local Employment Bill 1969** / 1/625,000

| [N] | A/ | 1969hd | 1966 | - | k | Ow2 | fouo | 30.6.69 | - | - | SOrm |
| [S] | A/ | 1969hd | 1966 | - | k | Ow2 | fouo | 30.6.69 | - | - | SOrm |

The additional title is on the North Sheet only

■Department of Employment and Productivity Local Office Areas: **Assisted Areas as defined by the Department of Trade and Industry at 5.8.1971** [green] / ■Administrative Areas>: Ordnance Survey North (or South)

| N | J11/003 | nd:jf | 1971 | - | k | Ow4 | fouo | 5.8.71: 30.6.71 | - | - | Cg,Lse,NTg |
| S | J11/003 | nd:jf | 1971 | - | k | Ow4 | fouo | 5.8.71: 30.6.71 | - | - | Cg,Lse,NTg |

Compiled by DOE. Made and printed for DOE by OS. With inset map at 1:250,000 of London

1	2	3	8	10	11	12	■13	■14	15	16	17	18	19

■**Department of Employment and Productivity "Travel to Work" Areas** / 1/625,000

| [N] | A/ | | 1969hd | 1966 | - | k | Ow2 | fouo | 30.6.69 | - | - | Lse,SOrm |
| [S] | A/ | | 1969hd | 1966 | - | k | Ow2 | fouo | 30.6.69 | - | - | Lse,SOrm |

Prepared by MHLG in consultation with SDD from data supplied by DEP

Two 1:625,000 **Digital Datasets** went on sale in 1986 (OSPR 7/86). The first contained Coastline, Rivers and Canals, Lakes and Reservoirs, County Boundaries, District Boundaries, the second Cities, Towns, Villages, Built-up Areas, Roads.

■**Farmhouse Auto-Tours in Great Britain Ltd** / Routeplanner, ["T" edition]

| NS | trimmed | | 1980rt | ■1981u | - | k | Rb1 | nd | 6.80 | - | - | SOos |

Overprinted with 1:50,000 sheet lines in green. With a Great Britain diagram relating to "Phone 'n Roam" vouchers

1:50,000 Sheet Lines: Layout No 7

| N | B | | | - | k | Ow1 | fiuo | | 50:30.4.70 | - | - | not found |
| S | B | | | - | k | Ow1 | fiuo | | 50:30.4.70 | - | - | not found |

Recorded in Job File R1000/28

■**[Finally proposed] 1:50,000 Sheet Lines: Layout No 8** ["Finally proposed" on 8.1970 revision only]

N	B			-	k	Ow1	fiuo		12:19.6.70	-	-	not found
N	B			-	k	Ow1	fiuo		3:10.8.70	-	-	not found
N	B			-	k	Ow1	fiuo	8.70	41:25.8.70	-	-	not found
S	B			-	k	Ow1	fiuo		12:19.6.70	-	-	not found
S	B			-	k	Ow1	fiuo		3:10.8.70	-	-	not found
S	B			-	k	Ow1	fiuo	8.70	41:25.8.70	-	-	not found

Recorded in Job File R1000/4. The next issue (12.1970) of this was at 1:1,250,000 (qv)

■**Finally proposed 1:50,000 Sheet Lines: Layout No 8 and One inch Seventh Series**

N	B			-	k	Ow2	fiuo	8.70	21:25.8.70	-	-	not found
N	B			-	k	Ow2	fiuo		36:10.6.71	-	-	not found
S	B			-	k	Ow2	fiuo	8.70	21:25.8.70	-	-	not found
S	B			-	k	Ow2	fiuo		36:10.6.71	-	-	not found

Recorded in Job Files R1000/4 and R1000/28

■**General Accident: Details of United Kingdom Areas and Office Locations (at 1st January 1979)** / Outline, "C" edition

| N | C | | nd:km | 1974u | - | jh | Ow1 | nd | 1.1.79 | - | - | SOrm |
| S | C | | nd:km | 1975u | - | jh | Ow1 | nd | 1.1.79 | - | - | SOrm |

■**Geological Map of the United Kingdom:** North (or South) (North (or South) of National Grid Line 500 Km N)

| ■N | C:■45000/79 | - | 1979u | - | k | Ob4 | 1979 | nd | | 203.2 | note | CPT, on sale |
| ■S | C:■50000/79 | - | 1979u | - | k | Ob4 | 1979 | nd | | 203.2 | note | CPT, on sale |

Third Edition Solid, 1979. Published by OS for IGS. With an index of 1:63,360 and 1:50,000 sheets in red. That part of Northern Ireland on the North Sheet is also overprinted. Recorded in GSPR 1980/1

■**Index to the Ordnance Survey Maps at Scales of 1:10,000 or 1:10,560 (six inches to the mile) on National Grid Sheet Lines** / Outline "B" edition

| N | B:■C | | 1969rf | ■1972 | - | k | Ow1 | nd | nd | - | - | Bg,Lrgs,RH |
| S | B:■C | | 1969rf | ■1972 | - | k | Ow1 | nd | nd | - | - | Bg,Lrgs,RH |

With inset maps at 1:250,000 of London, Birmingham, Manchester, Edinburgh, Glasgow and Newcastle upon Tyne

Index to 1:25 000 Pathfinder Maps and Outdoor Leisure Maps

[NS]A		-	1990d	-	k	Ow2	nd	1.1.90		304.1c	3/90	CPT
[NS]A1		-	1991d	-	k	Ow2	nd	1.1.91		304.1c	-	CPT
[NS]A2		-	1992d	-	k	Ow2	nd	1.1.92		304.1c	-	CPT

Folded integrally. The sheet size is enlarged for indexing reasons, and names and insets have been moved. National Grid letters are on the face of the map. Rivers and places are mapped and named, and National Trails and National Parks are shown, but other features, such as roads, railways, boundaries and contours, are excluded

1	2	3	8	10	11	12	■13	■14	15	16	17	18	19

Metalliferous Mineral Resources Excluding Energy Minerals / [Outline Map, "C" edition]
N C   nd:km   1974u   -   jf   0-0   nd   7.74   -   -   Ebgs
S   not found
On permatrace film

Ministry of Labour titles - See Department of Employment

■**Ministry of Transport Traffic Areas** / RPM, "G" edition
N G   1969rg   1969   -   k   3b1   nd   9.69   -   -   SOrm
S G   1969rg   1969   -   k   3b1   nd   9.69   -   -   SOrm

■Ordnance Survey **Monastic Britain**: North (or South) Sheet / [Outline Map, "C" edition]
N proof:B   -   ■1975u -   k   3g2   nd   nd   21.11.75   -   -   SOos,RH
N B   -   ■1976u -   k   3g2   nd   nd   11881:7.76   178   10/78 CPT
S proof:B   -   ■1975u -   k   3g2   nd   nd   21.11.75   -   -   SOos,RH
S B   -   ■1976u -   k   3g2   nd   nd   12328:26.7.76 178   10/78 CPT

Third Edition, compiled by R.N.Hadcock. With an inset map entitled **Religious Houses in London**. With letterpress. The proof copies have different colour hillshading. Lit: Hellyer (1987),(1988),(1989a),(1989d)

■**National Health Service, Scotland Health Board Boundaries** / ■Administrative Areas and Local Government Changes: Ordnance Survey North
N 81900c:■iic   nd:j-   1970   -   k   0w4   1973   1.11.73; 1.11.73   -   -   SOrm
National Health Service boundaries drawn by Graphics Group SDD, May 1974

■**Pilgrim's Progress Heritage Vacations** / RPM, "Z" edition
NS Z   nd:rz   1986d   -   k   1b1   nd   11.86   -   -   SOos
Overprinted with 1:50,000 series sheet lines in green for Pilgrim's Progress Tours

■**Quaternary Map of the United Kingdom:** North (or South) (North (or South) of National Grid Line 500 Km N)
■N ■20000/77   -   1977u   -   k   0b5   1977   nd   203.1 -   on sale
■S ■20000/77   -   1977u   -   k   0b5   1977   nd   203.1 -   on sale

Based on the work of IGS and other published data. Northern Ireland is overprinted on both sheets. With an index of 1:63,360 and 1:50,000 sheets in magenta. With a diagram of Distribution of sea-floor Quaternary sediments

■Ordnance Survey **Roman Britain:** North (or South) Sheet: [Fourth Edition] (or Fourth Edition (Revised))
■N proof 1   -   ■1977u- -   b   3b4   nd   nd   18.4.77   -   -   SOos,CCS
■N proof 2   -   ■1978u- -   b   3b4   nd   nd   22.12.77   -   -   SOos,CCS   1
■N ■A   -   ■1978u- -   b   3b4   nd   nd   20430:19.10.78 84   3/79 CPT   2
■S proof 1   -   ■1977u- -   b   0b4   nd   nd   4.1.77   -   -   CCS,RH
■S proof 2   -   ■1978u- -   b   3b4   nd   nd   16.12.77   -   -   SOos,CCS,RH   1
■S proof 3   -   ■1978u- -   b   3b4   nd   nd   3.2.78   -   -   SOos,CCS,RH
■S ■A   -   ■1978u- -   b   3b4   nd   nd   20462:19.10.78 84   3/79 CPT   2
 NS A   -   1991d- -   kj   6b4   nd   nd   407.8 12/90 CPT, on sale 3

The base map for the 1978 edition was the blue plate of the RPM, "P" edition, the 1991 edition the RPM, "AA" edition
 1. Versions with yellow and buff hill layers
 2. Compiled by J.Fox, the Archaeology Officer, and his staff. With letterpress. Ellis Martin's initials were restored to the cover in 1989, having been deleted in 1956. Lit: Hellyer (1987),(1988),(1989d)
 3. Published in conjunction with RCHM England, RCAHM Wales and RCAHM Scotland. For the first time on a topographical base map. With lists and photographs. Railways are omitted. All RPM elements are krystal black except inland open water, sea and sea names. Thus "Bristol Channel" is blue, "Mouth of the Severn" krystal black. The National Grid (in black dots) again extends across Irish and French mapping. County boundaries and names are added in green

■**Scotland: Sub-regions for Economic Planning (By Administrative Areas)** / ■Administrative Areas North>: 1/625,000
■N A/:■bbb   1966hd   1966   -   k   0w4   1967   1.4.67   -   -   Cg
The extra overprint is blue: overprinted by the Scottish Office, July 1967 (Ad/N/2/9)

1	2	3	8	10	11	12	■13	■14	15	16	17	18	19

■Scottish Development Department: **Reform of Local Government in Scotland: Government Proposals for New Authorities** / ■Administrative Areas>: Ordnance Survey North
N no code   nd:jf 1970   -   k Ow3 1972   1.72; 1.4.71   -   -   En
   The green overlay (Scotland only) drawn by Technical Support Services SDD. With inset maps at 1:250,000 of Glasgow, Edinburgh and Tyneside

---

Supplement 1. Military issues, published by D Survey, Ministry of Defence

1	2	3	6	8	10	11	12	■13	■14	15	16	17	18	19

■Roads
1 B:■A   p1958ga   -   CCR   1958 k 7b3 1958   5.56   -   -   ABn,Lrgs,Ob
2 A/:■A   p1956ga   1958   CCR   1958 k 7b3 1958   5.56   -   -   ABn,Lrgs,Ob
   M322 (ex GSGS 4813) Edition 2-GSGS, 1958. Print codes 5000/1/59/5834/OS. On the 1956 **Roads** map. Edition 3-GSGS is on the RPM "A" edition, and new editions have appeared annually at least as far as Edition-25 on RPM "Y" edition, including supplementary Editions 3 on "A/", 4 on "B/*", and 12 on "K/*", together with accompanying North Sheet reprints

Ordnance Survey "Ten Mile" Map of Great Britain Sheet 1 (or Sheet 2)
1 B:note   1958ga   -   CCR   1958 jg 9b0 1957   -   -   -   Lrgs,Ob   1
1   not found   2
2 nc:note 1955ga   -   CCR+ 1955 jg 9b0 1957   -   -   -   Lrgs,Ob   1
2 B:note   1962gc   -   1962 1962 jg 9b0 1962   -   -   -   Lrgs,Ob   2
   1. M325 (ex GSGS 4676) Edition 1-GSGS, 1957. Print codes 5000/2/58/5690/OS. Sheet 2 on the 3971 printing
   2. M325 (ex GSGS 4676) Edition 2-GSGS, 1962. Print code 6000/7/62/6324/OS

Ordnance Survey "Ten Mile" Map of Great Britain Sheet 1 (or Sheet 2)
1 C:note   1963gc   -   1963 6.63+ gb 8b0 1964   -   -   -   Ob
2 B:note   1962gc   -   1962 1962 gb 8b0 1964   -   -   -   Ob
   GSGS (Misc.) 1820 Edition 1-GSGS, 1964. With print codes 750/9/64/6654/OS. On the coloured "Ten Mile" Map

---

Supplement 2. Editions not published by the Ordnance Survey

Assessment of Climatic Conditions in Scotland: Based on Accumulated Temperature above 5.6°C and Potential Water Deficit
Assessment of Climatic Conditions in Scotland: Based on Exposure and Accumulated Frost
   North Sheets only, published for the Soil Survey of Scotland by the Macauley Institute for Soil Research (Aberdeen, 1969 and 1970). The National Grid is in 100km squares. Copies in Gg
Water Resources in Scotland
   North Sheet only, published by SDD, 1973
Average Annual Rainfall (in Millimetres) International standard period 1941-70: Northern (or Southern) Britain
   North and South Sheets published by the Meteorological Office in 1977 (print codes G16, G17). With a "km" legend, and the 7.1974 road revision, with three coloured layers and a blue grid. With an inset map of the Channel Islands (with UTM Grid). There is a companion map of the North of Ireland (HMSO, 1976)
Hydrogeological Map of Scotland (or **England and Wales**)
   Northern and southern sheets drawn and printed by Cook, Hammond & Kell, published by BGS (ex-IGS) London (NERC Copyright reserved 1988, 1977). A grey base with reduced topographical detail. Cartography by J.L.Meikle (BGS, Edinburgh)
Regional Gravity Map of the British Isles
   The northern sheet only. Printed by Bartholomew, and published in: R.G.Hipkin and A.Hussain *Regional gravity anomalies. 1. Northern Britain* (London, HMSO, 1983: BGS (IGS) Report 82/10 (ISBN 0 11 884235 8)
Bouguer Anomaly Map of the British Isles
   Coastal outline only, under milligal contours in three colours. Northern and southern sheets published by BGS (CC University of Edinburgh 1981, 1986). Northern sheet printed by Bartholomew, and published in: R.G.Hipkin and A.Hussain *Regional gravity anomalies. 1. Northern Britain* (London, HMSO, 1983: BGS (IGS) Report 82/10 (ISBN 0 11 884235 8)

Supplement 3. Special sheets

1 2   3   8   10 11 12 ■13   ■14   15   16   17   18   19

    Population Change 1951-1961 by Wards and Civil Parishes / "Ten Mile" ["B" edition, Sheet 2 section]
EW   B   nd:gc   CCR   -   g   Ow3   1963   1951-61   -   -   Cg
    Compiled and drawn by MHLG from the 1961 Census. A special pre-publication map, restricted to MHLG, printed by PRB/
MOD (Air). The sheet is non-standard, being narrower to the west, with the base map title at the foot

    Administrative Areas: England & Wales
EW   L1579   1966   1966   -   j   Ow0   -   -   -   -   SOrm
    A diagram, often used for overprinted information as below

    ?Administrative Areas: Scotland
    ?L1580
    Not found, and as yet I have no proof of its existence

    ■Assizes (Circuits and Towns) / Administrative Areas: England & Wales
EW   L1579   1966   1966   -   j   Ow1   nd   nd   -   -   SOrm
    Printed by OS for MHLG. The overprint (blue) supplied by the Lord Chancellor's office

    ■Cases for Trial at Courts of Quarter Sessions in 1965 / Administrative Areas: England & Wales
EW   L1579   1966   1966   -   j   Ow1   nd   1965   -   -   SOrm
    ■Civil Actions tried and Judgment Summonses heard in County Courts 1965 / Administrative Areas: England & Wales
EW   L1579   1966   1966   -   j   Ow1   nd   1965   -   -   SOrm
    ■Criminal Cases for Trial and Civil Actions tried at Assizes in 1965 / Administrative Areas: England & Wales
EW   L1579   1966   1966   -   j   Ow1   nd   1965   -   -   SOrm
    These three printed by OS for MHLG. The overprints (red) from the Royal Commission on Assizes and Quarter Sessions

    ■East Anglia Economic Planning Region
-    no code   nd   6.66   -   jg 4b1   nd   nd   -   -   book
    Output information supplied by MHLG. Printed by OS June 1966. On the coloured topographical map base (4 layers).
In *East Anglia: A Study: A first Report of the East Anglia Economic Planning Council* (London, HMSO, 1968)

    General Reference Map / From **A Strategy for the South East**
-    no code   1967re 1967   -   k   <u>3b2</u> nd   nd   -   -   book
    Showing the London area only. In *A Strategy for the South East* (London, HMSO, 1967)

    1/625,000 Map of Great Britain / [RPM, "B" edition]
EW   A   1967hd 1965   -   k   Ow0   -   -   -   -   SOrm
EW   A/   not found
EW   A//   1969hd 1966   -   k   Ow0   -   -   -   -   SOrm
    A special issue, extending to 660km N, prepared for MHLG. With inset maps at 1:250,000 of Manchester, Birmingham,
Newcastle-upon-Tyne and London

    ■Regional Hospital Areas / Administrative Areas: England & Wales
EW   L1581a   1969   1969   -   k   Ow1   nd   nd   -   -   SOrm
    Printed by OS. Prepared by MHLG for DHS

    Local Government Bill: **Government Proposals for New Counties in England:** with the Proposed Names 4th November 1971 /
Administrative Areas
EW   no code   -   1971   -   k   Ow1   nd   1.4.71   -   -   Cg
    Department of the Environment Circular 84/71. A diagram made and printed by OS for DOE

    Grassland Poaching Map of England and Wales
EW   no code   -   1976u   -   k   Ow2   nd   nd   -   -   Mg
    Compiled by P.M. Patto. Published by the Grassland Research Institute and Agricultural Development and Advisory
Service, drawn by Survey Section ADAS, MAFF

V.3.Supplement 3   173

1	2	3	8	10	11	12	■13	■14	15	16	17	18	19

■**Soil Survey of England and Wales: Bioclimatic Classification**
EW C   nd:np   ■1978u -   j   Ob4   nd   nd   -   -   Lkg,Og
    Made and published by the OS for the Soil Survey of England and Wales. With an inset map of the Scilly Isles

Distribution of Cattle in Great Britain
EW no code   -   1978u -   k   Ow1   nd   nd   -   -   SOos
Distribution of Pigs in Great Britain
EW no code   -   1978u -   k   Ow1   nd   nd   -   -   SOos
Distribution of Sheep in Great Britain
EW no code   -   1978u -   k   Ow1   nd   nd   -   -   SOos
    In each case the number per 50 hectares of agricultural land. Drawn by Survey Section, ADAS, printed by OS, fouo

Agricultural Land Classification of England and Wales
EW no code   -   1979u -   k   Ow3   nd   nd   -   -   SOos
    Based on 1977 OS map. Published by MAFF Welsh Office Agriculture Dept, drawn by ADAS, printed by OS

Ministry of Agriculture, Fisheries and Food: Offices and Centres
EW no code   -   1979u -   k   Ow1   nd   nd   -   -   SOos
    Based on 1977 OS map. Published by MAFF, drawn by ADAS, printed by OS

Mean Annual Excess Winter Rainfall
EW no code   -   1979u -   k   Ow1   nd   1941-70   -   -   LEg
    Based on 1977 OS map. Published by MAFF, map prepared by Survey Section, Land Service, ADAS

■**Potential Cycleroutes and disused Railways in England & Wales 1981**
EW no code   -   CCR   -   k   Ob2   nd   1981   -   -   Eu
    Showing Great Britain north to Edinburgh. With lists, photographs and an illuminated border. Scotland also overprinted. With the National Grid, but no values. Printed by the OS. Based on the 1981 1:625,000 map with overprint information supplied by John Grimshaw & Associates. "*In 1981 the Government published a consultation paper on cycling and the Secretary of State for Transport commissioned a study of disused railway lines in England & Wales to see which could be converted into routes for cyclists. This map shows the disused railways, existing railway paths and the potential cycleroutes examined in the study. A full report of this study is available from H.M.S.O. and separately published annexes on detailed routes from the Department of Transport.*"

■**National Health Service Boundaries 1982** / Administrative Areas: England & Wales
EW no code   -   1982u -   k   Ow1   nd   1.4.82   -   -   SOrm
    Made and printed by OS, and prepared for DHSS by OS

Ordnance Survey **Maritime England** / [RPM, "U" edition, outline state]
EW no code   -   1982u -   kj 1b4   nd   nd   401.2   4/82   CPT
    With copious lists and photographs. Published by OS in conjunction with English Tourist Board

---

    V.4. 1:1,250,000 Great Britain, "E" edition, 1975

2	3	8	11	12	■13	■14	15	17	18	19
proof:E	nd:ue:np	1975u-	j	Ow0	-	-	16.4.75	-	SOos	
proof:E	nd:ue:np	1975u-	J	Ow0	-	-	16.4.75	-	SOos	
E	nd:ue:np	1975u-	j	Ow0	-	-	10220:7.75	-	SOos,SOrm,RH	
E	nd:ue:np	1975u-	J	Ow0	-	-	?do	-	En,SOrm	

    Derived from the RPM "M" edition, so showing the new county boundaries and names throughout Great Britain. See also pp.151,154. Entitled **Ordnance Survey Great Britain**. *NB*: Editions to "D" are in section IV.4 (p.139)

■**Index to the 1:25 000 First & Second Series**: Index to the 1:25 000 Outdoor Leisure Maps
AN   nd:ue:np   1984u- jf 0-0   nd   1.10.84   -   Ng

INDEX MAPS

Main sheet lines only are depicted; insets and extrusions are not indicated. The only overlaps are in the Great Britain three sheet series

Projection - Transverse Mercator

Scale - 1:5,600,000 or 88 miles to 1 inch

Origins of Projections and Grids

▲ Trigonometrical station
● Graticule intersection
⊙ False origin

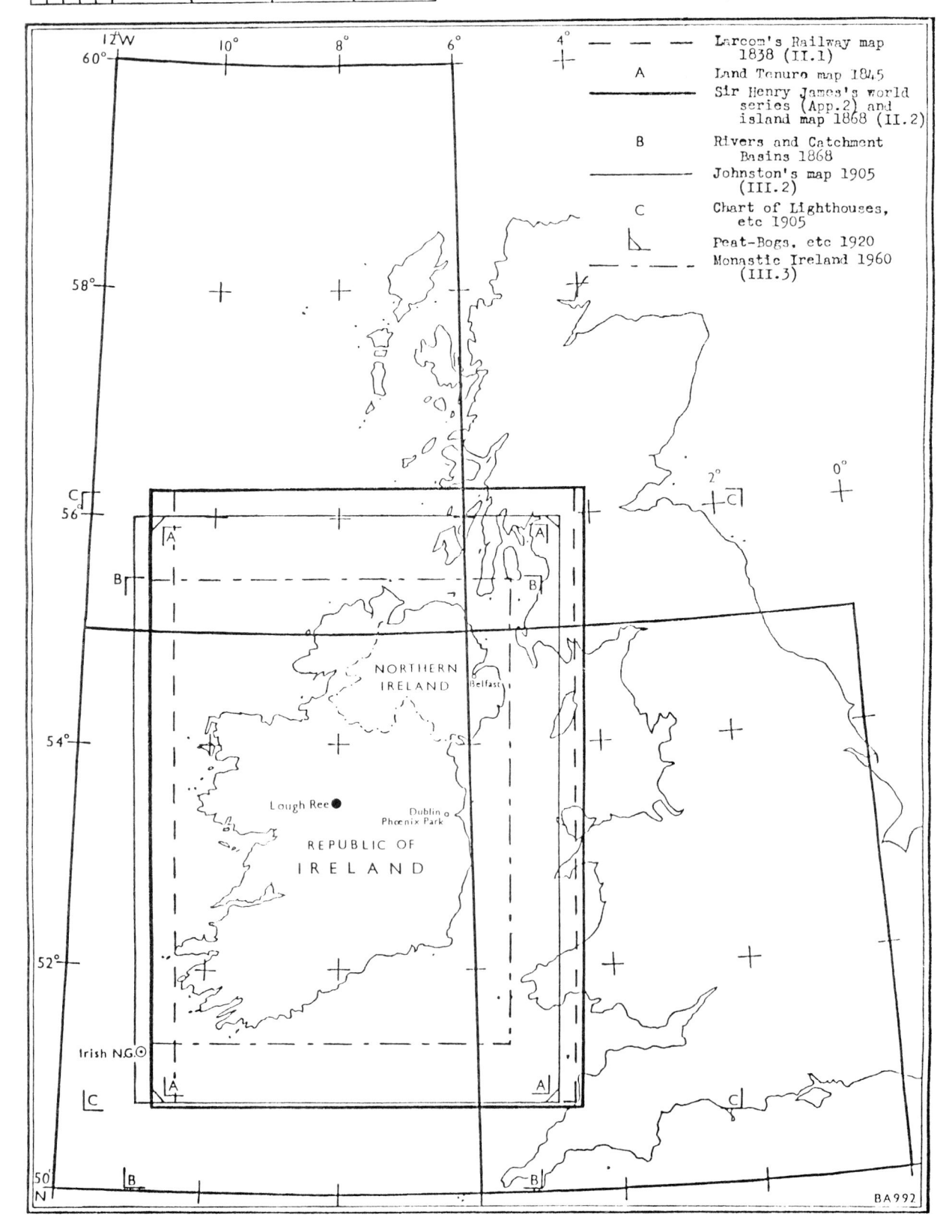

— — —	Larcom's Railway map 1838 (II.1)
A	Land Tenure map 1845
———	Sir Henry James's world series (App.2) and island map 1868 (II.2)
B	Rivers and Catchment Basins 1868
———	Johnston's map 1905 (III.2)
C	Chart of Lighthouses, etc 1905
	Peat-Bogs, etc 1920
—·—·—	Monastic Ireland 1960 (III.3)

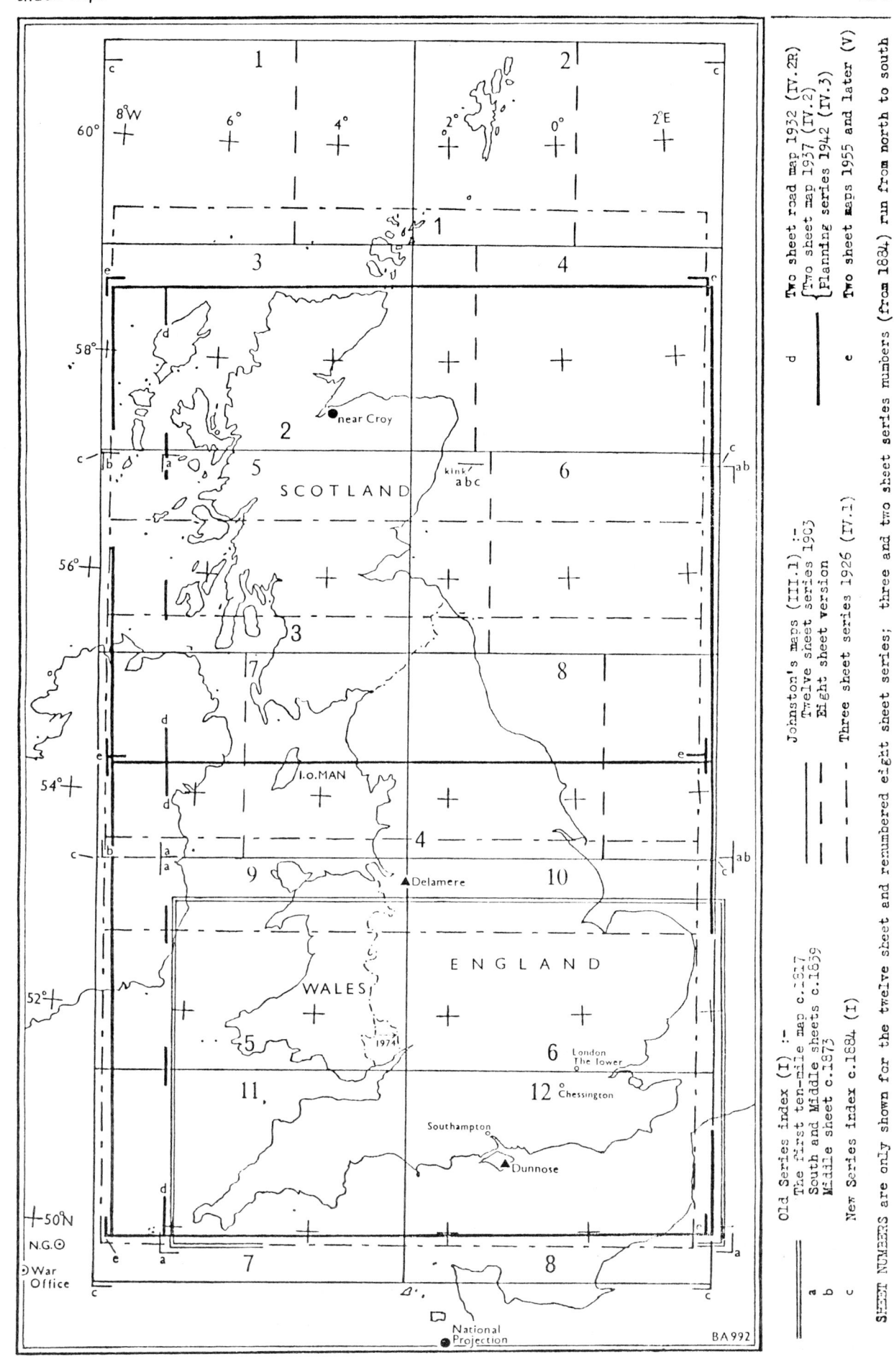

Index Maps 175

Old Series index (I) :-
═══════ The first ten-mile map c.1817
─────── South and Middle sheets c.1839
─ ─ ─ ─ Middle sheet c.1873

c New Series index c.1884 (I)

Johnston's maps (III.1) :-
═══════ Twelve sheet series 1903
─────── Eight sheet version

─ · ─ · Three sheet series 1926 (IV.1)

d Two sheet road map 1932 (IV.2R)
   { Two sheet map 1937 (IV.2)
   { Planning series 1942 (IV.3)

e Two sheet maps 1955 and later (V)

SHEET NUMBERS are only shown for the twelve sheet and renumbered eight sheet series; three and two sheet series numbers (from 1884) run from north to south

APPENDICES

Appendix 1. Projections of the Ordnance Survey Ten-mile Maps. (Contributed by Brian Adams)

"Begin at the beginning, and go on till you come to the end; then stop." This simple rule, enunciated gravely by the King of Hearts, may perhaps seem unusually straightforward for the Wonderland of Alice; I find it altogether inappropriate for the wonderland of the Ordnance Survey ten-mile map projections. Instead I propose to start in the middle, go on to the end, then backwards to the beginning, and then across the sea to Ireland; and somewhere I need to fit in a projection whose use was very largely mythical. Reader, take heed! But first:-

---

1. Ordnance Survey Fundamental Elements

The projections of all Ordnance Survey maps of Great Britain, of regular series commenced after 1830 and on scales of 1:633,600 and larger, are calculated on the Airy spheroid (Airy's figure of the Earth) defined in terms of the foot of Bar O_1. All such projections prior to the adoption of the National Projection had a scale factor of unity. Maps of Ireland on scales from 1:63,360 to 1:253,440 inclusive, and all scales on the Transverse Mercator Projection of Ireland until 1965, are also on the Airy spheroid, and all these have a scale factor of unity.

The reader will observe that several of the ten-mile maps fall outside the above-mentioned ranges, either of date or in Ireland of scale, and further details of these maps will be found in the relevant sections following. All subsequent references to feet, including yards proportionally, are to feet of Bar O_1.

2. The Properly Projected Series

---

1. Three-sheet series, 1926 (IV.1)

This was the first series of ten-mile maps of Great Britain to be properly projected throughout; it was constructed on Cassini's Projection on the origin of Delamere, latitude 53°13'17".274N, longitude 2°41'03".562W, the unit of calculation being the foot. This specific projection had originally been adopted for the northern sheets of the one-inch Old Series maps of England & Wales, followed by the one-inch New Series and quarter-inch and half-inch maps of those countries. But it was only after the first world war that its use was extended to maps of Scotland, on the one-inch and quarter-inch scales, thereby facilitating its adoption for an all Great Britain smaller scale series.

I have described elsewhere (Adams 1989a,1990) how the two mile squares on one-inch Popular Edition maps were formed by rectangular co-ordinate lines of this same Cassini projection, and so also are the squares of the two inch, twenty scale mile, reference grid on this ten-mile series. Thus each two inch interval represents 105,600 feet on the projected spheroid, north-south distances on the ground usually being slightly less due to the effects of the projection. According to Ordnance Survey records[1] the co-ordinates of the sheet lines of the series are as follows:-

Common west neat line	1,006,890 feet west
Common east neat line	999,510 feet east
Horizontal neat lines in order:-	
Sheet 1 north	2,263,727 feet north
Sheet 2 north	1,207,727 feet north
Sheet 1 south	890,927 feet north
Sheet 3 north	151,727 feet north
Sheet 2 south	165,073 feet south
Sheet 3 south	1,221,073 feet south

---

1. Box TL 23, item *Cassini Co-ords One Inch Qtr In. Half In.*

Appendix 1

The Delamere origin is at the point marked on this series by the spot height 575, nine miles E by N from Chester. In accordance with the above limits, the co-ordinates of the sides of the twenty mile square containing Delamere, M10 on Sheet 2, B10 on Sheet 3, are:-

46,127 feet north
56,490 feet west   49,110 feet east
59,473 feet south

Can these be right, or are we already following the White Rabbit down the rabbit-hole?; let me apply some elucidation. The lines forming the two inch, eight scale mile, squares on the quarter-inch Third Edition maps coincide with every fourth line of the Popular Edition reference grid, and the ten-mile map squares might be expected to coincide with every tenth. But the north-south lines are displaced by one mile, falling centrally down Popular Edition squares; when it is realised that this brings the western neat line of the ten-mile series just 0.17 inch (4.4mm) outside Sròn an Dùin, the western point of Mingulay, and the eastern neat line 0.16 inch (4.0mm) outside Lowestoft Ness, it can be seen that this implies some accurate scheming[2]. The east-west lines, however, are a different matter; they are displaced southwards from such lines on the Popular Edition by 2813 feet, or half a mile plus 173 feet, which latter increment is an insignificant 0.003 inch (0.008mm) on the map.

There appears no rational explanation for this shift in the horizontal lines of the ten-mile reference grid, which automatically brings an equal shift in the north and south limits of the map sheets. In fact, it brings a worsening in the positioning of these limits in that the clearance north of Seal Skerry, northern Orkney, is reduced from 0.18 to 0.13 inch (4.6 to 3.3mm), whilst that south of Gilstone Ledges, southern Scilly, is increased from 0.27 to 0.32 inch (6.8 to 8.1mm); a half mile shift <u>northward</u> in the limits would have evened these clearances to 0.23 and 0.22 inches (5.8 and 5.5mm) respectively! On the other hand, it may be no coincidence that with the strange recorded shift the latitude of the north-west corner of Sheet 3 works out at 53°32'50".000N, but there again seems no rational reason for this one of twenty-four geographical co-ordinates being made a round figure.

2. Two-sheet **Road Map**, 1932 (IV.2R), and topographical map, 1937 (IV.2)

The National Projection, already in embryo at the time the three-sheet series was published, was used in its original form for the construction of the succeeding pair of series. The National Projection was a Transverse Mercator Projection with its true origin at 49°N, 2°W, and as originally computed in imperial terms had a scale factor of 0.9996. This was the first Ordnance Survey projection to be manifested to the purchasing public by the incorporation of a numbered co-ordinate grid, initially appearing on the one-inch Fifth Edition in 1931. Although then known as the national grid, I use the full title National Yard Grid throughout to distinguish it clearly from its metric successor. As may be presumed, the unit of this grid was the yard, and the true origin was given the false co-ordinates 1,000,000 yards east, 1,000,000 yards north. The sheet line co-ordinates were:-

West neat line (1937)	620,000 yards east
West neat line (1932)	680,000 yards east
East neat line	1,300,000 yards east
North neat line Sheet 1	2,180,000 yards north
Joining line Sheets 1,2	1,650,000 yards north
South neat line Sheet 2	1,120,000 yards north

The reader will hardly need me to point the difference between this orderly array of round figures, and the disorderly array which had resulted from the true origin of the projection being at an eccentric point within a grid square (to apply the term anachronistically to an earlier equivalent). I would add, however, that the earlier practice was in line with that used on the large scale, six-inch and twentyfive-inch, projections.

On the new projection the meridian of Delamere slopes at an angle of 0°33', the convergence of the meridians, to the central meridian of 2°W and all north-south grid lines. As the topography from the former three-sheet series was transferred, presumably by manipulation of photography, on to the new projection it can clearly be seen, particularly in longer names such as Inverness-shire, Yorkshire and Irish Sea, that the "horizontal" work slopes downward to the right at the same angle; this is even more evident on the succeeding series with its closer spacing of grid lines. Being re-lettered, this problem did not afflict the **Road Map**.

3. 1:625,000 Ministry of Town and Country Planning Series, 1942 (IV.3), and 1:1,250,000 sheet, 1947 (IV.4)

Long before the appearance of the next ten-mile series, and indeed well before the Davidson Committee had completed its deliberations, the Survey had decided that it had no option but to adopt the international metre as the unit for its retriangulation. This necessitated recalculation of the National Projection on the same unit, making it very easy to fall in with the later Davidson recommendations that a national grid should be introduced with that unit.[3] The geodetic standards of the retriangulation required that a very accurate conversion factor was determined for feet of Bar O_1 to international metres, but as the time was still that of logarithmic computation this factor was defined by the absolute

---

2. This is the technical term, and no deviousness is implied.
3. *Final Report...* (1938) 5.

value of an eight-figure logarithm. In natural terms it is equivalent to 1 foot = 0.304,800,749,1 metres. The scale factor of this metricated National Projection was also defined by an eight-figure logarithm, equivalent to a natural value of 0.999,601,271,7.

The redefined National Grid now introduced on the modified projection, in addition to having its unit as the metre, had the false co-ordinates of the true origin amended to 400,000 metres east, -100,000 metres north. The formulae for conversion between National Yard Grid co-ordinates e,n and (metric) National Grid co-ordinates E,N are:-

$$E = 0.914,403,41\ e - 514,403.41$$
$$N = 0.914,403,41\ n - 1,014,403.41$$
$$e = 1.093,609,22\ E + 562,556.31$$
$$n = 1.093,609,22\ N + 1,109,360.92$$

Applying these formulae to the limits of the 1937 two-sheet series above, we obtain:-

West neat line	52,526.70 metres east
East neat line	674,321.02 metres east
North neat line Sheet 1	978,996.02 metres north
Joining line Sheets 1,2	494,362.22 metres north
South neat line Sheet 2	9,728.41 metres north

These when rounded to the nearest metre are seen to be the same as the neat line co-ordinates stated on the 1:625,000 and 1:1,250,000 maps, which as recorded in Chapter IV were direct conversions from the 1937 series. The figures also demonstrate how the mainland of Great Britain falls neatly between the northings of 0 and 1,000,000 on the revised grid, one of the facts which the Ordnance Survey used to promote this metric intrusion into the imperial atmosphere of the time.

4. 1955 and subsequent maps (V.1,2,3,4)

These series were constructed from the start on the National Projection with National Grid as currently in use. The defining elements of these have been recorded above, but for convenience I tabulate them here in the standard form:-

Projection	Transverse Mercator
Spheroid	Airy*
Unit	International Metre*
True Origin	49°N, 2°W.
False Co-ordinates	400,000 m.E, -100,000 m.N.
Scale Factor	0.999,601,271,7

    * 1 foot of Bar O_1 = 0.304,800,749,1 metre

The sheet line co-ordinates are:-

West neat line	50,000 metres east
East neat line	670,000 metres east
North neat line Sheet 1	990,000 metres north
Joining line Sheets 1,2	500,000 metres north
South neat line Sheet 2	10,000 metres north

3. The Mixed Parentage Series

Twelve-sheet series, 1903, later eight-sheet (III.1)

As described in Chapter III, this series was directly reduced from the then current series of quarter-inch maps, themselves derived from one-inch maps. The problem was that these maps of England & Wales were on a different projection from those of Scotland, and the two projections could in no way be accurately fitted together at the border, or anywhere else; whilst east-west distances were broadly the same on the two projections, the curvature of the projected parallels of latitude was different.

The one-inch and quarter-inch maps of England & Wales were on Cassini's Projection on the origin of Delamere, as specified in section 2.1 above. The corresponding maps of Scotland at this time were constructed on Bonne's Projection on the origin of 57°30'N, 4°W, with the foot as unit. The radius of projection of the initial parallel of latitude, 57°30'N, was 13,361,612.2 feet, and the position of the origin was in the vicinity of Croy, some eight miles E by N from Inverness and on original Scotland one-inch Sheet 84; the co-ordinates on Bonne's Projection of the neat lines of this map were:-

                  41,957 feet north
44,349 feet west   82,371 feet east
                  53,083 feet south

In another place (Adams 1991) I distinguished as (a) and (b) the two intertwined problems I found in looking into the projection of a particular map series. I now define these items in more general terms, and with a marked lack of humility I christen them:-

Adams's Components

In the study of the projections of maps, as distinct from the study of map projections, two separate components are sometimes present; they should not then be confused, though their joint presence may confuse the issue. Such dichotomy is largely, but not entirely, restricted to maps from the era of national surveys, and on topographic scales. I label these components:-

(A) The projection upon which the sheet lines, graduation and graticule (where present) of the map or series are constructed; and

(B) The method used in laying down the topographical or other detail upon that framework.

Considering component (A) for the series under review, the sheet lines for the whole series were set out on the projection of the English larger scale maps; on this projection the central meridian of the Scottish projection was a very shallow curve, running at an angle to the north-south sheet lines varying from 1°04' to 1°09'. Turning to component (B), the quarter-inch maps of

Appendix 1   179

England & Wales were correctly reduced in on their own projection, but how were those on the discordant projection laid down? We are enabled to examine this question by the hindsight presented by the subsequent re-publication of the one-inch and quarter-inch maps of Scotland on Cassini's Projection on Delamere. From these we can trace where the detail on the ten-mile maps ought to be, and compare this with where it was actually drawn.

But first we need to obtain the co-ordinates of the sheet lines of the ten-mile series. I have not been able to find any extant documentation giving these figures, neither at the Ordnance Survey Office nor at the Public Record Office, so we have to resort to the maps themselves, remembering after the revelations in section 2.1 above that no certain values can be obtained therefrom. The sheet lines running east-west across England can be visually identified quite precisely with lines of the Popular Edition reference grid, and it is reasonable to suppose that these do coincide. But the central north-south join does not coincide with such lines, and tests show it to be running about one mile east of the meridian of Delamere; whilst it is only possible to estimate the position to some 250 scale feet, there is no reason to depart from the guess of exactly one mile east. Consequently, we arrive at the estimated co-ordinates of the sheet lines of the original twelve-sheet series as:-

Common west neat line	1,050,720 feet west
Central north-south joining line	5,280 feet east
Common east neat line	1,061,280 feet east
North neat line Sheet 1,2	2,826,220 feet north
Joining line Sheets 1,2/3,4	2,139,820 feet north
Joining line Sheets 3,4/5,6	1,453,420 feet north
Joining line Sheets 5,6/7,8	767,020 feet north
Joining line Sheets 7,8/9,10	80,620 feet north
Joining line Sheets 9,10/11,12	632,180 feet south
South neat line Sheet 11,12	1,344,980 feet south

The estimated co-ordinates of those sheet lines which were altered for the subsequent eight-sheet versions (which running in the sea, away from confirmatory land features, must be regarded as more suspect than those above) were:-

Sheet 1.2.(later 1) west neat line	401,280 feet west
Sheet 1.2.(later 1) east neat line	549,120 feet east
Sheet 3.4.(later 2) east neat line	216,480 feet east
Sheet 5.6.(later 3) east neat line	269,280 feet east
Sheet 7.8.(later 4) west neat line	546,480 feet west
Sheet 7.8.(later 4) east neat line	667,920 feet east

Returning to component (B) for the area of Scotland, I have traced the above co-ordinate lines across the one-inch Popular Edition maps of that country; from these it can be seen that, within the limits of accuracy of the ten-mile scale, the greater part of Scotland is laid down in its true longitude position, but is mainly placed nearly half a mile too far north, with slight variations attributable to the Bonne's Projection. The main shift is, however, much greater than can be accounted for by the different projection, whilst in the vicinity of the Anglo-Scottish border what should be straight co-ordinate lines are found to take on snake-like forms. These discrepancies cannot be fully explained, as no evidence has been found to show exactly how the reduced Scottish quarter-inch sheets were laid down; but although it is a commonplace that independently drawn maps of adjoining areas will not join precisely, it seems to me that the Scottish maps could have been positioned more accurately, and if so would have needed less fudging at the border.

4. The Original (One-inch Index) Series

Before considering the projection, or indeed any other aspect, of the very first Ordnance Survey map on the scale of ten miles to one inch, the reader should attempt to put him/herself into the state of knowledge of the shape of England and Wales in 1817; it is not particularly easy. I do not detract at all from the better efforts of the county surveyors, but there was no totally accurate knowledge of the disposition of any given area of the country until the relevant one-inch sheet was published. Nor was there any framework, beyond the area currently reached by the Principal Triangulation, upon which a small-scale map could be based. Thus one early map showing the progress of the survey[4] has some northern coasts badly displaced, and although north-east Norfolk is only two miles too far west this brings it wholly within a standard sized Old Series sheet 68; no doubt a similar map was used for scheming the main body of the sheet lines, with the result that sheet 68 is the only known entity to consist of six quarters.

It also needs to be explained that no knowledge survives of the actual method of construction of many of the early maps, and it is not generally possible to determine the projection of such a map by inspection; the difference between the various projections used for constructing maps in the scale range we are concerned with is often less than the distortion which may occur in the paper during the printing process (or in fact in the drawing process in those days). It may however be possible to say from examination of the graticule, for example, which projections were not used, or it may be possible to make deductions from other items present on the map such as the sheet line pattern on an index sheet.

My main regret in compiling this appendix is that I

---

4. PRO MPHH 239.

have had to tackle this section without benefit of a detailed study of the projections, or lack of them, of the Old Series one-inch maps themselves, but those who have attempted to look into this matter, myself included, have been daunted by the sheer magnitude of the investigations which would be required. Nevertheless, an adequate assessment of the index sheets can still be made.

1. South sheet (I.1)

Remembering the deprecating way in which this map was referred to as merely an index, etc, on several occasions as described in Chapter I, it comes as no great surprise when the first thing to be discovered about its construction is that the "rectangle" is not rectangular. In fact there are several features which lead me to suggest that the initial engraving was entrusted to an apprentice. Notwithstanding this, it would appear that the map was intended to be drawn on an accurate projection for, whilst the south-west corner was 3mm higher than the south-eastern, the measurements from each corner to equivalent latitude degree markings were the same on both sides. All degrees of both latitude and longitude except one were marked by prick-holes in the copper plate, which show as dots on printed copies, and all but three of these had ticks added outside the neat line; of varying lengths on the initial state and altered to a fairly uniform half-inch (13mm) on the next state, they were nearly all engraved at the wrong slope, usually shallower than that of the meridian or parallel they were supposed to be lying along.

Concentrating first on the initial state of this sheet, almost certainly dated 1817 as explained in Chapter I, it could well be thought that the one-inch sheets indexed on this state and the topography contained within them, mainly south of the line of the River Kennet plus the county of Essex, were laid down on the same projection as the border (with the 3mm mis-plotting being somehow "accommodated"). There being no graticule on the copper, there was presumably one on a paper drawing upon which the one-inch limits were plotted before being transferred to the plate for engraving. This very first known proof is on a firm piece of paper showing little evidence of distortion other than direct paper shrinkage, and it is clear that the one-inch sheet rectangles do not come out as quite rectangular on the Index, and that the longer horizontal lines across groups of sheets on the same meridian exhibit very slight upward curvature thereon. These effects imply that the Index was on a different projection from that of the one-inch maps, with two equally unlikely possibilities; that the Index was on a conical-type projection with its origin in the south of France, or that there was something strange and hitherto unsuspected in the projections of the one-inch maps. But examination of the relative disposition of the western, central and eastern areas indicates a quite different projection again.

An attempt to trace the course of a latter-day straight line, a National Grid northing, across the face of the map produces some marked zig-zags; this is not unduly surprising as central southern England was surveyed and mapped after Essex and Kent to the east and Devon and Cornwall to the west, and had to be inserted between those areas. But a similar operation with a parallel of latitude makes it clear that its curvature is greater than would be the case if the Index were on Cassini's Projection, and implies a conical-type projection, possibly Bonne's, with its origin in the centre of England or of Great Britain. Returning to the degree ticks on the neat line, it is seen that the south-west corner lay on the meridian of 6°W and the south-east on that of 2°E so that the central meridian of this first ten-mile map was 2°W, giving shades of *déjà vu* to the National Projection of over a century later. From the 1820 state of the Index it is seen that the two northern corners were precisely on the parallel of 53°N, and from these various figures I calculate that the latitude of the mid-point of the south neat line would be latitude 49°54'35"N, if the Index were on Cassini's Projection, but 49°55'00" if it were on a Bonne's Projection with its origin on 53°N; the round figure seems the more in keeping with the exact degree values at the corners as noted above but, since a shift of the origin to 55°30'N would only drop the south neat line by 0.1mm, nothing positive about the origin is suggested.

On the next few states of the south sheet new one-inch sheets were added piecemeal but presumably on the original projection. However, when the plain neat line came to be deleted and a piano key border added farther out, c.1839-41, something most peculiar also happened; an attempt seems to have been made to change the projection in mid-stream, not of course that such a thing is possible, at least for the material already *in situ*. On the early states of this sheet carrying the new border, a score-mark of a revised meridian can be seen just over 1mm to the west of the 2°E meridian in the south-east corner; the new border itself carried no graduation whatsoever. I should also add that the neat line of the piano key border in the south-west corner was engraved just over 3mm farther below the old neat line than it was in the south-east corner, thus slightly over-correcting the lack of rectangularity in the original.

From further checks on sheet lines and the paths of National Grid lines and parallels of latitude across the sheet, it does seem that the material added with and later than the new border was on a different projection to the earlier material, very possibly Cassini's. Once again, however, discontinuities occur across Lincolnshire, surveyed and mapped in advance of the adjoining areas. Examining now the New Series index version of this map, it would appear that all the sheet lines of this series were engraved as though the entire basic map was on Cassini's Projection, even though the southern counties had almost certainly been drawn on a different projection. Added to the distortion which produced the zig-zag effects already referred to, this meant that the sheet lines, particularly in the south, were badly

Appendix 1

placed in relation to the topography they cut through. The neat lines of the new border which was engraved on the New Series version were realigned to be parallel and perpendicular to the meridian of Delamere, the central meridian of this series as mentioned in section 2.1.

2. Middle sheet (I.2), north sheet (I.3)

There being no graduation on these sheets from the start, all that is possible here is to examine the sheet patterns, bearing in mind that the situation was akin to that already examined in section 3 above - the one-inch sheets of England were on Cassini's Projection whilst those of Scotland at the time were on Bonne's Projection. As the sheet lines in both countries appear as true rectangles, it seems that we again have the case that the two areas were drawn on the two different respective projections, both appearing together on the middle sheet; in this instance, however, the two countries seem to have been accurately laid down relative to each other. As with the south sheet (New Series), and with the subsequent twelve-sheet series, the neat lines of the middle and north sheets of the New Series index were parallel and perpendicular to the meridian of Delamere.

---
5. The Mainly Mythical Projection

G.B.Airy, later Sir George Airy, Astronomer Royal 1835-81, made a number of notable contributions directly and indirectly to the scientific work of the Ordnance Survey. His name appears very early in this appendix as having calculated the figure of the Earth, or spheroid, which has been used for the vast majority of all OS surveys and mapping; it was first published on 17 August 1830 when Airy was still Plumian Professor of Astronomy at Cambridge, but is thought to have been communicated to the Survey a little before that.

During the summer of 1861 Airy's active mind thought up an idea for an improved map projection[5] and this was published by him (1861) with the name of a Projection by Balance of Errors; amended slightly by James and Clarke (1862) it is known today as Airy's Projection. Plainly described by Airy as for maps of a very large extent of the Earth's Surface, and recommended by more than one authority as deserving of more attention, it has in fact received very little use, no doubt due to the rather complex computations necessary to construct it.

However, the then Major C.F.Close stated (1901) that it had been used for the Ordnance Survey ten-mile map of England (note specifically England) and that statement has been repeated by many authors right to the present day; I have not been able to trace any prior such reference. But the only ten-mile map of England published between 1861 and 1901 was that of **Rivers and their Catchment Basins** 1868 (p.51) and whilst, as remarked in 4 above, it is not possible to determine absolutely the projection of such a map by inspection in the absence of any details of its construction, it is inconceivable that this map was on Airy's Projection. For, as shown in Chapter I, Sir Henry James was personally involved in the conception of the 1868 map, and Sir Henry knew very well that Airy's Projection was only intended for very small scale maps since he had invented a similar projection himself and he and Airy had corresponded about the two projections, each defending his own as the better.[6] Furthermore, the form of Close's reference in 1901 seems to imply a regular map rather than a particular thematic issue.

Nor was Close's remark apposite to a map then in the course of preparation, even had we not known that the 1903 map was on a different projection for England (and another for Scotland), so to what then was Close referring, for we know he would not have made such a statement lightly. It would seem that he must have been repeating something garnered during his first spell of duty at the OS Office in 1897-98. I now move forward to 1934 when Brigadier H.St J.L.Winterbotham as Director General wrote:[7]

It has always been, in my time, a tradition that the ten mile map of Great Britain (sic) *was originally upon Airy's projection. At the end of my time as OTT here I began to be extraordinarily doubtful,* and requested the Research Officer, Harold L.P.Jolly, to "*substantiate*". Jolly submitted a lengthy paper effectively coming to similar conclusions to those above, but also quoting a minute by Major Alan Wolff in 1919:

There are very old drawings in the MS store of an index of old series 1-inch maps on 10 mile scale which is quite different from any of the existing copper plates and it seems very probable that this may be on Airy's projection but there is nothing on the drawings to indicate the central point or the circular area included in the projection".

Jolly dismissed Wolff's suggestion, but without adequate reason apparent from this distance of time, and the "*very old drawings in the MS store*" may well be the source for Close's 1901 statement; no other possible source is currently evident.

However, as to the basic "tradition", Winterbotham accepted that "*The myth of the Airy Projection is now laid*", but also handed down the decision that Jolly's memorandum "*washes too much soiled linen for publication.*" So it is only now that it appears in print for the first time that no Ordnance Survey map on the ten-mile scale was ever published on Airy's Projection. I have to add that particular thanks are due to Dr Richard Oliver for first bringing to light the PRO document detailing this matter.

---

5. Cu RGO 6/435,475.
6. Cu RGO 6/472.
7. PRO OS 1/144.

6. The Irish Ten-mile Maps

1. Larcom's map, 1838 (II.1)

Professor John Andrews[a] has not been able to find any record of the actual method of construction of this map, and the chances of anyone else so doing must be regarded as microscopic; he does however suggest, no doubt rightly, that the mathematical basis would have been set out by Captain Joseph Portlock. The only positive thing that can be said about the projection is that the central meridian was 7°50'W; otherwise, even less can be learned from this map than were a complete graticule present. Speculation as to the projection used can only be based on what was fashionable at the time, and even the projections used for related scales were not particularly relevant at this date; its place is therefore in the realm of day-dreams, and not in this appendix.

2. Sir Henry James's world series (Appendix 2)

The projection used by Sir Henry for this ambitious project had the best contemporary documentation (see Appendix 2) of any of the early projections employed by the Ordnance Survey, but the material has to be read with circumspection and translated into modern terms, the main trap being that two slightly different projections were involved, whilst a third had also been suggested by Sir Henry for world-wide use for larger scale maps. The latter, which he distinguished (James 1868) as "the purely Tangential Projection", was that known today as the Simple Polyconic Projection, and from this James O'Farrell evolved an alternative form of construction to produce the Rectangular Tangential Projection, today's Rectangular Polyconic Projection; these names reflect the feature that the meridians and parallels cut at right angles, but the projection is not conformal.

It was first used for very small scale maps of whole continents and the world, but when Sir Henry James proposed its adoption for a systematised scheme of sheets to cover the world at the ten-mile scale it was again O'Farrell who calculated a modification, claimed to reduce the average distortion of projected distances in both principal directions to zero; this Modified Rectangular Polyconic Projection is not to be confused with one written up by G.T.McCaw in 1921. O'Farrell's published computations only permitted the construction of sheets conforming with the standard scheme of sheet lines detailed in Appendix 2; they were calculated on the Clarke 1858 spheroid, and the centre line of each sheet was the central meridian of its projection. Across each sheet the meridians of longitude were drawn straight, so that the convergence of a particular meridian on that sheet was constant, but on the adjoining sheet towards the Equator the angle would be smaller and on the next sheet away from the Equator it would be greater.

3. Sir Henry James's island map, 1868 (II.2)

This single sheet map of Ireland was not on a proper projection; it was formed by butting together the three Irish sheets of Sir Henry's world scheme, each <u>individually</u> on the projection described above, leaving the north-east corner to be filled in separately. Consequently there were changes in the directions of the meridians (except for the original central meridian of 9°W) across the sheet join along 55°N, producing slight dog-legs in those lines. The maximum kink occurred in the sheet edge meridian of 6°W, and although the change of direction was only 0°09', this can be clearly seen with the aid of a straight-edge on those maps which retain a graticule.

A new rectangular border was constructed around the island of Ireland with its west and east neat lines parallel to the meridian of 8°W, adopted in the previous decade as the central meridian for the one-inch series of Ireland (on a different projection, see below). A graduation was established in this border, presumably derived from its cuts across the graticule, and this was completed, without any change in the spacing of the ticks, round the initially blank north and east of 55°N, 6°W; by joining the relevant ticks to the existing graticule on the rest of the map, a rather distorted framework was produced within which the Firth of Clyde, Sound of Jura and adjoining areas were drawn in skeleton form.

4. Johnston's map, 1905 (III.2)

This map was constructed on what had become the regular national projection for Ireland, originally adopted for the country's one-inch maps, almost certainly on the recommendation of Captain William Yolland. These were constructed on Bonne's Projection on the Airy spheroid, on the origin of 53°30'N, 8°W, the unit of calculation being the foot; the radius of projection of the initial parallel of latitude, 53°30'N, was 15,516,209.8 feet, and the position of the origin was just in the waters of Lough Ree on Ireland one-inch Sheet 98. The co-ordinates on Bonne's Projection of the neat lines of the latter were:-

```
                35,720 feet north
27,040 feet west              68,000 feet east
                27,640 feet south
```

The above elements specify component (A) for the Ireland one-inch series, and in my 1991 article referred to in section 3 above I explained the incorrect use of a component (B) for compiling that series from the six-inch maps. However, the maximum error resulting from this misuse was 0.004 inch (0.10mm) on the one-inch scale, which is almost invisible, and becomes totally invisible when reduced again to the ten-mile scale. Johnston's map may therefore be said without reservation to be drawn on the Bonne's Projection specified above.

---

8. Personal discussion.

Appendix 1

5. 1:625,000 Monastic Ireland, 1960 (III.3)

This was the first ten-mile map of Ireland to be presented on the metric version of the scale, on the Transverse Mercator Projection of Ireland, and carrying the Irish National Grid; there was however no mention of the latter fact on the face of the map and numbering of the grid lines was limited to the 10km figures, repeating every 100km across the map. The attached pages of text contained brief notes on the lettering of the 100km squares (known as Sub-zones) and on grid references; for the second edition the notes were longer but rather convoluted. The defining elements of the projection and grid as first introduced were:-

Projection	Transverse Mercator
Spheroid	Airy*
Unit	International Metre*
True Origin	53°30'N, 8°W.
False Co-ordinates	200,000 m.E, 250,000 m.N.
Scale Factor	1

* 1 foot of Bar O_1 = 0.304,800,749,1 metre

The actual sheet line co-ordinates are:-

West neat line	10,000 metres east	(numbered 1)
East neat line	370,000 metres east	(numbered 7)
North neat line	470,000 metres north	(numbered 7)
South neat line	10,000 metres north	(numbered 1)

When the retriangulation of Great Britain was carried out, from 1935 onwards, it was decided early on to avoid disturbance to existing material as far as possible, and steps were taken to carry the scale and orientation of the old Principal Triangulation through to the new Primary Triangulation and new mapping based thereon. But after completion of the retriangulation of Ireland in 1964, and its adjustment in the following year, it was decided between the Irish surveys to alter the scale of their triangulation and mapping to bring distances determined therefrom into line with electronically measured true distances. This was effected by altering the scale factor of the projection and grid from unity to 1.000,035 with an equal and opposite alteration in the dimensions of the fundamental spheroid, in order to avoid recalculation of the projection tables, co-ordinates etc. Reduced by 0.999,965,001,2 it was christened the Airy (Modified) spheroid.

Subsequently, it appears that Northern Ireland had second thoughts on the latter part of these alterations, and by 1983 the six counties had reverted to the use of the true Airy spheroid.[9] As the new scale factor could not be interfered with, the projection was concurrently re-christened the "Modified Transverse Mercator Projection"; this appears to me to be nonsense, there is no modification in the projection at all. I do not think either of the Irish surveys' solutions was right, and they were surely unnecessary today when software is available which can instantaneously calculate distances on the Earth, on in terms of other national or international datums. I therefore leave the listing of the projection elements to "as first introduced" above.

7. Military Grids

With the exception of the National Grid, which in this context may be regarded as a dual-purpose grid, all the military grids which appear on OS ten-mile maps are on different projections or different origins from the maps on which they are superimposed. An attempt to explain this apparent contradiction to the uninitiated has been made by me (Adams 1989b).

1. War Office Cassini Grid (military title: English Grid)

Projection	Cassini's
Spheroid	Airy (1 foot of Bar O_1 =
Unit	Metre 0.304,799,73 metre)
True Origin	Dunnose (50°37'03".748N, 1°11'50".136W)
False Co-ordinates	500,000 m.E, 100,000 m.N.
Scale Factor	1
Grid colour	Purple

Co-ordinates on this grid are known as War Office False Origin (familiarly Woffo) co-ordinates.

2. War Office Irish Grid (military title: Irish Grid)

Projection	Cassini's
Spheroid	Airy (conversion factor as in
Unit	Metre 7.1, the Benoit & Chaney relationship)
True Origin	53°30'N, 8°W.
False Co-ordinates	199,990 m.E, 249,975 m.N.
Scale Factor	1
Grid colour	Red

The false co-ordinates of the origin were originally 200,000 m.E., 250,000 m.N., but were altered after a War Office adjustment of trigonometrical station co-ordinates.

3. Continental grids

All of these grids of which portions fall in the sea on ten-mile maps of Great Britain are on the Lambert Conformal Conical Projection and have the metre as their unit; they have a variety of spheroids and scale factors and a full listing of their elements does not seem relevant here. The grids concerned and their colours are:-

French Lambert Zone I	Red
Nord de Guerre Zone	Blue
Northern European Zone III	Blue

---

9. *The Irish Grid*, OSNI Leaflet No 1, July 1983.

Appendix 2. Sir Henry James's Map of the World

Those who seek information on the plans of Sir Henry James to cover the land masses of the globe with maps at the ten-mile scale are referred to James (1860), James (1868), Mumford and Clark (1985) and Appendix 1 above. But a brief resumé here might prove useful. It was probably in 1859 that James began to investigate the possibility of a uniform system of mapping throughout the world at specific scales, coupled to a general map of the world at a single scale. A rectangular polyconic projection had been invented in February 1858, which James referred to as the Rectangular Tangential Projection. The formula was first published in James (1860), a paper in fact written by Captain A.R.Clarke. James chose for his general map a modified form of this projection, the technical aspects of which were written up in James (1868), probably mainly by James O'Farrell. Therein James listed Canada, New Zealand, Cape of Good Hope, India, China, Abyssinia and England as places where he made his general maps of the world using this projection and the ten-mile scale. To date only 21 sheets of North America (United States and Canada) and Europe are known to have been printed, with no reference to any others in OSR. Three Irish sheets were in active preparation, and there is evidence that a British plate was made, or at least started, entitled **London**. The sheet lines were graticule based, and were planned as 4° wide by 5° high from the equator to 50°, the width increasing to 6° thence to 70° and 12° to 85°, with common sheet limits throughout at 6°, 18° etc west and east of Greenwich. The increase to 6° above 50° was not applied to the continental European sheets, though it was to the Irish. Portrait format sheets minimised the distortions which distance from the central meridians magnified.

The known North American and European sheets were prepared by TDWO. Their reproduction was by the OS. The North American sheets should probably be seen as provisional editions, since several of them lack apparently standard features. Place and county names may be lacking or superficially engraved. Some have rouletted rather than finished boundaries. No known sheet has waterlining, but in time that was to be added at Southampton. Of particular relevance here is their common specification with James's 1868 map of Ireland. This is discussed on p.57. For comparison with the Old Series Index, see the specification table on p.9.

---

1. The North American group

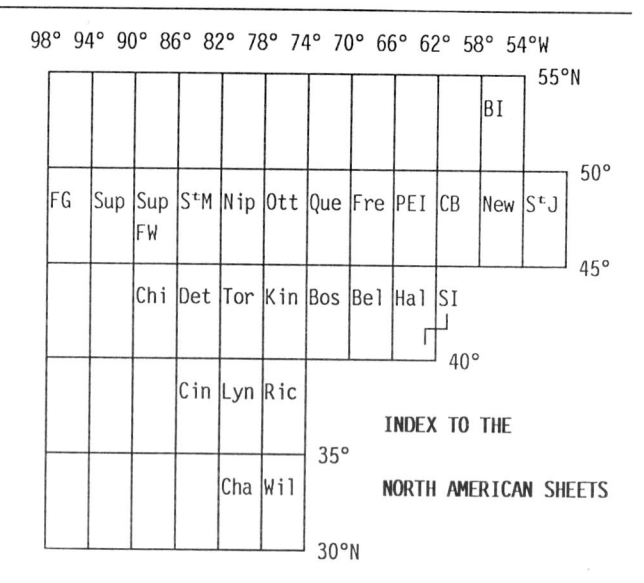

"*America, North, incl. Canada, Engraving, Drawing &c, Map of Part of, 10m to 1in*" (OSR/TDWO 1865). OSR 1866 notes 200 as the total number of copies printed, but it is unclear whether this was of each sheet. The number was unchanged in 1870. Sheets printed were not named.

98° 94° 90° 86° 82° 78° 74° 70° 66° 62° 58° 54°W

								BI			
FG	Sup	Sup FW	SᵗM	Nip	Ott	Que	Fre	PEI	CB	New	SᵗJ
			Chi	Det	Tor	Kin	Bos	Bel	Hal	SI	
				Cin	Lyn	Ric					
					Cha	Wil					

55°N
50°
45°
40°
35°
30°N

INDEX TO THE NORTH AMERICAN SHEETS

Representative dimensions (from **Quebec**, PRO WO 78/4649)
  Paper: 690mm W-E by 1000mm+ S-N
  Plate: 584mm W-E by 975mm S-N

Standard marginal information:
Top centre: [sheet title] / bottom centre: Scale Ten Miles to One Inch [scale-bar 10+100 miles] Constructed and Engraved at the Ordnance Survey Office Southampton in 1864 Colonel Sir Henry James R.E.,F.R.S.,M.R.I.A.&c. Director. Elsewhere: titles of adjacent sheets, name extensions carried over from the sheet

There is a graticule at 1° intervals, and a border graduated at 10' intervals

Standard features present (see col."E" on p.9)
B.1   Boundaries
  a   Rouletted, provisionally engraved
  b   State border
  c   County boundary (many linking rivers)
  d   Province border
  e   National border
B.2   Topographic features
  a   River (pecked when unsurveyed)
  b   Lake
  e   Canal
  h   Road (pecked if under construction)
  k   Track
  l   Built-up area
  q   County town (perhaps within built-up area)
  s   Small town
  t   Village, settlement (eg mission)
  w   Fortification
  nn   Railway
  vv   Site of battle
  ww   Bridge
  xx   Cliff

Appendix 2

B.5		Hydrographic features
	b	Sandbank, mudbank
	c	Underwater rock, often surrounded by B.5p
	f	Lighthouse (usually with L.H., L.Ho., L.Ho^s.)
	g	Lighthouse
	p	Danger area
C.1		Roman capitals
	b	City
	c	Large island
	g	Province
	h	State
C.2		Roman open capitals
	a	Marine name
C.3		Roman
	b	Minor island
	c	Principal headland
	d	Minor bay, harbour, basin
	g	Small town
	m	Important port (Quebec harbour, basin)
	n	Fortification
C.4		Roman sloping capitals
	b	Principal bay, channel, harbour, navigable river
	c	Large forest, moor, marsh, swamp
	h	Principal lake
C.6		Egyptian capitals
	b	Railway company name
	d	County
	f	County town
C.7		Egyptian
	a	Range of hills, peak
C.8		Egyptian sloping capitals
	a	Canal
	b	Important mountain range
	c	Extensive district
C.9		Italic
	b	River, including Source
	c	Canal
	d	Minor lake
	e	Minor headland
	f	Small island, rock
	g	Village, settlement (eg mission)
	k	Lighthouse
	m	Minor bay, channel, harbour, cove, strait, sound
	q	Important lowland topographic feature, eg swamp
	s	Road name
	cc	Industrial location: eg mine, salt work, mill
	dd	Reservoir
	ff	Railway junction
	gg	Date of battle
	hh	Military establishment: eg U.S. Arsenal
	jj	Portage
	kk	Inland harbour
	ll	Falls, rapids, spring, waterfall
	mm	Ferry
	nn	Railway tunnel

NB: All sheets known in two states have an identical variant: State 2 adds 1/633600 below scale-bar, 22mm bnl. All Lbl sheets carry this feature.

The following sheets appear on printed indexes: "+" sheets are in **Index to the Ten Mile Map of North America**

Belle Isle+ (58°-54°W, 50°-55°N)
James's schema would have had this sheet 6° wide, but the **Index** does not show this.   ?Not printed

Fort Garry (98°-94°W, 45°-50°N)   ?Not printed

Superior (94°-90°W, 45°-50°N)   ?Not printed

Superior+ / Fort William (90°-86°W, 45°-50°N)
Renamed by c.1872.   ?Not printed

St Mary+ (1864) (86°-82°W, 45°-50°N)
Graticule values lack degree and minute symbols
State 1: C-On NMC 51486 (EPD May 1868)
State 2: Lbl 69915 (99) (EPD Jan.1880)

Nipissing+ (1864) (82°-78°W, 45°-50°N)
No county names given
State 1: C-On NMC 51487 (EPD May 1868)
State 2: Lbl 69915 (99) (EPD Jan.1880)

Ottawa+ (1864) (78°-74°W, 45°-50°N)
No county names given
State 1: C-On NMC 51488 (EPD May 1868)
State 2: Lbl 69915 (99) (EPD Jan.1880)

Quebec+ (1864) (74°-70°W, 45°-50°N)
No county names given
State 1: C-On NMC 51489 (EPD May 1868)
State 2: Lbl 69915 (99), Lpro WO 78/4649 (EPD Jan. 1880)

Frederickton+ (1864) (70°-66°W, 45°-50°N)
No county names given
?State 1 not found.
?State 2: Lbl 69915 (99) (EPD Jan.1880)

Prince Edward Island+ (1864) (66°-62°W, 45°-50°N)
No county names given, rouletted county boundaries
?State 1 not found.
?State 2: Lbl 69915 (99) (EPD Jan.1880)

Cape Breton+ (1864) (62°-58°W, 45°-50°N)
No county names given, rouletted county boundaries
?State 1 not found.
?State 2: Lbl 69915 (99) (EPD Jan.1880)

Newfoundland+ (58°-54°W, 45°-50°N)   ?Not printed

St John+ (54°-50°W, 45°-50°N)   ?Not printed

Chicago+ (90°-86°W, 40°-45°N)   ?Not printed

Detroit+ (1864) (86°-82°W, 40°-45°N)
No county names given, no national border shown
State 1: C-On NMC 51490 (EPD May 1868)
State 2: Lbl 69915 (99) (EPD Jan.1880)

Toronto+ (1864) (82°-78°W, 40°-45°N)
Provisional (rouletted) national border shown
State 1: C-On NMC 51491 (EPD May 1868)
State 2: Lbl 69915 (99) (EPD Jan.1880)

Kingston+ (1864) (78°-74°W, 40°-45°N)
Manhattan Island, Philadelphia and the New Jersey coastline extrude into the borders. With county names, but rouletted county boundaries
State 1: C-On NMC 51492 (EPD May 1868)
State 2: Lbl 69915 (99) (EPD Jan.1880)

Boston+ (1864) (74°-70°W, 40°-45°N)
Cape Cod, Nantucket Island extrude into the border
State 1: C-On NMC 51493 (EPD May 1868)
State 2: Lbl 69915 (99) (EPD Jan.1880)

Belfast+ (1864) (70°-66°W, 40°-45°N)
No county names given. Cape Cod and Nantucket Island are omitted
?State 1 not found.
?State 2: Lbl 69915 (99) (EPD Jan.1880)

Halifax+ (1864) (66°-62°W, 40°-45°N)
No county names given. Sable Island is inset
?State 1 not found.
?State 2: Lbl 69915 (99) (EPD Jan.1880)

Lynchburg+ (1864) (82°-78°W, 35°-40°N)
The North Carolina section is largely blank
?State 1 not found.
?State 2: Lbl 69915 (99) (EPD Jan.1880)

Richmond+ (1864) (78°-74°W, 35°-40°N)
No county names given
State 1: Lpro FO 925/1837; PC (TJH 1867)
State 2: Lbl 69915 (99) (EPD Dec.1872)

Cincinnati, Charlestown and Wilmington are given in the margins as adjacent sheet titles, but it is unlikely that they were printed.

All Lbl sheets in 69915 (99) were acquired on 14 July 1923: they (and **Quebec** in Lpro WO 78/4649) are stamped "For War Department Purposes Only"

Relevant indexes:

Index to the Ten Mile Map of North America

Showing ten-mile sheet lines, outline, names and railways. The sheets named are marked "+" above

Lbl 69915 (99); Lpro WO 78/4676 (TJH 1867)

Index to the Ten-mile Maps of the World. (North America) [Plate] N° IV / Scale:- Four Equatorial Degrees to an Inch or = 1:17,531,000 [scale-bar 100+1000 miles] Constructed on the Rectangular Tangential Projection of the Spheroid by Mr J.O'Farrell of the Ordnance Survey, under the Direction of Colonel Sir Henry James,R.E., F.R.S.,&c. Director General of the Ordnance Survey

Engraved. Dimensions: paper 645mm by 825mm, plate 561mm by 704mm, neat line 451mm by 560mm
The North Pole is at the centre of the north neat line. The south neat line is the Equator from 130°W to 58°W. Meridians are drawn, where falling within the west and east neat lines, from 174°W to 18°W; parallels are completed only within these meridians
Showing coastal outline and main river systems

State 1 (?1868)

Changes to ten-mile sheets indexed, compared with the **Index to the the Ten Mile Map of North America**
1. Belle Isle lacking
2. Lynchburgh to Lynchburg

In James (1868)

State 2 (?1872)

Changes
1. Geographical and placenames added
2. Name of sheet diagram added, top right
3. Fort Garry (98°-94°W) sheet added
4. Superior (94°-90°W) sheet added
5. Superior (90°-86°W) renamed Fort William

Lpro OS 5/26 (THSL) (EPD Dec.187?2)

Index to the Ten-mile Maps of the World / Scale:- Four Equatorial Degrees to an Inch or = 1:17,531,000 [scale-bar 100+1000 miles] Constructed on the Rectangular Tangential Projection of the Spheroid by Mr J. O'Farrell of the Ordnance Survey, under the Direction of Colonel Sir Henry James,R.E.,F.R.S.,&c. Director General of the Ordnance Survey

Appendix 2

Engraved. Dimensions: paper 666mm by 847mm, plate 568mm by 704mm, neat line 528mm by 562mm.

The North Pole is at the centre of the north neat line. The south neat line is the Equator, covering 84° longitude. Parallels are completed to all neat lines, meridians shown covering 200° longitude. No topography.

Lpro OS 5/26, OS 5/100 (both THSL, EPD Aug.1872)

---

2. The Irish group

The drawing for these maps was a topographical layer which William Harvey added to his **Plan of the Catchment Basins of the Rivers of Ireland.** (for details of the original, see pp.56,69). Harvey's model for the specification of the additional material was one of the North American sheets listed above. Detail includes outline, roads, railways, canals, county boundaries, the Curragh, lighthouses and location dots for towns and villages. The military origins of the map are reflected in the inclusion of security installations. Harvey's additional lettering also followed the North American map, with Gothic lettering introduced for the "Seven Churches" antiquity. Counties are not named. The railways drawn on the plates were as Harvey depicted, except that the Ballingrane-Newcastle West line was omitted. This system is identical to State 1 of the island map (II.2), where they are listed.

Harvey's drawing carries no border. The North American pattern came to be used for the Irish sheets. Harvey also drew no waterlining, but that was added to the Irish sheets in advance of the North American ones (where it seems in the event never to have been drawn), modelled on OS one-inch practice. The three sheets covering Ireland were destined never to be published. Instead the plates were electrotyped together to form the 1868 island map of Ireland (see p.57). Initially this carried identical characteristics to these sheets.

Additions to standard features (compared with the North American sheets)
B.2 Topographic features
  p Medium town (in Great Britain only)
  v Barracks: The Curragh Camp
B.4 Waterlining
C.10 Gothic
  a Antiquity: Seven Churches (Co. Wicklow)

Changes from standard features of North American sheets
1. Province boundaries and names absent at first
2. National, state boundaries unused
3. Concentric circles county town symbol not used
4. Egyptian sloping capitals (for mountains) not used
5. Site of battle symbol not used
6. Track symbol not used
7. Underwater rock, heavy dot for lighthouse, not used
8. These plates are 6° wide, not 4°

Changes from Harvey's drawing
1. Graticule sheet lines used, and border added
2. County names are included
3. Lettering used for islands and headlands changed
4. Ballingrane-Newcastle West railway is missing
5. The Scottish coastline west of 6°W is added

Documentary source: NA OS 5/3007

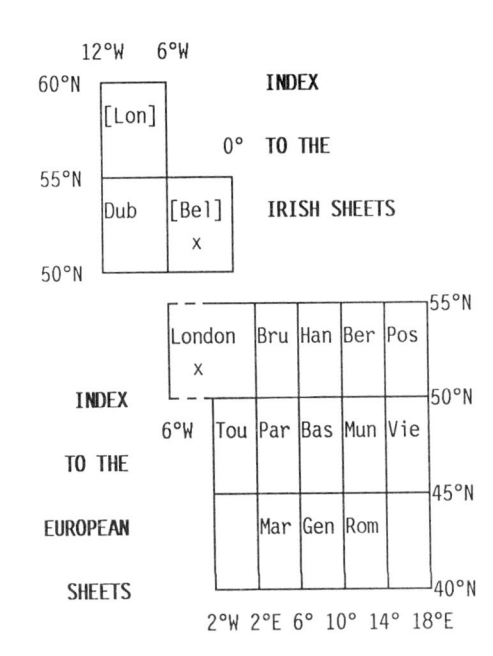

```
        12°W   6°W
   60°N ┌─────┬─────┐     INDEX
        │[Lon]│     │
        │     │     │  0°  TO THE
   55°N │─────┼─────│
        │ Dub │[Bel]│     IRISH SHEETS
        │     │  x  │
   50°N └─────┴─────┘
```

```
                                                    55°N
           ┌──────┬───┬───┬───┬───┐
           │London│Bru│Han│Ber│Pos│
   INDEX   │  x   │   │   │   │   │
           ├──────┼───┼───┼───┼───┤ 50°N
           │      │Tou│Par│Bas│Mun│Vie
   TO THE  │ 6°W  │   │   │   │   │
           ├──────┼───┼───┼───┼───┤ 45°N
   EUROPEAN│      │   │Mar│Gen│Rom│
           │      │   │   │   │   │
   SHEETS  └──────┴───┴───┴───┴───┘ 40°N
            2°W 2°E 6°  10° 14° 18°E
```

[?Londonderry (no title)] (12°-6°W, 55°-60°N)

Plate dimensions (1): 710mm W-E by 1030mm S-N
                 (2): 706mm W-E by 1028mm S-N

The Ireland area, a complete graticule and border was engraved in 1867 in Dublin. Some five miles of the Antrim coast lying east of 6°W appears as an extrusion. With waterlining, but no marginalia. The first copy has the coastal outline of the Hebrides added in Southampton in 1869. This is not present on the second plate. OSI files refer to it as Sheet 1.

Plates (two copies): Dna OS 106. No print recorded

Dublin (12°-6°W, 50°-55°N)
Bottom centre, 16mm bnl: Scale_Ten Miles to One Inch [scale-bar 10+100 miles (21mm bnl)]; 28mm bnl: Constructed and Engraved at the Ordnance Survey Office Phoenix Park Dublin in 1867, under the direction of Captain Wilkinson,R.E. Colonel Sir Henry James,R.E.,F.R.S.,M.R.I.A. &c. Superintendent. 36mm bnl: The Outline by John West, the Writing by John F.Ainslie, the Water by James Jones

Plate dimensions: 708mm W-E by 1026mm S-N (copper)

With a 1° graticule and a border graduated at 10' intervals. Waterlining breaks the eastern border for 188mm off Dublin. OSI files refer to it as Sheet 2.

Plate: Dna OS 106. A pull from the plate is in Dos: it is planned to transfer it to Dna. Not published

---

[?Belfast (no title)] (6°W-0°, 50°-55°N)

Plate dimensions: 706mm W-E by 1027mm S-N (copper)

The Ireland area only, a complete graticule and border, engraved 1867. With waterlining, but no marginalia. OSI files refer to it as Sheet 3.

Plate: Dna OS 106. No print recorded

---

3. The European group

"*Europe, Map of Part of, compiling, 10m to 1in*" (OSR/TDWO 1866). For sheet index, see p.187.

Standard marginal information:
Top centre: [sheet title] / bottom centre: Scale Ten Miles to One Inch [scale-bar 10+100 miles] Constructed on a general projection of the Sphere and Photozincographed at the Ordnance Survey Office, Southampton. Major General Sir Henry James R.E.,F.R.S.,&c. Director 1870. Elsewhere: titles of adjacent sheets, name extensions carried over from the sheet.

With a graticule at 1° intervals and values in the border, drawn to a revised specification

Brussels (1870) (2°-6°E, 50°-55°N)

Hanover (1870) (6°-10°E, 50°-55°N)

Berlin (1870) (10°-14°E, 50°-55°N)

Paris (1870) (2°-6°E, 45°-50°N)

Basle (1870) (6°-10°E, 45°-50°N)
Additional wording, above the title: Rectangular Tangential Projection of the Spheroid 1/633600

Munich (1870) (10°-14°E, 45°-50°N)

All known copies are in Lbl 1030 (368). These are hand coloured copies, acquired 12 March 1921. Five are date stamped 10.8.1870 by the Geographical Department War Office: the date on **Basle** is 1.9.1870.

NB: London, Posen, Tours, Vienna, Marseilles, Genoa and Rome are given in the margins as the titles of adjacent sheets, but whether any of them were printed is unknown. **Brussels**, **Hanover** and **Berlin** are not drawn to the specification given in James (1868), which suggests that sheets above 50°N have 6° of longitude. How London was to have accorded with the unnamed Belfast sheet, which was drawn to 6° of longitude, 6°W-0°, is unclear. There may be some explanation in a letter that Cameron in Southampton wrote to Wilkinson in Dublin on 1 November 1869 concerning the Ireland plates:

....*for the other containing the Down & Antrim coast I have no use, as the part of England which comes in the Sheet has been engraved in another plate.*[10]

It is possible therefore that at Southampton a plate extending west from 2°E covering all Great Britain between 50°N and 55°N was engraved, presumably entitled London. James (1868) also refers to the existence of this "England" sheet, but nothing is known of it.

---

10. NA OS/5 3007.

Appendix 3. County and other indexes

1. England and Wales

South Eastern Counties of England (OSPR 5/1882)
Counties of Gloucester, Hereford, Monmouth, Somerset, Warwick, Wilts, Worcester, South Wales (1/1883)
South Eastern Counties of England (9/1883)
Counties of Nottingham, Leicester, Lincoln, Rutland, Northampton, Bedford, Huntingdon, Cambridge, Norfolk and Suffolk (4/1884)
Counties of Chester, Derby, Shropshire, Stafford and North Wales (3/1885)
Counties of Cornwall, Devon and Dorset (3/1885)
None found. *NB*: Other counties appear at other scales.

2. Scotland

1. **Index Shewing the State of the Ordnance Survey of Argyleshire and Buteshire** Published on a Scale of Six Inches to a Mile on the... / Scale 10 Miles to One Inch [Scale-bar 10+20 miles]. Copy: Eu
2. Outer Hebrides (existence uncertain)
Not found.
3. **Index Shewing the State of the Ordnance Survey of the Shetland and Orkney Islands** Published on a Scale of Six Inches to one Mile: Scale of this Index [Scale-bar 10+30 miles]. Copy: Eu

Undated and presumably unpublished. Coastlines are sketched. The Argyleshire and Buteshire map also shows road alignments. Named triangulation points, county and parish boundaries, county and parish names, and some important geographical names are shown.

These indexes show one-inch and six-inch sheet lines, and were probably printed for internal use during the survey of the six-inch map, c.1860-70s. There are tables of reference listing the several compilation and production procedures of the one-inch, six-inch and twentyfive-inch map, with boxes to be filled in showing the month or quarter of the completion of each stage.

The Argyleshire and Buteshire map has only lists for Detail Survey and Contouring procedures: that of the Shetland and Orkney Islands is undoubtedly later and altogether more complicated, with sections for Hill Sketching &c., Levelling & Contouring, Detail Survey, Boundaries, Triangulation, Publication One Inch Scale, Publication Six Inch Scale and Publication 1/2500 Scale. Mainland counties are known treated at four-mile and seven-mile scales. Perhaps there are others, but it is quite probable that the Outer Hebrides was also covered at the ten-mile scale. Lit: Hellyer (1991a)

Ten Mile Index to the Ordnance Survey of Scotland, on the one inch, six inch & 1/2500 Scales / Photozincographed and Published at the Ordnance Survey Office, Southampton 1889 (OSPR 2/1889)
  Section I   ARRR 6d. Not found
  Section II   ARRR 6d. Not found
  Section III ARRR 6d. Copies: Lse,NTg
Outline diagrams, no roads, no rivers. With county boundaries and names, principal placenames, sheet lines and numbers. With a table showing the areas of counties

NB: See also IV.2 Supplement 4

3. Ireland

"Small indexes 10 miles to an inch are published at 2d each of counties Antrim, Clare, Cork, Donegal, Galway, Kerry, Kilkenny, Limerick, Fermanagh, Mayo" (OSC 1898). Additional counties listed in OSC 1900 were Armagh, Down, Dublin, Kildare, King's County, Londonderry, Monaghan, Queen's County, Tipperary, Tyrone, Waterford, Wexford, Wicklow. OSC 1901 advertised "of all counties"

Combined Indexes to the OS Maps of County........, on the 1-inch and 6-inch scales
  *Cork (OSPR 9/05)
  +Clare, Down, Galway, Limerick, Mayo (1/06)
  +Antrim, Donegal, Kilkenny, Tipperary, Wexford (1/07)
  *Armagh, Carlow, Dublin, Fermanagh, King's County, Leitrim, Londonderry, Louth, Queen's County, Roscommon, Sligo, Tyrone, Waterford, Westmeath, Wicklow (12/07)
  *Cavan, Kildare, Longford, Meath, Monaghan (1/08)
  Price 1d each. *NB*: Kerry is not recorded

+   In Dtc with ARRR
*   Most in Dtc with Phoenix Park embossed stamp

4. Other indexes

Index to Map of Frontier of Canada East Shewing divisions of work Surveyed & Revised by Officers for the Quarter Master General's department in Canada / Scale Ten Miles to One Inch [scale-bar 10+100 miles] Zincographed at the Topographical Department of the War Office Southampton Colonel Sir Henry James R.E.,F.R.S.,&c. Director. 1866

Sheet limits: 75°-70°W, 45°-47°N. With a graticule at 30' intervals. Nineteen of the graticule sectors are numbered. Mapping as on **Ottawa** and **Quebec** (see p.186), without the outer border and state names. Boundaries of areas surveyed 1863-1866 are marked in red

Lpro WO 78/4676

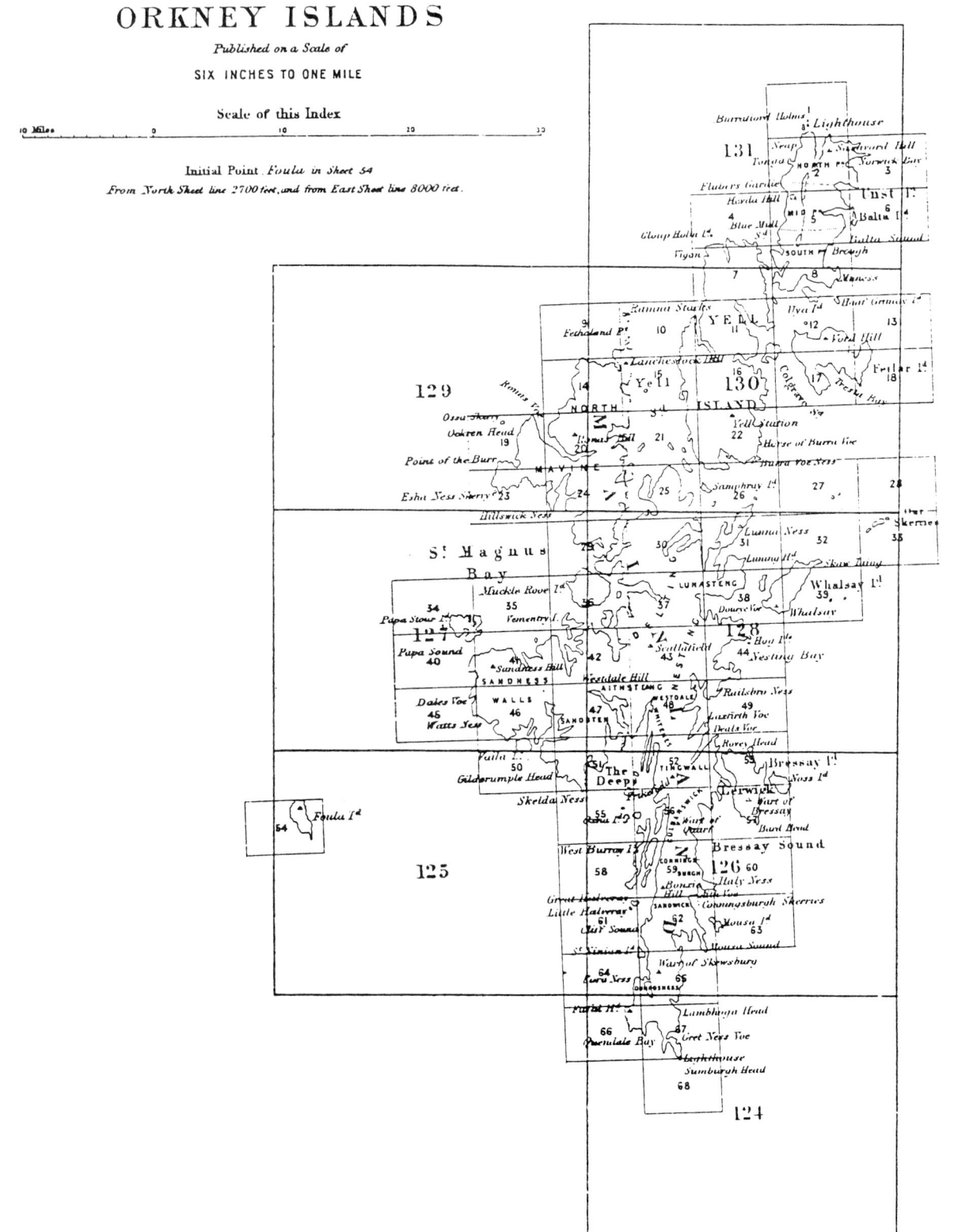

Plate 15. Extract from **Index Shewing the State of the Ordnance Survey of the Shetland and Orkney Islands.** Reproduced by courtesy of Edinburgh University Library.

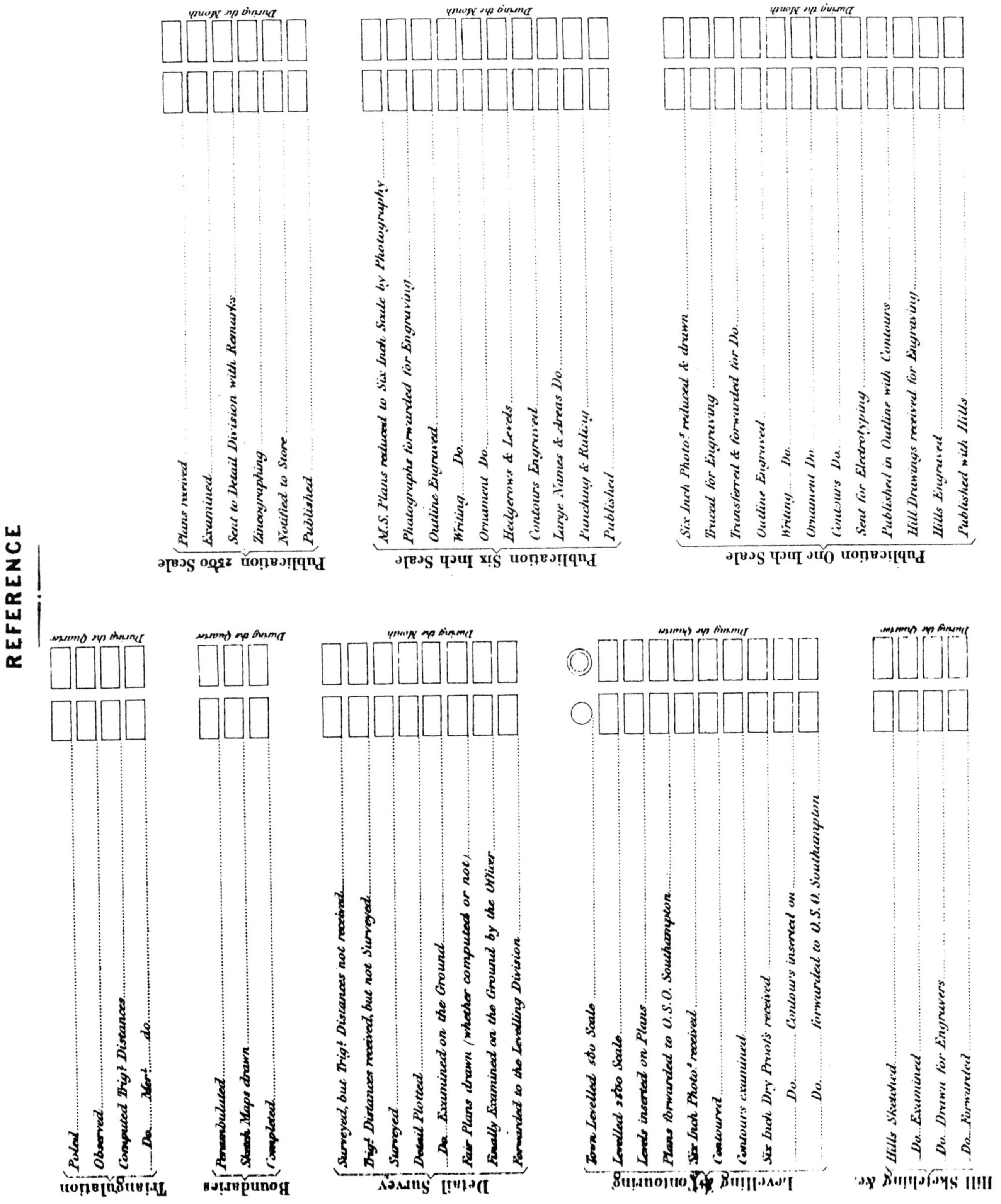

REFERENCE

Triangulation
Poled
Observed
Computed Trig! Distances
Do. Mer! do.
During the Quarter

Boundaries
Perambulated
Sketch Maps drawn
Completed
During the Quarter

Detail Survey
Surveyed, but Trig! Distances not received
Trig! Distances received, but not Surveyed
Surveyed
Detail Plotted
Do. Examined on the Ground
Fair Plans drawn (whether computed or not)
Finally Examined on the Ground by the Officer
Forwarded to the Levelling Division
During the Month

Levelling & Contouring
Town Levelled 500 Scale
Levelled 2500 Scale
Levels inserted on Plans
Plans forwarded to O.S.O. Southampton
Six Inch Photo! received
Contoured
Contours examined
Six Inch Dry Proofs received
Do. Contours inserted on
Do. forwarded to O.S.O. Southampton
During the Quarter

Hill Sketching &c.
Hills Sketched
Do. Examined
Do. Drawn for Engravers
Do. Forwarded
During the Quarter

Publication 2500 Scale
Plans received
Examined
Sent to Detail Division with Remarks
Zincographing
Notified to Store
Published
During the Month

Publication Six Inch Scale
M.S. Plans reduced to Six Inch Scale by Photography
Photographs forwarded for Engraving
Outline Engraved
Writing Do.
Ornament Do.
Hedgerows & Levels
Contours Engraved
Large Names & Areas Do.
Punching & Ruling
Published
During the Month

Publication One Inch Scale
Six Inch Photo! reduced & drawn
Traced for Engraving
Transferred & forwarded for Do.
Outline Engraved
Writing Do.
Ornament Do.
Contours Do.
Sent for Electrotyping
Published in Outline with Contours
Hill Drawings received for Engraving
Hills Engraved
Published with Hills
During the Month

Plate 16. The legend from plate 15. Reproduced by courtesy of Edinburgh University Library.

Appendix 4. Unique numbers

Maps are at 1:625,000 unless otherwise noted. Short form titles are used. Number allocated but unused: ≡

1. MTCP Series numbers

Allocated, probably on initiation of the system in 1947
- 3801 Great Britain (base map) 1 1948
- 3802 Great Britain (base map) 2 1948
- 3803 Administrative Areas 1 1947
- 3804 Administrative Areas 2 1947
- 3805 Population Density, 1931 1 1949
- 3806 Population Density, 1931 2 1948
- 3807 Land Utilisation 1≡
- 3808 Land Utilisation 2≡
- 3809 Physical 1≡
- 3810 Physical 2≡
- 3811 Types of Farming 1≡
- 3812 Types of Farming 2 1948
- 3813 Types of Farming (north strip)≡
- 3814 Railways 1≡
- 3815 Railways 2≡
- 3816 Roads 1≡
- 3817 Roads 2 1948
- 3818 Land Classification 1≡
- 3819 Land Classification 2≡
- 3820 Coal and Iron 1≡
- 3821 Coal and Iron 2≡
- 3822 Topography 1≡
- 3823 Topography 2 not used until 1951
- 3824 Population of Urban Areas 1≡
- 3825 Population of Urban Areas 2≡
- 3826 Iron and Steel 1≡
- 3827 Iron and Steel 2≡
- 3828 Electricity: Statutory Supply Areas 1≡
- 3829 Electricity: Statutory Supply Areas 2≡
- 3830 Vegetation 1≡
- 3831 Vegetation 2≡
- 3832 Population Total Changes, 1921-31 1 1949
- 3833 Population Total Changes, 1921-31 2 1949
- 3834 Population Changes by Migration, 1921-31 1 1949
- 3835 Population Changes by Migration, 1921-31 2≡
- 3836 Population Total Changes, 1931-39 1 1949
- 3837 Population Total Changes, 1931-39 2 1949
- 3838 Population Changes by Migration, 1931-39 1 proof
- 3839 Population Changes by Migration, 1931-39 2 proof
- 3840 Gas and Coke 1 1947 proof only
- 3841 Gas and Coke 2 1947 proof only
- 3842 Rainfall Annual Average 1881-1915 1 1949
- 3843 Rainfall Annual Average 1881-1915 2 1949
- 3844 Limestone 1≡
- 3845 Limestone 2≡
- 3846 1:1,250,000 National Parks 1947
- 3847 National Parks extract maps 1947
- 3848 1:253,440 Greater London Region Housing Diagram≡
- 3849 Ministry of Labour Local Office Areas 1
- 3850 Ministry of Labour Local Office Areas 2
- 3851 1:1,250,000 Conservation of Nature 1947

2. MTCP Series numbers: extended list

- 3852 Rainfall Annual Average 1 reprinted 1951
- 3853 Rainfall Annual Average 2≡
- 3854 Gas and Coke 1 1951
- 3855 Gas and Coke 2 1951
- 3856 Administrative Areas 1 reprinted 1951

3. Miscellaneous sequence

Begun at 3201 in 1947, and reached 38xx area in 1953-54
- 3252 1:1,250,000 Land required 1947 proof
- 3363 1:1,250,000 Great Britain reprinted 1948
- 3370 Areas of Gas Boards 1 (MFP issue) 1948
- 3371 Areas of Gas Boards 2 (MFP issue) 1948
- 3382 Areas of Gas Boards 1 (published issue) 1949
- 3383 Areas of Gas Boards 2 (published issue) 1949
- 3393 1:1,250,000 Index: NP, 5th, Popular Edns 1950
- 3394 Monastic Britain 2 1950
- 3395 Monastic Britain 1 1950
- 3618 Monastic Britain 1 reprinted 1951
- 3624 Ancient Britain 1 1951
- 3625 Ancient Britain 2 1951
- 3748 Jurisdiction of the Durham Palatine Court 1952
- 3781 Large Scale Field Programme 2 1953
- 3794 1:1,250,000 Index: Popular, NP, 7th Series 1952
- 3802 Large Scale Field Programme 1 1953
- 3897 Population Changes by Migration, 1939-47 1 1954
- 3898 Population Changes by Migration, 1939-47 2 1954
- 3899 Population Total Changes, 1939-47 1 1954
- 3900 Population Total Changes, 1939-47 2 1954
- 3926 Population of Urban Areas, 1951 2 1954
- 3927 Monastic Britain (Second Edition) 2 1954
- 3944 Local Accessibility 1 1955
- 3945 Local Accessibility 2 1955
- 3965 Large Scale Field Programme 1 reprinted 1955
- 3966 Large Scale Field Programme 2 reprinted 1955
- 3970 "Ten-Mile" Map 1 1955 (coloured and outline)
- 3971 "Ten-Mile" Map 2 1955 (coloured and outline)
- 3977 Limestone 1 1955
- 3978 Limestone 2 1955
- 3979 Types of Farming 1 reprinted 1955
- 3980 Monastic Britain (Second Edition) 1 1955

Appendix 4   193

4063	Land Utilisation 2 reprinted 1956	105	1:1,250,000 Six-inch NG Series 1.7.1966
4125	Limestone 2 reprinted 1957	106	1:1,250,000 Six-inch NG Series 1.1.1967
4156	Population of Urban Areas 1951 1 1957	107	1:1,250,000 Six-inch NG Series 1.7.1967
4194	Large Scale Field Programme 1 1958 (1955 base)	108	1:1,250,000 Six-inch NG Series 1.1.1968
4195	Large Scale Field Programme 2 1958 (1955 base)	109	1:1,250,000 Six-inch NG Series 1.7.1968
4224	Ancient Britain 1 reprinted 1959	110	1:1,250,000 Six-inch NG Series 1.1.1969
4225	Ancient Britain 2 reprinted 1959	111	1:1,250,000 Six-inch NG Series 1.7.1969
4266	Population Density 1951 2 1960 (1955 base)	112	1:1,250,000 Six-inch NG Series 1.1.1970
4288	Population Density 1951 1 1961 (1955 base)	113	1:1,250,000 1:10,000-1:10,560 NG Series 1.7.1970
4353	Land Utilisation 1 reprinted 1962	114	1:1,250,000 1:10,000-1:10,560 NG Series 1.1.1971
4354	Coal and Iron 2 reprinted 1962	115	1:1,250,000 1:10,000-1:10,560 NG Series 1.7.1971
4355	Iron and Steel 2 reprinted 1962	116	1:1,250,000 1:10,000-1:10,560 NG Series 1.1.1972
		117	1:1,250,000 1:10,000-1:10,560 NG Series 1.7.1972
		118	1:1,250,000 1:10,000-1:10,560 NG Series 1.1.1973

4. Indexes to large scale maps

42	Index to 6-inch maps on NG Sheet Lines 1,2 1950	119	1:1,250,000 1:10,000-1:10,560 NG Series 1.7.1973
69	1:1,250,000 Six-inch NG Series 1.7.1956	120	1:1,250,000 1:10,000-1:10,560 NG Series 1.1.1974
70	1:1,250,000 Six-inch NG Series 1.10.1956	121	1:1,250,000 1:10,000-1:10,560 NG Series 1.7.1974
71	1:1,250,000 Six-inch NG Series 1.1.1957	122	1:1,250,000 1:10,000-1:10,560 NG Series 1.1.1975
72	1:1,250,000 Six-inch NG Series 1.4.1957	123	1:1,250,000 1:10,000-1:10,560 NG Series 1.7.1975
73	1:1,250,000 Six-inch NG Series 1.7.1957	124	1:1,250,000 1:10,000-1:10,560 NG Series 1.1.1976
74	1:1,250,000 Six-inch NG Series 1.10.1957	125	1:1,250,000 1:10,000-1:10,560 NG Series 1.7.1976
75	1:1,250,000 Six-inch NG Series 1.1.1958		
76	1:1,250,000 Six-inch NG Series 1.4.1958		
77	1:1,250,000 Six-inch NG Series 1.7.1958		

NB: 85, 95 and two between 89 and 92 had other uses, or the quarterly sequence was broken. See p.142

78	1:1,250,000 Six-inch NG Series 1.10.1958
79	1:1,250,000 Six-inch NG Series 1.1.1959

5. Indexes to medium scale maps

768	Index to 1:25,000 on NG Sheet Lines 1 1954
769	Index to 1:25,000 on NG Sheet Lines 2 1954
773	1:1,250,000 Index to 1:25,000 20km x 15km sheets

80	1:1,250,000 Six-inch NG Series 1.4.1959
81	1:1,250,000 Six-inch NG Series 1.7.1959
82	1:1,250,000 Six-inch NG Series 1.10.1959
83	1:1,250,000 Six-inch NG Series 1.1.1960
84	1:1,250,000 Six-inch NG Series 1.4.1960
85	
86	1:1,250,000 Six-inch NG Series 1.7.1960
87	1:1,250,000 Six-inch NG Series 1.10.1960
88	1:1,250,000 Six-inch NG Series 1.1.1961
89	
90	
91	
92	

6. "L" sequence of unique numbers

L54	Boundary Commission 1 1947
L55	Boundary Commission 2 1947
L58	Boundary Commission 1 1947
L59	Boundary Commission 2 1947
L182	Retriangulation, Secondary Blocks 1 1950
L316	1:633,600 Diagram of Scotland 4 1950
L525	Retriangulation, Primary Observations 1 1952
L527	Retriangulation, Primary Observations 2 1952
L536	1:633,600 Diagram of Scotland 4 1953
L615	Boundary Commission 1,2 1954
L709	Boundary Commission 1 1954
L1579	Administrative Areas Diagram: E&W 1966
?L1580	Administrative Areas Diagram: Scotland ?1966/?≡
L1581	1:1,250,000 Administrative Areas Diagram 1966
L1581	Regional Hospital Areas, 1969
L1619	Administrative Areas 1 1970 (1955 base)
L1620	Administrative Areas 2 1970 (1955 base)

93	1:1,250,000 Six-inch NG Series 1.10.1961
94	1:1,250,000 Six-inch NG Series 1.1.1962
95	
96	1:1,250,000 Six-inch NG Series 1.4.1962
97	1:1,250,000 Six-inch NG Series 1.7.1962
98	1:1,250,000 Six-inch NG Series 1.1.1963
99	1:1,250,000 Six-inch NG Series 1.7.1963
100	1:1,250,000 Six-inch NG Series 1.1.1964
101	1:1,250,000 Six-inch NG Series 1.7.1964
102	1:1,250,000 Six-inch NG Series 1.1.1965
103	1:1,250,000 Six-inch NG Series 1.7.1965
104	1:1,250,000 Six-inch NG Series 1.1.1966

Appendix 5. Planning Maps: Explanatory Texts

Texts were issued by MTCP and DHS, and published by OS

1. *Land Classification*
    1950   K24 4/50, 9d (OSPR 7/50)

2. *Average Annual Rainfall based on Observations during 1881 to 1915*
    1950   K8 4/50, 9d (OSPR 7/50)
           400 1/60, 1/-
           300 5/64, 2/6d (OSPR 8/64)

2A. *Average Annual Rainfall 1916-1950*
    1967   1967 2/6d (OSPR 12/67)

3. *Population*
    1950   K40 4/50, 9d (OSPR 7/50)

4. *Limestone*
    1957   K8 2/57, 9d (OSPR 2/57)
           1M 7/57, 9d

5. *Vegetation: The Grasslands of England and Wales*
    1952   K10 8/52, 9d (OSPR 10/52)
           500 5/64, 2/6d (OSPR 8/64)

6. *Local Accessibility*
    1955   1000 8/55, 9d (OSPR 9/55)
           4000 4/56, 9d

7. Igneous and Metamorphic Rocks
    Not published (listed in Explanatory Texts 4,8)

8. *Vegetation: Reconnaissance Survey of Scotland*
    1958   K8 4/58, 1/- (OSPR 5/58)

Appendix 6. Unpublished maps at 1:625,000

There are two important sources for details of the MTCP ten-mile maps which were not published: the minutes of the Maps Advisory Committee meetings, to which John Dower's 1941 lists of possible map subjects given on p.105 provides a suitable starting point, and the unpublished MTCP internal catalogue. A second edition of this was issued on 1 May 1947 and kept up to date by accession lists, initially every other month, at least until 1951. Few of the unpublished titles could genuinely be claimed in the event as in active preparation towards OS publication, and in fact only **Igneous and Metamorphic Rocks** (the only one even to reach proof stage), Ports, Drift Geology, and Electricity Supply at one time or another reached OS publicity material.

The OS only published those deemed worthy of a popular market. They printed some others for official purposes only, but most titles were created by MTCP by one of several means, including outline drawing, lithography or dyelining the OS base map and adding an overprint. Many maps were MTCP preparations of England and Wales only. With DHS collaboration others were of Great Britain on two sheets. Several evolved from one form to the other, and some titles passed through initial stages of this kind to ultimate publication. Such titles do not appear below. Few of these maps now survive, and almost always only if they escaped official ownership prior to the mass destruction of material in the mid-1970s following the Rayner investigation of Government departments. This list is thus limited to those few maps which have been located, and in addition those which were intended for publication. For those interested in investigating the hundred or so further titles, they should consult the MTCP internal catalogue at present with the author and which in due course may be passed to the archive of the Charles Close Society.

Trade by Ports 1936-38 (Limits of Customs Ports; value of imports and exports and re-exports at each, distinguishing the main classes of commodities. From the Customs and Excise Department's *Annual Statement of the trade of the United Kingdom with British Countries and Foreign Countries*, Vol. 4.)
 EW: MTCP print, June 1944. Copy: NTg
 NB: Intended for publication on two sheets: listed in DOSSSM 1947, and PRO OS 1/432 3 (7.1949)

Distribution and Intensity of Unemployment in June 1937 (From MOL's Local Unemployment Index). The areas of depression delimited by the Special Areas Act 1934
 EW: MTCP coloured print 1944. Copy: NTg

Fireclay, Silica Stone and Ganister Mines and Quarries (Location of Mines 1941 and Quarries 1938 and exposed coal measures)
 EW: MTCP print 1945; GB: MTCP dyeline compiled 9.1945. Copies (Sheets 1,2): Ng,NTg

Sandstones and Sands (Geological outcrops and quarries 1938, provisional edition)
 EW: MTCP ms 1944. Listed as in preparation for OS publication in the press release on 6 September 1945

Brickworks: Classified by geological formation of clay and shale used
 EW: MTCP print 1945; GB: MTCP dyeline compiled 9.1945. Copy (Sheet 2): Ng

Manufacture of Fireclay and Silica Refractories 1944
 GB: MTCP dyeline compiled 9.1945. Copies: Ng, NTg

Forestry Commission Areas and Conservancy Boundaries in Great Britain 1947
 GB: Forestry Commission print 1947
Kept up to date on one-inch and ten-mile scales. Four later 1:625,000 issues recorded: on 1946 "Roads" Map (to 1950, overprinted September 1951 (copies: Ob), on 1956 "Roads" Map (copies: Ob), on 1958 "Ten Mile" Map (copy: Eg (N)), and 1965 "Outline" Map (copies: Eg)

Bomb Damage: Towns with over 100 Dwellings destroyed or damaged beyond repair by 30th June 1943
 EW: MTCP ms 1944
On Administrative Areas Map Sheet 2 proof. Copy: Lrgs

Suggested Names for Physical Map
 GB: MTCP print 1944
NB: A dyeline outline draft map was produced by MTCP in 8.1943 for the use of Fellows of the RGS who responded to Willatts (1943a). Copies (Sheets 1,2): Bg

Electricity Generating Stations
 EW: MTCP ms 1947
NB: As Electricty Transmission Lines in DOSSSM 1947, and still in preparation 7.1949. Not printed

Spheres of Influence as Determined by Motor Bus Services (published as "Local Accessibility")
 EW: MTCP ms 1948
A dyeline of Sheet 1 was prepared by DHS in 4.1951 on the Great Britain Outline Base Map, 3801 edition.
Copy: Ng

Conservation and Recreation in England & Wales: Proposed Areas of Outstanding Natural Beauty
 Published by the Countryside Commission 1984
Copies (Dyelines): as at 1.1.86: Bg, as at 1.9.86: Bg

Countryside Commission: Countryside Recreation Service 1984. On 1975 EW map. Copy: Lse

Geology (Drift Edition)
 GB: Still in preparation for publication in 1949

Hydrometric Areas
Various titles. Four editions known: on an unpublished "Physical Features" Map of 1948 (copy: Ob), then overprinted on various "Ten Mile" maps in 1960 (copies: Ob), 1.1965 (copies: NTg) (these prepared by MHLG), and 1.4.1974 (copies: Lrgs, Ob) (this prepared by the Central Water Planning Unit Drawing Office, Reading 1974)

Major Post-War Developments
EW: Compiled and lithographed by MHLG, 1958. On the 1955 "Ten Mile" Outline Map, with the west border at 100km E. Copy: Lse

Ministry of Labour Local Area Offices
GB: Several (?later annual) editions were produced between the printed maps of 1947 and 1967. Copies: Lse (overprinted 30.6.1961), LDg (overprinted 30.6.1962). Both lithographed on the 1955 "Ten Mile" base map: A Edn (N), A/* Edn (S) (print code CBH 33905)

National Parks
EW: Lithographed by MHLG. On the Outline Map, 1969 Copy: Lkg

River Authority Areas, Statutory Catchment Areas and Administrative Areas in England and Wales
Prepared by Survey Section for Land Drainage, Water Supply and Machinery Division, MAFF, 1966 (print code CBH 6416 11/66). EW, on the outline map. Copy: Og

NB: There were in addition many regional maps at the ten-mile scale, not listed here. See in particular *Conurbation: a Survey of Birmingham and the Black Country* (Birmingham, MTCP, 1948)

A much more ephemeral group of maps was published in January 1942 under a general heading "Ground Plan of Britain" by "The 1940 Council", a non-elected body concerned about future planning and chaired by Lord Balfour of Burleigh. Dudley Stamp was a member. The maps were prepared by Miss J.Tyrwhitt of the Association for Planning and Regional Reconstruction. Reductions from the 1:625,000 scale to 40 miles to the inch were published, and dyelined working copies at 1:625,000 could also be obtained. A set of these is in Bg. They are of mainland Great Britain, and carry the 100km squares of the National Grid over the sea areas. They are worth examining if only for their titles!

Speed the Plough & Milk the Cow
Save the best Farmland
Foundations of Industry: Coal, Iron, Salt
The Mosaic of Local Authorities
Chief Built up Areas
Population Movement: The Uncontrolled Shift 1931-37
Pivots of Industry: Power & Transport
Upland, Hill & Vale
Climates, Coming of Spring
Rural Solitudes & Urban Sprawl
Trees & Trusts
Raw Material of Cities
Wheels of Industry
Seek the Sun, Shun the Fog

NB: I am indebted to Christie Willatts for much of the information in this appendix.

BIBLIOGRAPHY AND DOCUMENTARY SOURCES

1. Bibliography

Adams,B., 1989a,1989b,1990. 198 Years and 153 Meridians, 152 defunct. *Sheetlines*, 25: 3-6; 25: 6-7; 27: 8-9.
Adams,B., 1991. The Projection of the original One-Inch Map of Ireland (and of Scotland). *Sheetlines*, 30: 12-15.
Airy,G.B., 1861. Explanation of a Projection by Balance of Errors for Maps applying to a very large extent of the Earth's Surface; and Comparison of this projection with other projections. *Philosophical Magazine*, 22: 409-421.
Andrews,J.H., 1974. *History in the Ordnance Map. An introduction for Irish Readers*. Dublin, Ordnance Survey.
Andrews,J.H., 1975. *A Paper Landscape: the Ordnance Survey in nineteenth century Ireland*. Oxford University Press.
Andrews,J.H., 1983. *The Thematic Maps of Henry Harness* (Unpublished paper read to the ICA Commission on the History of Cartography, Dublin, 29 August 1983) (Copy Dtf).
Anon., 1933. Road Map of Great Britain (Review). *Geographical Journal*, 81: 438-439.
Atthill,R., 1967. *The Somerset & Dorset Railway*. Newton Abbot, David & Charles.
Bolger,P., 1984. *An illustrated History of the Cheshire Lines Committee*. Merseyside, Heyday Publishing Co.
Borley,H.V., 1982. *Chronology of London Railways*. Oakham, Railway and Canal Historical Society (RCHS).
Bradshaw's Railway Manual, Shareholders' Guide and Directory (annual editions from 1863 to 1923). London, G.Bradshaw.
Browne,J.P., 1991. *Map Cover Art*. Southampton, Ordnance Survey. (With Hellyer list of OS covers)
Christiansen,R. and Miller,R.W., 1967,[1968]. *The Cambrian Railways* (Vols 1,2). David & Charles, Newton Abbot.
Christiansen,R. and Miller,R.W., 1971. *The North Staffordshire Railway*. Newton Abbot, David & Charles.
Clark,R.H., 1964. *A Southern Region Record*. Lingfield, Oakwood Press.
Clark,R.H., 1967. *A Short History of the Midland & Great Northern Joint Railway*. Norwich, Goose & Son.
Clarke,R.V., 1969. The use of watermarks in dating Old Series One-Inch Ordnance Survey maps. *Cartographic Journal*, 6: 114-129.
Clinker,C.R., 1954. *Railways of the West Midlands: A Chronology 1808-1954*. London, Stephenson Locomotive Society.
Clinker,C.R., 1960. *The Railways of Northamptonshire (including the Soke of Peterborough) 1800-1960*. Rugby, Author.
Clinker,C.R., 1961. *London and North Western Railway: A chronology of opening and closing dates of lines and stations, including joint, worked and associated undertakings 1900 to 1960*. Dawlish, David & Charles.
Clinker,C.R., 1963. *The Railways of Cornwall 1809-1963: A list of authorising acts of Parliament, opening and closing dates, with other information, compiled from original and contemporary sources*. Dawlish, David & Charles.
Close, Major C.F., 1901. *A Sketch of the Subject of Map Projections*. London, HMSO.
Close, Colonel C.F., 1913. *Ordnance Survey Maps of the United Kingdom. A Description of their Scales, Characteristics, &c*. London, HMSO.
Close, Colonel Sir Charles, 1926. *The Early Years of the Ordnance Survey* (Reprinted from the *Royal Engineers Journal*). Chatham, Institution of Royal Engineers (reprinted 1969, Newton Abbot, David & Charles).
Collinson, General T.B., 1903. *General Sir Henry Harness K.C.B., Colonel Commandant Royal Engineers*. London, Royal Engineers Institute Committee.
Cook,R.A., 1974. *Lancashire & Yorkshire Railway: Historical Maps* (Second Edition, 1976). Caterham, RCHS.
Cook,R.A. and Hoole,K., 1975. *North Eastern Railway: Historical Maps*. Caterham, RCHS.
Cook,R.A., 1977. *Great North of Scotland Railway and Highland Railway: Historical Maps*. Caterham, RCHS.
Cox,C.H., 1924. *Exercises on Ordnance Maps* (Second and subsequent editions, 1928 etc). London, G.Bell & Sons.
Dow,G., 1959,1962,1965. *Great Central* (Vols 1-3). London, Locomotive Publishing Co.
Final Report of the Departmental Committee on the Ordnance Survey, 1938. London, HMSO. (ie Davidson Committee Report)

Gough,J., 1989. *The Midland Railway: A Chronology listing in geographical order the opening dates of the lines and additional running lines of the company (with the powers under which they were constructed) together with the dates of the opening, re-naming, and closing of stations and signal boxes.* Mold, RCHS.

Greville,M.D., 1981. *Chronology of the Railways of Lancashire and Cheshire* (Revised and Combined Edition). London, RCHS.

Hajducki,S.M., 1974. *A Railway Atlas of Ireland.* Newton Abbot, David & Charles.

Harley,J.B., 1969-71. *The Old Series of Ordnance Survey One-inch Maps* (notes). Newton Abbot, David & Charles.

Harley,J.B., 1975. *Ordnance Survey Maps: a descriptive manual.* Southampton, Ordnance Survey.

Harrison,J.H. 1949. Maps of the Biological Subdivisions of Ireland (review). *Irish Naturalists' Journal* 9: 331-334.

Hellyer,R., 1987,1988,1989a. The archaeological and historical maps of the Ordnance Survey. *Sheetlines*, 20: 2-8; 21: 6-15; 24: 14-16 (Addenda & Corrigenda).

Hellyer,R., 1989b. Ordnance Survey Aviation Maps (note). *Sheetlines*, 24: 22.

Hellyer,R., 1989c. Edition codes on OS Maps (note). *Sheetlines*, 24: 23.

Hellyer,R., 1989d. The archaeological and historical maps of the Ordnance Survey. *Cartographic Journal*, 26: 111-133.

Hellyer,R. and Oliver,R.R., 1990. The earliest 'Ten-Mile' Map. *Sheetlines*, 27: 11-12.

Hellyer,R., 1991a. Another early large-scale survey progress index (note). *Sheetlines*, 31: 54-55.

Hellyer,R., 1991b. Ordnance Survey covers: the extended list. *Supplement to Sheetlines*, 31: 19-23.

Hellyer,R., 1992a. The Physical Maps of the Ordnance Survey. *Sheetlines*, 32: 18-32.

Herries Davies,G.L., 1983. *Sheets of many colours: The Mapping of Ireland's rocks 1750-1890.* Royal Dublin Society.

Hinks,A.R., 1913. *Maps and Survey* (Fourth Edition, 1942). Cambridge University Press.

Hodson,Y., 1989. *Ordnance Surveyors' Drawings 1789-c.1840.* Reading, Research Publications.

Holmes,P.J., [1975]. *The Stockton and Darlington Railway 1825-1975.* Ayr, First Avenue Publishing Co.

James, Colonel Sir Henry, 1860. Description of the Projection used in the Topographical Department of the War Office for Maps embracing large portions of the Earth's Surface. *Journal of the Royal Geographical Society*, 30: 106-111.

James, Colonel Sir Henry and Clarke, Captain A.R., 1862. On Projections for Maps applying to a very large extent of the Earth's Surface. *Philosophical Magazine*, 23: 306-312.

James, Colonel Sir Henry, 1868. *On the Rectangular Tangential Projection of the Sphere and Spheroid, with tables of the quantities requisite for the construction of maps on that projection; and also for a Map of the World on the Scale of Ten Miles to an Inch, with Diagrams and Outline Map.* Southampton, Ordnance Survey.

James, Lieut-General Sir Henry, 1875. *Account of the Methods and Processes adopted for the Production of the Maps of the Ordnance Survey of the United Kingdom drawn up by Officers of Royal Engineers.* London HMSO (revised 1902 by Colonel Duncan A.Johnston).

James,L., 1983. *A Chronology of the Construction of Britain's Railways 1778-1855.* London, Ian Allan.

Land Utilisation Survey of Britain, The, [?1943]. [?London, Land Utilisation Survey of Britain].

London & North Eastern Railway Company, 1926. *List of Lines (not including those jointly owned) with the Acts of Parliament authorising them, and the Dates of Opening* (reprinted 1987, Weston-super-Mare, Avon-Anglia).

MacDermot,E.T., 1927,1931. *History of the Great Western Railway* (Vols 1,2). London, Great Western Railway (revised 1964 by C.R.Clinker, London, Ian Allan).

Maggs,C.G., 1967. *The Midland & South Western Junction Railway.* Newton Abbot, David & Charles.

Margary,H. (ed.), 1975,1977,1981,1986,1987,1992,1989,1991. *The Old Series Ordnance Survey Maps of England and Wales* (Vols 1-8). Lympne Castle, Harry Margary.

Marshall,J., 1969,1970. *The Lancashire & Yorkshire Railway* (Vols 1,2). Newton Abbot, David & Charles.

Ministry of Agriculture and Fisheries, 1941. *Types of Farming Map of England & Wales.* London, Geographical Publications.

Mumford,I., 1974. Lithography, Photography and Photozincography in English Map Production before 1870. *Cartographic Journal*, 11: 19-33.

Mumford,I. and Clark,P.K., 1985. Tomorrow the World: Sir Henry James' Map of the World at Ten Miles to an Inch. *Eleventh International Conference on the History of Cartography.* Ottawa, National Map Collection, Public Archives of Canada.

Nicholson,T.R., 1988a. An Introduction to the Ordnance Survey Aviation Maps of Britain 1925-39. *Sheetlines*, 23: 5-18.

Nicholson,T.R., 1988b. The Ordnance Survey and Smaller Scale Military Maps of Britain 1854-1914. *Cartographic Journal*, 25: 109-127.

Nicholson,T.R., 1991. The Ordnance Survey and a nineteenth century environmental 'crisis'. *Sheetlines*, 31: 12-18.

Nock,O.S., 1967. *History of the Great Western Railway: Volume Three 1923-1947.* London, Ian Allan.

Oliver,R.R., 1985. *The Ordnance Survey in Great Britain 1835-1870* (Unpublished D.Phil. thesis at University of Sussex).

Oliver,R.R., 1988. Edition codes on Ordnance Survey Maps. *Sheetlines* 22: 4-7. Oliver,R.R., 1991. *A Guide to the Ordnance Survey One-inch Seventh Series.* London, Charles Close Society.

Owen,T. and Pilbeam,E., 1992. *Ordnance Survey: Map Makers to Britain since 1791.* Southampton, Ordnance Survey, and London, HMSO.

Paar,H.W., 1963. *The Severn & Wye Railway.* Dawlish, David & Charles, and London, MacDonald.

Bibliography

Paar,H.W., 1965. *The Great Western Railway in Dean*. Dawlish, David & Charles, and London, MacDonald.
Pilkington White,T., 1888. The Romance of Mapping. *Blackwood's Edinburgh Magazine*, 144: 548-564, esp.555.
Praeger,R.L., 1950. The Biological Subdivision of Ireland. *Irish Naturalists' Journal* 10: 29-30.
Railway Clearing House, 1867-1939 (occasional). *Railway Junction Diagrams*. London.
Railway Clearing House, 1867-1956 (occasional). *Hand Book of the Stations* (with title variants). London.
Report of the Departmental Committee appointed by the Board of Agriculture to inquire into the present Condition of the Ordnance Survey. BPP(HC) 1893-94, [C 6895, C 6895-I], LXXII. (ie Dorington Committee Report)
Robinson,A.H., 1955. The 1837 Maps of Henry Drury Harness. *Geographical Journal*, 121: 440-450.
Seymour,W.A. (ed.), 1980. *A History of the Ordnance Survey*. Folkestone, Dawson.
Stamp,L.D., 1941a. *Fertility, Productivity and Classification of Land in Britain*. London, Geographical Publications.
Stamp,L.D., 1941b. *Map of Predominant Farming Types in Scotland*. London, Geographical Publications.
Stamp,L.D., 1948. *The Land of Britain: its use and misuse* (Third Edition, 1962). London, Longmans, Green and Co.
Stephenson Locomotive Society, 1950. *The Glasgow and South Western Railway 1850-1923*. London.
Stephenson Locomotive Society, 1954. Great North of Scotland Railway 1854-1954. *Journal of the Stephenson Locomotive Society*: 30, 277-320 (revised and published in book form 1972, London, Stephenson Locomotive Society).
Stephenson Locomotive Society, 1955. *The Highland Railway Company and its Constituents and Successors 1855-1955*. London.
Taylor,E.G.R., 1940. Plans for a National Atlas. *Geographical Journal*, 95: 96-108.
Thomas,J., 1969,1975. *The North British Railway* (Vols 1,2). Newton Abbot, David & Charles.
Tomlinson,W.W., [1915]. *The North Eastern Railway: Its Rise and Development*. Newcastle upon Tyne, Andrew Reid & Co, and London, Longmans, Green and Co (reprinted 1967, Newton Abbot, David & Charles).
War Office, 1889. *Catalogue of Maps in the War Office* (with supplements 1890-1914). London.
War Office, 1892. *Report of Committee on a Military Map of the United Kingdom*, A.237. London.
War Office, 1936. *Catalogue of Maps not on sale to the public, published by the Geographical Section of the General Staff* (October 1936). London. (Copy RGS).
War Office, 1944. *Numerical Catalogue of Maps published by the Directorate of Military Survey and Geographical Section General Staff* (January 1944). London. (Copies PRO,RGS)
Willatts,E.C., 1943a. The Names of the Physical Features of Britain. *Geographical Journal*, 102: 20-26.
Willatts,E.C., 1943b. Physical Names for the Map of Britain. *Geographical Journal*, 102: 145-169.
Willatts,E.C., 1950. A New Series of Maps of Britain. *Compte Rendu du XVIe Congrès International de Géographie, Lisbonne 1949*: 390-398.
Willatts,E.C., 1963. Some Principles and Problems of Preparing Thematic Maps. *Conference of Commonwealth Survey Officers*, Paper No. 36.
Willatts,E.C., 1971. Planning and Geography in the last three Decades. *Geographical Journal*, 137: 311-338.
Willatts,E.C., 1987. Geographers and their involvement in planning. Chapter 8 in R. Steel *British Geography*. Cambridge University Press.
Willatts,E.C., 1988. The Maps of the Land Utilisation Survey of Britain 1931-49. *Sheetlines*, 23: 2-5.
Winterbotham,H.St J.L., 1936. *A Key to Maps*. London and Glasgow, Blackie & Son.
Wrottesley,A.J.F., 1970. *The Midland & Great Northern Joint Railway*. Newton Abbot, David & Charles.
Wrottesley,A.J.F., 1979a,1979b,1981. *The Great Northern Railway* (Vols 1-3). London, B.T.Batsford.

NB: Other railway dates (especially Great Eastern, Caledonian, London & North Western, railways of South Wales, Westmorland and Cumberland) from the various volumes of *A Regional History of the Railways of Great Britain* (London, Phoenix House, then Newton Abbot, David & Charles). In search of information on railway chronology I have examined well over one hundred sources: only the principal ones are listed here.

British Parliamentary Paper (House of Commons) (BPP(HC)), or (House of Lords) (BPP(HL))
Irish Parliamentary Paper (IPP)
Geological Survey Publication Reports (GSPR)
A Description of the Ordnance Survey Small Scale Maps (with slight title variants). Editions published in [1919],1920, 1921,1923,1925,1927,[1930],[1935],[1937],1947,1947 Reprinted with Addendum 1951,1957 (DOSSSM)
Ordnance Survey Annual Reports: various titles (OSR)
Ordnance Survey Catalogues: various titles until 1924, from 1967 (OSC)
Ordnance Survey Indexes to the 1/2500 and 6-inch scale Maps (of England & Wales, Scotland, Ireland), c.1905-06
Ordnance Survey Leaflet OSO 3000 16.2.27 re the coloured and outline 1926 maps
Ordnance Survey Leaflet OSO 3000 12.5.27 re corrections to Map of the Solar Eclipse 29th June 1927 (copy: PRO OS 1/15/3)
Ordnance Survey Leaflet No 42/34 (OSO 5000/3/34) re air editions (copy: PRO OS 1/456)
Ordnance Survey Office (Dublin) Quarterly Reports
Ordnance Survey Pamphlet OSO 1047 10000 12.63, re RPM
Ordnance Survey Publication Reports: various titles (OSPR)

2. Documentary sources

Ordnance Survey files in the Public Record Office, Kew (= PRO):
Eclipse Map	OS 1/15/3
Design of new 10 mile or 1/500,000 flying maps	OS 1/55, 1/456
Ten-mile map 1927-43	OS 1/144
Planning maps 1938-42, 1942-43, 1943-48, 1948	OS 1/155, 1/156, 1/157, 1/376
MTCP series 1943-49, 1949-50, 1950-57, 1957-69	OS 1/162, 1/432, 1/433, 1/999
Roman Britain 1955-57	OS 1/327
Monastic Britain 1935-52, 1952-63	OS 1/352, 1/353
Limestone 1957-72	OS 1/400
Ancient Britain 1950-59	OS 1/440
MHLG 1:625,000 series 1943-45, 1946, 1947-50	OS 1/556, 1/557, 1/558
Outline map 1965-69	OS 1/646
1:625,000 series 1954-60, 1964-68	OS 1/705, 1/706
Roads (RPM) 1955-56, 1957-64, 1964-65	OS 1/709, 1/1136, 1/1291
Iron age Britain 1958-64	OS 1/719
Rainfall 1959-68	OS 1/721
Topography 1948-61	OS 1/789
Period map production 1952-54	OS 1/825
Administrative areas 1956-66	OS 1/1135
Physical 1955-57	OS 1/1137
Gravel 1961-66	OS 1/1224
Maps on scales smaller than 1:625,000 1951-67	OS 11/5
1:625,000 series 1949-64, 1964-69	OS 11/42, 11/43
Period maps 1951-69, 1970-75	OS 11/54, 11/55
Specification files of the "Ten Mile" map	OS 36/4, 36/5, 36/6

Register of GSGS Maps 1-2299, 2300-4795 (beginning 15.2.1881). A photocopy of the original is in the Map Department, PRO
Other PRO Ordnance Survey, Treasury, and War Office documents as quoted in footnotes
Job files and control cards, now in the custody of the Charles Close Society. Almost all the Planning Map job files survive (not Roads [1946], Railways or Administrative Areas), also some period map files, and the 1:1,250,000 map

Board of Works documents in National Archives, Dublin (= NA):
Irish Railway Commissioners minute book	2D 59 51
Irish Railway Commissioners correspondence	2D 59 52, 2D 59 53

OSI files in National Archives, Dublin (= NA):
Land Tenure	OSLR 10997 (1844)
Board of Works	OS 5/2934
Sir Henry James's ten-mile maps	OS 5/3007
Bog maps	OS 5/3514, 6/18000
Ten-mile index	OS 5/3616
Poor Law Unions	OS 5/4239
Irish Land Commission	OS 5/4525
Railway revision	OS 5/8409
New ten-mile map	OS 6/8540, 6/9070, 6/9254
Urban and District Boundaries	OS 6/8603
Rivers and Catchment Basins	OS 6/8633
County and Rural District Boundaries	OS 6/11727
Barony Boundaries	OS 6/15664
Murders in Ireland	OS 6/19806
Six-inch index	OS 6/20612

INDEX

NB: Only the text areas of this book have been indexed, for the names of persons and institutions

Adams,B.W. xii
Advisory Committee on Surveys and Mapping 149
Agriculture, Board of 1,74
Agriculture and Technical Instruction (Ireland),
    Department of 79
Agriculture and Fisheries, Ministry of 96,101,106,107
Air Ministry 96,98
Airy,G.B. 73,176,181,182,183
Andrews,J. 182
Ansell,G.K. 76,98,99
Automobile Association 149

Bacon,G.W. xi,100,106
Bailey, Sir E. 103
Baker,B. 2
Bonne 5,74,178,179,180,181,182
Brigstoke,C.R. 96
British Association for the Advancement of Science 100,
    101,103
British Railways 111
Brown,R.L. 146
Butler,A.S. 103,112

Cameron,J. 56,57,188
Cary,J. 2
Cassini 5,74,98,176,178,179,180,181,183
Central Planning Authority 101,102,103,145
Champion Jones,D. 77
Cheetham,G. 112
Clarke,A.R. 181,182,184
Close,C.F. 181
Colby,T. 3,54,55
Cole,G. 79
Cox Son & Barnet 2
Crook,H.T. 1,73
Crowley 70

Davidson, Viscount 102,146,177
Davis,H.L. 101,103,104,106
Dawson,R.K. 55
Defence, Ministry of x
De Vitre 77
Dorington, Sir J. 1,5,73,74
Dower,J. 102,103,104,105,110,112,195
Dowson,A.H. 149
Drummond,T. 55
Duncan,J. 56

Elliot,J.H. 1,74
Environment, Department of the 109

Faden,W. 3
Fawcett 101
Fryer 98

Gardiner 150
Gardner,J. 54,55
Geological Survey 103
Geological Survey of Ireland 79
Griffith,R. 54,55,79

Harness,H.D. 54,55
Harvey,W. 56,57,58,187
Health for Scotland, Department of 103,106,108,111,112,
    145,148,149,194,195
Health, Ministry of 111
Hellard,R.C. 77
Heseltine,M. x
Hill 79
Hodges and Smith 56
Holford 101,102,103,104,105
Housing and Local Government, Ministry of 108,109,110,
    147,148,149

James, Sir H. xi,7,56,57,58,63,181,182,184
Jellicoe 101
Jerrard,R.A. 94
Johnston,D.A. 74,77,182
Jolly,H.L.P. 73,181
Jones,H.D. 54,55

Land Tenure Commissioners (Ireland) 56
Land Utilisation Survey of Britain xi,100,101,103,109
Larcom,T.A. 54,55,56,182
London Midland & Scottish Railway Co 111
Lyons,W.J. 70

MacLeod,M.N. 100,101,102,103,104,106,111
Malings 87
Manktelow,A.R. 101
Maps Advisory Committee 103,104,106,112,195
Martin,E 96,98,99
McCaw,G.T. 182
Messenger,K.G. x
Mill,H.R. 70
Mudge,W. 2

National Atlas Committee 100,101,102,103,112
Nelson 150

O'Farrell,J. 182,184
Oliver,R.R. x,181
Ordnance Survey *passim*
Ordnance Survey of Ireland xi, Chapter II, Chapter III.2
    passim

Place,C.O. 98
Portlock,J. 182

Railway Commissioners (Ireland) 54,55,56,79
Reith, Lord 101,103
Royal Air Force xi,96,98
Royal Automobile Club 149

Scientific and Industrial Research, Department of 103
Scott, Sir L. 100
Scottish Development Department 109
Shaw,T.B. 57
Sim 58,77
Stamp,L.D. 103,106,196
Standidge 62
Stanford,E. 76,98,99,110,150

Taylor,E. 100,101,102,103
Town and Country Planning, Ministry of xi,xii,103,106,
    107,110,111,112,145,194,195
Town Planning Institute 100
Transport, Ministry of x,96
Treasury 1,73,74,77,93,101,103,104,106,107,108,110,146

Vincent,H.G. 101,102,103,104

War Office xi,1,58,79,100,149,150,183,184
Wilkinson,B.A. 56,57,58,188
Willatts,E.C. 101,103,104,106,107,108,109,111,112,114,
    147,148,149,150
Willis,J.C.T. 108,147,148
Wilson, Sir C. 1,5,73,74
Winterbotham,H.St J.L. 5,73,76,93,96,98,99,100,181
Wolff,A. 181
Works, Ministry of 101,103,104

Yolland,W. 182

CGS 05

THE CHARLES CLOSE SOCIETY

for the Study of

Ordnance Survey Maps

AIMS OF THE SOCIETY

Ordnance Survey maps and plans, from their first publication in 1801 to the present day, provide a fascinating graphic portrayal of the changing face of Britain, and are a prime source of information for many users such as local historians, planners, engineers and walkers.

Founded in 1980, the Charles Close Society aims to bring together all those with any interest in the maps, plans and related materials of the Ordnance Survey, to promote the exchange of information, and to encourage and co-ordinate research.

Membership is open to all upon payment of an annual subscription.

PUBLICATIONS

The Society issues **Sheetlines**, three times a year, in which the results of members' research are published together with reviews of recent OS publications and news of current OS activities. A detailed monograph on the Third Edition (Large Sheet Series) has been published. Other publications include guides to the Seventh Series, the New Popular and Fifth editions of the one-inch map. In preparation are works on the 10 mile map and the methods and processes of nineteenth century OS map production.

ACTIVITIES

The Society arranges several meetings a year, providing a varied programme in a geographically wide selection of venues such as Birmingham, Sheffield, Leicester, Bradford Southampton and Manchester. Visits are made to some of the country's leading map collections and have included those of the British Library, National Library of Scotland and Public Record Office. A popular event is the annual map market, where members have an OS 'bring and buy' sale.

Further details may be obtained from:

The Hon. Secretary, David Archer,
c/o The Map Library,
British Library,
Great Russell Street,
London WC1B 3DG.

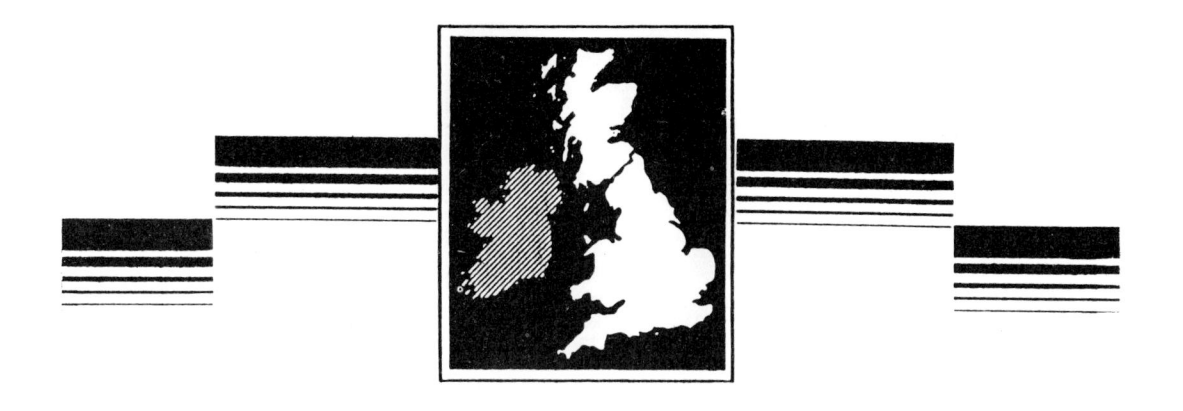

A Guide to the
ORDNANCE SURVEY
ONE-INCH
SEVENTH SERIES

by

RICHARD OLIVER

THE
CHARLES
CLOSE
SOCIETY

for the study of Ordnance Survey maps

A Guide to the

ORDNANCE SURVEY
ONE-INCH MAP OF SCOTLAND

THIRD EDITION IN COLOUR

by

GUY MESSENGER

THE CHARLES CLOSE SOCIETY

for the study of Ordnance Survey maps

SHEETLINES

Nos 19 to 21　　　　　ISSN 0962-8207

Ordnance Survey Leaflets: No. 4.

THE
ORDNANCE SURVEY
AND
ARCHÆOLOGY

G R
Ordnance Survey Office,
Southampton.

Stonehenge
Ellis Martin

The CHARLES CLOSE SOCIETY for the Study of Ordnance Survey Maps

The CHARLES CLOSE SOCIETY
for the Study of Ordnance Survey Maps

A Preliminary List of
Ordnance Survey "One-Inch"
DISTRICT and TOURIST MAPS
AND SELECTED PRECURSORS IN
The British Library

by Karen Severud Cook and Robert P. McIntosh

AN INCH TO THE MILE

The Ordnance Survey One-Inch Map 1805-1974

Catalogue of an exhibition to commemorate the bicentenary of the Ordnance Survey

Edited by

YOLANDE HODSON

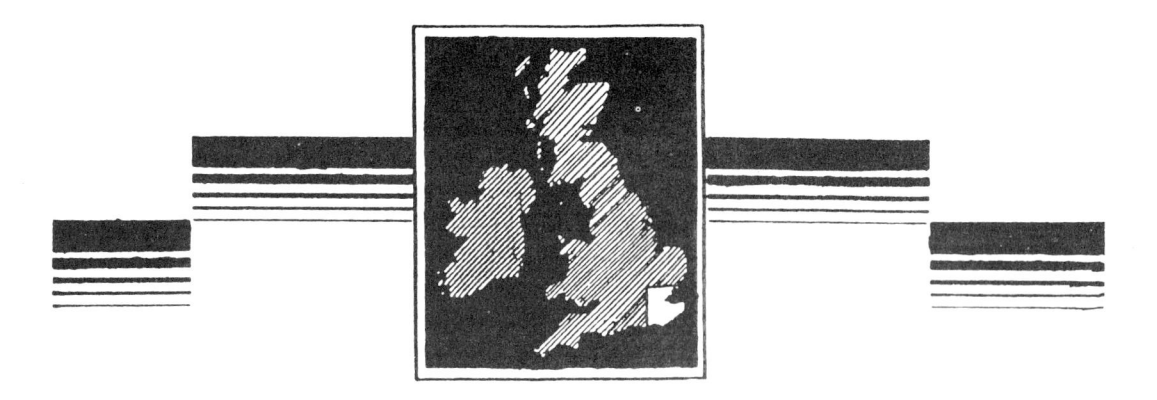

The Sheet Histories of the

ORDNANCE SURVEY

One-Inch

OLD SERIES MAPS

of

ESSEX and KENT

A Cartobibliographic Account

by

Guy Messenger

THE
CHARLES
CLOSE
SOCIETY

for the study of Ordnance Survey maps